2013 International Conference on Indium Phosphide and Related Materials

(IPRM 2013)

Kobe, Japan
19 – 23 May 2013

IEEE Catalog Number: CFP13IIP-PRT
ISBN: 978-1-4673-6130-9

**Copyright © 2013 by the Institute of Electrical and Electronic Engineers, Inc
All Rights Reserved**

Copyright and Reprint Permissions: Abstracting is permitted with credit to the source. Libraries are permitted to photocopy beyond the limit of U.S. copyright law for private use of patrons those articles in this volume that carry a code at the bottom of the first page, provided the per-copy fee indicated in the code is paid through Copyright Clearance Center, 222 Rosewood Drive, Danvers, MA 01923.

For other copying, reprint or republication permission, write to IEEE Copyrights Manager, IEEE Service Center, 445 Hoes Lane, Piscataway, NJ 08854. All rights reserved.

***This publication is a representation of what appears in the IEEE Digital Libraries. Some format issues inherent in the e-media version may also appear in this print version.**

IEEE Catalog Number: CFP13IIP-PRT
ISBN 13: 978-1-4673-6130-9
ISSN: 1092-8669

Additional Copies of This Publication Are Available From:

Curran Associates, Inc
57 Morehouse Lane
Red Hook, NY 12571 USA
Phone: (845) 758-0400
Fax: (845) 758-2633
E-mail: curran@proceedings.com
Web: www.proceedings.com

2013 International Conference on Indium Phosphide and Related Materials

Table of Contents

Monday, May 20, 2013

MoPLN

Plenary Session

MoPLN-3 (plenary) 10:30 - 11:10

Advances in III-V Semiconductor Photonics: Nanostructures and Integrated Chips 1

*Osamu Wada

Kobe University, Japan

MoPLN-4 (plenary) 11:10 - 11:50

III-V MOS Technology: From Planar to 3D and 4D 3

*Peide D Ye

Purdue University, United States of America

MoC3

Optical Properties of Nanostructures

MoC3-1 (invited) 14:00 - 14:30

Reshaping the optical properties of quantum dots via strain and electric fields 5

*Armando Rastelli(*1,*2), Rinaldo Trotta(*1,*2), Eugenio Zallo(*2), Paola Atkinson(*2), and Oliver G. Schmidt(*2)

*(*1)1Institute of Semiconductor and Solid State Physics, Johannes Kepler University of Linz, Austria and (*2)Insitute for Integrative Nanosciences, IFW Dresden, Germany*

MoC3-2 14:30 - 14:45

Suppression of multi-photon emission in 1.5-μm 7

*Toshiyuki Miyazawa(*1), Kazuya Takemoto(*2), Yoshiki Sakuma(*3), Haizhi Song(*2), Motomu Takatsu(*2), Tsuyoshi Yamamoto(*1), and Yasuhiko Arakawa(*2,*4)

*(*1)Fujitsu Laboratories Ltd., Japan, (*2)Institute for Nano Quantum Information Electronics, The University of Tokyo, Japan, (*3)National Institute for Materials Science, Japan, and (*4)Institute of Industrial Science, The University of Tokyo, Japan*

MoC3-3 (invited) 14:45 - 15:15

Wurtzite Gallium Phosphide has a Direct Band Gap 9

Simone Assali, Ilaria Zardo, Sebastien Plissard, Marcel Verheijen, Jos Haverkort, and *Erik Bakkers
TU Eindhoven, Netherlands

MoC3-4 15:15 - 15:30

Self-aligned quantum-dot growth for single-photon sources 11

U. W. Pohl, *A. Strittmatter, J. H. Schulze, D. Quandt, T. D. Germann, W. Unrau, T. Heindl, O. Hitzemann, D. Bimberg, and S. Reitzenstein
Institut für Festkörperphysik, Technische Universität Berlin, Germany

MoC3-5 15:30 - 15:45

Single-photon emission in telecommunication band from an InAs quantum dot in a pillar structure 13

*Xiangming Liu(*1), Natsuko Kobayashi(*1), Kouichi Akahane(*2), Masahide Sasaki(*2), Hidekazu Kumano(*1), and Ikuo Suemune(*1)
*(*1)Research Institute for Electronic Science, Hokkaido University, Japan and (*2)National Institute of Information and Communications Technology, Japan*

MoD3   **Modulators and Detectors**

MoD3-1 (invited) 14:00 - 14:30

Travelling Wave Mach-Zehnder Modulators 15

*Kelvin Prosyk(*1), Abderrahmane Ait-Ouali(*1), Junfu Chen(*1), Michael Hamacher(*2), Detlef Hoffmann(*2), Ronald Kaiser(*2), Ron Millett(*1), Alessio Piratsu(*1), Marco Totolo(*1), Karl-Otto Velthaus(*2), and Ian Woods(*1)
*(*1)Cogo Optronics, Canada and (*2)Heinrich Hertz Institute, Germany*

MoD3-2 14:30 - 14:45

Novel planar structure single-RF drive MZ optical modulator on InP(110) substrate 17

*Yoshihiro Ogiso, Masakazu Arai, Eiichi Yamada, Hiromasa Tanobe, Yasuo Shibata, and Masaki Kohtoku
NTT Photonics Laboratories, NTT corporation, Japan

MoD3-3   14:45 - 15:00

Flat-top Optical Frequency Comb Block Generation using InP-based Mach-Zehnder Modulator 19

Takeaki Saikai(*1), Takahiro Yamamoto(*1), Eiichi Yamada(*2), and *Hiroshi Yasaka(*1)

*(*1)RIEC, Tohoku University, Japan and (*2)NTT Photonic Laboratories, Japan*

MoD3-4   15:00 - 15:15

A 1×4 MMI-Integrated High-Power Waveguide Photodetector 21

*Efthymios Rouvalis, Philipp Müller, Dirk Trommer, Jens Stephan, Andreas G Steffan, and Günter Unterbörsch

u2t Photonics AG, Germany

MoD3-5   15:15 - 15:30

Study of lowering onset gain for a high-speed InGaAs/InAlAs avalanche photodiode 23

*Masahiro Nada, Yoshifumi Muramoto, Haruki Yokoyama, Tadao Ishibashi, and Hideaki Matsuzaki

NTT Photonics Laboratories, NTT Corporation, Japan

MoD3-6   15:30 - 15:45

Monolithic Integration of InP-Based Waveguide Photodiodes with MIM Capacitors for Compact Coherent Receiver 25

*Ryuji Masuyama, Hideki Yagi, Naoko Inoue, Yutaka Onishi, Tomokazu Katsuyama, Takehiko Kikuchi, Yoshihiro Yoneda, and Hajime Shoji

Sumitomo Electric Industries, LTD., Japan

MoD3-7   15:45 - 16:00

Monolithic InP Receiver Chip with a 90° Hybrid and a Variable Optical Attenuator for 100GBit/s Colourless WDM Detection 27

*Patrick Runge(*1), Stefan Schubert(*1), Angela Seeger(*1), Tom Gärtner(*1), Klemens Janiak(*1), Jens Stephan(*2), Dirk Trommer(*2), and Mads Lønstrup Nielsen(*2)

*(*1)Fraunhofer Heinrich-Hertz-Institute, Germany and (*2)u²t Photonics AG, Germany*

MoC4

Nano Devices

MoC4-1 (invited)   16:30 - 17:00

Highly non-linear phenomena and coherent effects in 1500 nm QD lasers and amplifiers 29
*Gadi Eisenstein(*1), Amir Capua(*1), Ouri Karni(*1), and Johann Peter Reithmaier(*2)
*(*1)Technion, Israel and (*2)Kassel University, Germany*

MoD4

Advanced Epitaxial Growth

MoD4-1 (invited)   16:30 - 17:00

Selective area MOVPE of InGaAsP and InGaN systems as process analytical and design tools for OEICs 31
*Yukihiro Shimogaki(*1), Masakazu Sugiyama(*2), and Yoshiaki Nakano(*3)
*(*1)Department of Materials Engineering, The University of Tokyo, Japan, (*2)Institute of Engineering Innovation, The University of Tokyo, Japan, and (*3)Research Center for Advanced Science and Technology, The University of Tokyo, Japan*

MoD4-2   17:00 - 17:15

MOVPE growth of InAs/InP QDs on directly-bonded InP/Si substrate 33
Keiichi Matsumoto, *Xinxin Zhang, Yoshinori Kanaya, and Kazuhiko Shimomura
Department of Engineering and Applied Sciences, Sophia University, Japan

MoD4-3   17:15 - 17:30

From surface dimer orientations to bonds at the GaP/Si(100) heterointerface 35
*Oliver Supplie(*1,*2), Sebastian Brückner(*1,*3), Henning Döscher(*1,*3), Peter Kleinschmidt(*1,*3,*4), and Thomas Hannappel(*1,*3,*4)
*(*1)Helmholtz-Zentrum Berlin, Institute Solar Fuels, Germany, (*2)Humboldt-Universität zu Berlin, Institut für Physik, Germany, (*3)Technische Universität Ilmenau, Institut für Physik, Germany, and (*4)CiS Forschungsinstitut für Mikrosensorik und Photovoltaik, Germany*

MoD4-4   17:30 - 17:45

in-situ Characterization of MOCVD grown GaAs- and InP-based tunable VSCEL structures 37

Christian Grasse(*2), *Yuto Tomita(*1), Peter Wiecha(*2), Ralf Meyer(*2), Tobias Gruendl(*2), Michael Mueller(*2), and Markus Christian Amann(*2)

*(*1)LayTec AG, Germany and (*2)Walter Schottky Institut, Technical University of Munich, Germany*

MoD4-5   17:45 - 18:00

Preparation of single-domain Si(100) surfaces with in situ control in CVD ambient 39

Sebastian Brückner(*1,*2), *Oliver Supplie(*1,*3), Peter Kleinschmidt(*1,*2,*4), Anja Dobrich(*1), Henning Döscher(*1,*2), and Thomas Hannappel(*1,*2,*4)

*(*1)Helmholtz-Zentrum Berlin, Institute Solar Fuels, Germany, (*2)Technische Universität Ilmenau, Institut für Physik, Germany, (*3)Humboldt-Universität zu Berlin, Institut für Physik, Germany, and (*4)CiS Forschungsinstitut für Mikrosensorik und Photovoltaik, Germany*

MoPI   **Poster Session**

MoPI-1

Zn Diffusion in Ruthenium Doped InP with Annealing by Metalorganic Vapor Phase Epitaxy 41

*Harunaka Yamaguchi, Takashi Nagira, Zempei Kawazu, Kenichi Ono, and Masayoshi Takemi

High Frequency & Optical Device Works, Mitsubishi Electric Corporation, Japan

MoPI-2

The Measurement of EPD on InP Single Crystal Wafers 43

*Qingfang Huang(*1), Zhiguo Liu(*2), Ruixia Yang(*2), Xiaolan Li(*1), Qiang Wang(*1), Xiuwei Tian(*1), Jianye Yang(*1), Shuai Li(*1), Yanlei Shi(*1), Huimin Shao(*1), Xin Zhang(*1), Ning Li(*1), Yong Kang(*1), Huisheng Liu(*1), Tongnian Sun(*1), and Niefeng Sun(*1)

*(*1)Science and technology on ASIC Laboratory, Hebei Semiconductor Research Institute, China and (*2)School of Information Engineering, Hebei University of Technology, China*

MoPI-3

Optimizing of metamorphic buffer layer for extended-InGaAs/InP photodetectors 45

*Sten Seifert

Fraunhofer institute for telecommunications, Heinrich Hertz institute Einsteinufer 37, 10587 Berlin, Germany, Germany

MoPI-4

Liquid-Phase Electroepitaxy of GaN at atmospheric pressure using ammonia and Ga-Ge solution 47

*Daisuke Kanbayashi, Takeshige Hishida, Masafumi Tomita, Hiroyuki Takakura, Takahiro Maruyama, and Shigeya Naritsuka

Department of Materials Science and Engineering, Meijo University, Japan

MoPI-5

Wide energy level control of InAs QDs using double-capping procedure by MOVPE 49

*Masayuki Yamauchi, Yuto Iwane, Shohei Yoshikawa, Yuta Yamamoto, and Kazuhiko Shimomura

Department of Engineering and Applied Sciences, Sophia University, Japan

MoPI-6

Analysis of photoluminescent properties of InAs/InGaAsP/InP quantum dots structure 51

*Rie Sato, Mariya Nakamura, and Hajime Imai

Faculty of Science, Japan Women's University, Japan

MoPI-7

Nature of the optical transition in (In,Ga)As(N)/GaP quantum dots (QDs): effect of QD size, indium composition and nitrogen incorporation 53

*Cedric Robert(*1), Charles Cornet(*1), Katiane Pereira Da Silva(*2), Pascal Turban(*3), Samuel Mauger(*4), Tra Nguyen Thanh(*1), Jacky Even(*1), Jean-Marc Jancu(*1), Mathieu Perrin(*1), Hervé Folliot(*1), Tony Rohel(*1), Sylvain Tricot(*3), Andrea Balocchi(*5), Philippe Barate(*5), Xavier Marie(*5), Paul M Koenraad(*4), Maria Isabel Alonso(*2), Alejandro Rodolfo Goñi(*2,*6), Nicolas Bertru(*1), Olivier Durand(*1), and Alain Le Corre(*1)

*(*1)CNRS UMR 6082 FOTON-OHM,INSA Rennes, France, (*2)Institut de Ciencia de Materials de Barcelona-CSIC, Spain, (*3)UMR UR1-CNRS 6251 , Equipe de Physique des Surfaces et Interfaces, Institut de Physique de Rennes, France, (*4)Department of Applied Physics, Eindhoven University of Technology, Netherlands, (*5)LPCNO, INSA-CNRS-UPS, Université de Toulouse, France, and (*6)ICREA, Spain*

MoPI-8

Optimizing The Double-Cap Procedure for InAs/InGaAsP/InP Quantum Dots by Metal-Organic Chemical Vapor Deposition 55

Shuai Luo, Haiming Ji, Xiaoguang Yang, and *Tao Yang

Key Laboratory of Semiconductor Materials Science,Institute of semiconductors,Chinese Academy of Sciences., China

MoPI-9
Junction Field-Effect Transistor Based on GaAs Core-Shell Nanowires 57

Oliver Benner, Andrey Lysov, Christoph Gutsche, *Gregor Keller, Claudia Schmidt, Werner Prost, and Franz-Josef Tegude

Solid State Electronics Department , University Duisburg-Essen, Germany

MoPI-10
MOVPE-preparation of Si(111) surfaces for III-V nanowire growth 59

Matthias Steidl(*1,*2), Agnieszka Paszuk(*1,*2), Weihong Zhao(*1,*2), Sebastian Brückner(*1,*2), Anja Dobrich(*2), *Oliver Supplie(*2,*3), Johannes Luczak(*2), Peter Kleinschmidt(*1,*2,*4), Henning Döscher(*1,*2), and Thomas Hannappel(*1,*2,*4)

*(*1)Technische Universität Ilmenau, Institut für Physik, Germany, (*2)Helmholtz-Zentrum Berlin, Institute Solar Fuels, Germany, (*3)Humboldt-Universität zu Berlin, Institut für Physik, Germany, and (*4)CiS Forschungsinstitut für Mikrosensorik und Photovoltaik, Germany*

MoPI-12
Bandgap Wavelength Shift in Quantum Well Intermixing using Different SiO₂ masks for Photonic Integration 61

*Jieun Lee(*1), Yoshiaki Yamahara(*1), Mitsuaki Futami(*1), Takahiko Shindo(*2), Tomohiro Amemiya(*2), Nobuhiko Nishiyama(*1), and Shigehisa Arai(*1,*2)

*(*1)Department of Electrical and Electronic Engineering, Tokyo Institute of Technology, Japan and (*2)Quantum Nanoelectronics Research Center, Japan*

MoPI-13

Carrier-transport, Optical and Structural Properties of Large Area ELOG InP on Si Using Conventional Optical Lithography 63

*Himanshu Kataria, Wondwosen Metaferia, Mony Nagarajan, Carl Junesand, Yanting Sun, and Sebastian Lourdudoss

Laboratory of Semiconductor Materials, KTH- Royal Institute of Technology, Sweden

MoPI-14
Low Power Consumption Operation of Light Sources for Inter-chip Optical Interconnects 65

*Nobuaki Hatori(*1,*2), Takanori Shimizu(*1,*2), Makoto Okano(*2,*3), Masashige shizaka(*1,*2), Tsuyoshi Yamamoto(*1,*2), Yutaka Urino(*1,*2), Masahiko Mori(*2,*3), Takahiro Nakamura(*1,*2), and Yasuhiko Arakawa(*2,*4)

*(*1)PETRA, Japan, (*2)PECST, Japan, (*3)AIST, Japan, and (*4)Univ. of Tokyo, Japan*

MoPI-15
1540 to 1645 nm continuous VCSEL emission based on quantum dashes 67

*Cyril Paranthoen(*1), Christophe Levallois(*1), Jean-Philippe Gauthier(*2), Fethallah Taleb(*1), Nicolas Chevalier(*1), Mathieu Perrin(*1), Yoan Leger(*1), Olivier De Sagazan(*2), and Alain Le Corre(*1)

*(*1)FOTON, INSA, France and (*2)IETR, Université Rennes 1, France*

MoPI-16
InAs/InP Quantum dot mode-locked lasers grown on (113)B InP substrate 69

Kamil Klaime(*1), Cosimo Calo(*2), Rozenn Piron(*1), *Cyril Paranthoen(*1), Thomas Batte(*1), Olivier Dehaese(*1), Julie Le Pouliquen(*1), Slimane Loualiche(*1), Alain Le Corre(*1), Kamel Merghem(*2), Anthony Martinez(*2), and Abderrahim Ramdane(*2)

*(*1)UEB INSA-RENNES, CNRS UMR6082 FOTON, FRANCE, France and (*2)CNRS LPN MARCOUSSIS, FRANCE, France*

MoPI-17

Analysis of Uni-Traveling-Carrier Photodetectors (UTC-PDs) with Dipole-Doped Interface 71

*Qianqian Meng(*1), Chongyang Liu(*1), Hong Wang(*1,*2), Kian Siong Ang(*1), Manoj Kumar C M(*1), Tina Xin Guo(*1), and Bo Gao(*1,*3)

*(*1)Temasek Laboratories, Nanyang Technological University, Singapore, (*2)School of Electrical and Electronic Engineering, Nanyang Technological University,, Singapore, and (*3)School of Electronic and Information Engineering, Xian Jiaotong University, China*

MoPI-18

Multi-regrowth steps for the realization of buried single ridge and μ-stripes quantum cascade lasers 73

*Olivier Parillaud(*1), Guy-Maël De Naurois(*1), Bouzid Simozrag(*1), Virginie Trinite(*1), Grégory Maisons(*1), Michel Garcia(*1), Bruno Gerard(*1), Mathieu Carras(*1), Wondwosen Metaferia(*2), Carl Junesand(*2), Himanshu Kataria(*2), Yanting Sun(*2), and Sebastian Lourdudoss(*2)

*(*1)III-V Lab, France and (*2)KTH - Royal Institute of Technology, Sweden*

MoPI-19

Mid-infrared photodetectors with InAs/GaSb type-II quantum wells grown on InP substrate 75

*Hiroshi Inada(*1), Kouhei Miura(*1,*2), Yasuhirso Iguchi(*1), Yuuichi Kawamura(*2), Junpei Murooka(*3), Haruyoshi Katayama(*3), Shota Kanno(*4), Tomoko Takekawa(*4), and Masafumi Kimata(*3,*4)

*(*1)Transmission Devices R&D Laboratories, Sumitomo Electric Industries, Ltd., Japan, (*2)Frontier Science Innovation Center, Osaka Prefecture University, Japan, (*3)Japan Aerospace Exploration Agency (JAXA), Japan, and (*4)Department of Mechanical Engineering, Ritsumeikan University, Japan*

MoPI-20

Cryogenic DC Characterization of InAs/Al$_{80}$Ga$_{20}$Sb Self-Switching Diodes 77

*Andreas Westlund(*1), Giuseppe Moschetti(*1), Per-åke Nilsson(*1), Jan Grahn(*1), Ludovic Desplanque(*2), and Xavier Wallart(*2)

*(*1)Department of Microtechnology and Nanoscience, Chalmers University of Technology, Sweden and (*2)Institute of Electronics, Microelectronics and Nanotechnology, University of Lille, France*

MoPI-21

Cryogenic Ultra-Low Noise Amplification - InP PHEMT vs. GaAs MHEMT 79

*Joel Schleeh, Helena Rodilla, Niklas Wadefalk, Per-åke Nilsson, and Jan Grahn

Department of Microtechnology and Nanoscience (MC2), Chalmers University of Technology, Sweden

MoPI-22

High Performacne InAs/AlSb HEMT with Refractory Iridium Schottky Gate Metal 81

*Wen-Yu Lin(*1), Chao-Hung Chen(*1), Hsien-Chin Chiu(*1), Wei-Jen Hsueh(*2), Yue-Ming Hsin(*2), and Jen-Inn Chyi(*2)

*(*1)Chang Gung Univ., Taiwan and (*2)National Central Univ., Taiwan*

MoPI-23

Influence of gate-channel distance in low-noise InP HEMTs 83

*Per-Ake Nilsson, Helena Rodilla, Joel Schleeh, Niklas Wadefalk, and Jan Grahn

Department of Microtechnology and Nanoscience, Chalmers University of Technology, Sweden

MoPI-24

Terahertz Oscillators using Resonant Tunneling Diodes with InAlGaAs/InP Composite Collector 85

*Riku Sogabe(*1), Kaoru Shizuno(*1), Hidetoshi Kanaya(*1), Safumi Suzuki(*1), Masahiro Asada(*1), Hiroki Sugiyama(*2), and Haruki Yokoyama(*2)

*(*1)Graduate School of Interdisciplinary Science and Engineering, Tokyo Institute of Technology, Japan and (*2)NTT Photonics Laboratories, NTT Corporation, Japan*

MoPI-25

120nm AlSb/InAs HEMT without gate recess : 290GHz f_T and 335GHz f_{max} 87

*Cyrille Gardès, Sonia Marcelle Bagumako, Ludovic Desplanque, Nicolas Wichmann, François Danneville, Sylvain Bollaert, Xavier Wallart, and Yannick Roelens

Institut d'Electronique de Microélectronique et de Nanotechnologie (IEMN), France

MoPI-26

Monte Carlo Simulation of InAlAs/InGaAs HEMTs with Buried Gate 89

*Akira Endoh(*1,*2), Issei Watanabe(*1), Akifumi Kasamatsu(*1), and Takashi Mimura(*1,*2)

*(*1)National Institute of Information and Communications Technology, Japan and (*2)Fujitsu Laboratories Ltd., Japan*

MoPI-27

Terahertz GaAs Schottky diode mixer and multiplier MICs based on e-beam technology 91

*Vladimir Drakinskiy(*1), Peter Sobis(*2), Huan Zhao(*1), Tomas Bryllert(*1,*3), and Jan Stake(*1)

*(*1)Department of Microtechnology and Nanoscience, Chalmers University of Technology, Sweden,*

*(*2)Omnisys Instruments AB, Sweden, and (*3)Wasa Millimeter Wave AB, Sweden*

MoPI-28

Simulation and Fabrication of InGaAs Planar Gunn Diode on InP Substrate 93

*Vasileios Papageorgiou, Ata Khalid, Chong Li, and David R. S. Cumming

School of Engineering, University of Glasgow, United Kingdom

MoPI-29

5 GHz Low-Power RTD-Based Amplifier MMIC With a High Figure-Of-Merit of 24.5 dB/mW 95

*Jongwon Lee, Jooseok Lee, Jaehong Park, and Kyounghoon Yang

Department of Electrical Engineering, KAIST, Republic of Korea

Tuesday, May 21, 2013

TuD1

Epitaxy for Advanced Devices

TuD1-2 (invited)   9:00 - 9:30

Light emission between 2 and 4 μm: Innovative active region designs for InP- and GaSb-based devices 97

Gerhard Boehm, *Stephan Sprengel, Kristijonas Vizbaras, Christian Grasse, Tobias Gruendl, Ralf Meyer, and Markus-Christian Amann

Walter Schottky Institut, Technische Universitaet Muenchen, Germany

TuD1-3   9:30 - 9:45

MOCVD Growth and Device Characterization of InP/GaAsSb/InP DHBTs with a GaAs Spacer 99

*Takuya Hoshi, Hiroki Sugiyama, Haruki Yokoyama, Norihide Kashio, Kenji Kurishima, Minoru Ida, and Hideaki Matsuzaki

NTT Photonics Laboratories, NTT Corporation, Japan

TuD1-4   9:45 - 10:00

High Growth Rate Gallium Phosphide for Red LEDs 101

*Stephen Farrell(*1), Chris Ebert(*2), and Devon Dyer(*3)

*(*1)Veeco Instruments, Inc., United States of America, (*2)Veeco Instruments, Inc., United States of America, and (*3)Veeco Instruments, Inc., United States of America*

TuD1-5   10:00 - 10:15

MOCVD growth of carbon-doped InGaAs layers using ethyl-base metal organic materials 103

*Hideo Yokohama(*1,*2), Kenji Shiojima(*1), and Gako Araki(*2)

*(*1)Graduate School of Electrical and Electronics Engineering, University of Fukui, Japan and (*2)OPTRANS Corporation, Japan*

TuD2   **Lasers**

TuD2-1 (invited)   11:00 - 11:30

High-speed directly modulated laser for applications beyond 100GbE 105

*Wataru Kobayashi, Takeshi Fujisawa, Toshio Ito, Takayuki Yamanaka, Yasuo Shibata, Takashi Tadokoro, and Hiroaki Sanjoh

NTT, Japan

TuD2-2   11:30 - 11:45

Simultaneous 40-Gbps Direct Modulation of 1.3-μm Wavelength AlGaInAs Distributed-Reflector Laser Arrays on Semi-Insulating InP Substrate 107

*Manabu Matsuda, Ayahito Uetake, Takasi Simoyama, Shigekazu Okumura, Kazumasa Takabayashi, Mitsuru Ekawa, and Tsuyoshi Yamamoto

Fujitsu Laboratories Ltd., Japan

TuD2-3   11:45 - 12:00

Static and dynamic characteristics of InAs/AlGaInAs/InP quantum dot lasers operating

at 1550 nm 109

*Johann Peter Reithmaier(*1), Vitalii Ivanov(*1), Vitalii Sichkovskyi(*1), Christian Gilfert(*1), Anna Rippien(*1), Florian Schnabel(*1), David Gready(*2), and Gadi Eisenstein(*2)

*(*1)Institute of Nanostructure Technologies and Analytics, University of Kassel, Germany and*

*(*2)Department of Electrical Engineering, Technion, Israel*

TuD2-4   12:00 - 12:15

Frequency-resolved optical gating measurements of sub-ps pulses from InAs/InP

quantum dash based mode-locked lasers 111

*Cosimo Calò(*1), Holger Schmeckebier(*2), Kamel Merghem(*1), Ricardo Rosales(*1,*2), François Lelarge(*3), Anthony Martinez(*1), Dieter Bimberg(*2), and Abderrahim Ramdane(*1)

*(*1)CNRS Laboratory for Photonics and Nanostructures, France, (*2)Institut für Festkörperphysik, Technische Universität Berlin, Germany, and (*3)III-V Lab, France*

TuD2-5   12:15 - 12:30

1480nm InGaAsP LOC Broad-Area-Lasers with >18W Pulsed Output Power at 20°C 113

*David Fendler, Martin Moehrle, Marc Spiegelberg, Wolfgang Rehbein, Wolfgang Passenberg, and Norbert Grote

Fraunhofer Institute for Telecommunications, Heinrich-Hertz-Institute, Germany

---

TuD3   Devices for Photonic Integration

TuD3-1 (invited)   14:00 - 14:30

Monolithically Integrated Optical Link Using Photonic Crystal Laser and Photodetector 115

*Shinji Matsuo(*1,*2)

*(*1)NTT Photonics Laboratories, NTT Corporation, Japan and (*2)Nanophotonics Center, NTT Corporation, Japan*

TuD3-2   14:30 - 14:45

Low Crosstalk and High Modulation Bandwidth 100GbE Optical Transmitter Using Flip-Chip Interconnects 117

*Shigeru Kanazawa, Takeshi Fujisawa, Kiyoto Takahata, Akira Ohki, Ryuzo Iga, and Hiroyuki Ishii
NTT Photonics Laboratories, Japan

TuD3-3   14:45 - 15:00

Non-blocking 4x4 InAlGaAs/InAlAs Mach-Zehnder-Type Optical Switch Fabric 119

Noriaki Koyama(*1), *Hiroki Kouketsu(*1), Shoko Kawasaki(*1), Aki Takei(*2), Takafumi Taniguchi(*2), Yuichi Matsushima(*3), and Katsuyuki Utaka(*1)
*(*1)Faculty of Science and Engineering, Waseda University, Japan, (*2)Central Research Laboratory, Hitachi Ltd., Japan, and (*3)Green Computing System Research Organization, Waseda University, Japan*

TuD3-4   15:00 - 15:15

Design of Multi-Functional GaInAsP/Si Hybrid Semiconductor Optical Amplifier Array with AlInAs-Oxide Current Confinement Layer 121

*Yusuke Hayashi(*1), Keita Fukuda(*1), Ryo Osabe(*1), Jun-Ichi Suzuki(*1), Joonhyun Kang(*1), Yuki Atsumi(*1), Nobuhiko Nishiyama(*1), and Shigehisa Arai(*1,*2)
*(*1)Department of Electrical and Electronic Engineering, Tokyo Institute of Technology, Japan and (*2)Quantum Nanoelectronics Research Center, Tokyo Institute of Technology, Japan*

TuD3-5   15:15 - 15:30

New Fabrication Method of Trapezoidal Polarization Converters for InP-Based Photonic Integrated Circuits 123

*Dzmitry O. Dzibrou, Jos J. G. M. van der Tol, and Meint K. Smit
Group of Photonic Integration, Eindhoven University of Technolgy, Netherlands

TuD3-6   15:30 - 15:45

Butt-Joint Built-in (BJB) Structure for Membrane Photonic Integration 125

*Daisuke Inoue(*1), Jieun Lee(*1), Takahiko Shindo(*2), Mitsuaki Futami(*1), Kyohei Doi(*1), Tomohiro Amemiya(*2), Nobuhiko Nishiyama(*1), and Shigehisa Arai(*1,*2)
*(*1)Department of Electrical and Electronic Engineering, Tokyo Institute of Technology, Japan and (*2)Quantum Nanoelectronics Research Center, Tokyo Institute of Technology, Japan*

TuD3-7LN   15:45 - 16:00

Tunable InP Photonic Integrated Circuit for Millimeter Wave Generation 127

*Marco Lamponi(*1), Mourad Chtioui(*2), François Lelarge(*1), Gaël Kervella(*1), Efthymios Rouvalis(*3), Cyril Renaud(*3), Martyn Fice(*3), Guillermo Carpintero(*4), Frederic van Dijk(*1)
*(*1)III-V Lab, a joint Laboratory of "Alcatel Lucent Bell Labs", "Thales Research & Technology" and "CEA-LETI", Palaiseau, France, (*2)Thales Air Systems, 91470 Limours, France, (*3)Department of Electronic and Electrical Engineering, UCL, Torrington Place, WC1E 7JE, United Kingdom, and (*4)Universidad Carlos III de Madrid, Av de la Universidad, 30 Leganes 28911 Madrid, Spain*

TuD4

Integrated Devices

TuD4-1 (invited)   16:30 - 17:00

AlGaInAs Selective Area Growth for high-speed EAM-based PIC Sources 129

*Jean Decobert(*1), Pierre-Yves Lagree(*2), Hugues Guerault(*3), and Christophe Kazmierski(*1)
*(*1)III-V lab, Route de Nozay, 91460 Marcoussis, France, (*2)CNRS, UPMC Univ Paris 06, IJLRA, 75005 Paris, France, and (*3)Bruker AXS GmbH, O. Rheinbrueckenstr. 49, 76187 Karlsruhe, Germany*

TuD4-2   17:00 - 17:15

56Gb/s PDM-BPSK Experiment with a Novel InP-Monolithic Source Based on Prefixed Optical Phase Switching 131

*Christophe Kazmierski(*1), Nicolas Chimot(*1), Fabrice Blache(*1), Jean Decobert(*1), Francois Alexandre(*1), Jorg Honecker(*2), Christoph Leonhardt(*2), Andreas Steffan(*2), Oriol Bertran-Pardo(*3), Haik Mardoyan(*3), Jeremie Renaudier(*3), and Gabriel Charlet(*3)
*(*1)III-V Lab, France, (*2)U2T Photonics, Germany, and (*3)Alcatel-Lucent, Bell Labs, France*

TuD4-3   17:15 - 17:30

InP-based Compact Reflection-Type Transversal Filter 133

*Yuta Ueda, Takeshi Fujisawa, Kiyoto Takahata, Masaki Kohtoku, Hiroshi Takahashi, and Hiroyuki Ishii
NTT Photonics Laboratories, NTT Corporation, Japan

TuD4-4   17:30 - 17:45

Transmitter PIC for THz Applications Based on Generic Integration Technology 135
*Norbert Grote
Fraunhofer Heinrich-Hertz-Institut, Germany

TuD4-5 17:45 - 18:00

Intermixng of Highly-Stacked InAs/InGaAlAs Quantum Dots Grown on InP(311)B

Substrate by SiO$_2$ Sputtering and Annealing Technique 137
*Asuka Matsushita(*1), Atsushi Matsumoto(*1), Kouichi Akahane(*2), Yuichi Matsushima(*3), and Katsuyuki Utaka(*1)
*(*1)Faculty of Science and Engineering, Waseda University, Japan, (*2)National Institute of Information and Communications Technology, Japan, and (*3)Green Computing System Research Organization, Waseda University, Japan*

Wednesday, May 22, 2013
WeD1   III-V MOSFETs

WeD1-1   8:30 - 8:45

High Transconductance Surface Channel In$_{0.53}$Ga$_{0.47}$As MOSFETs Using MBE

Source-Drain Regrowth and Surface Digital Etching 139
*Sanghoon Lee(*1), Cheng-Ying Huang(*1), Andrew D. Carter(*1), Jeremy J. M. Law(*1), Doron C. Elias(*1), Varistha Chobpattana(*2), Brian J. Thibeault(*1), William Mitchell(*1), Susanne Stemmer(*2), Arthur C. Gossard(*2), and Mark J. W. Rodwell(*1)
*(*1)Department of Electrical and Computer Engineering, UCSB, United States of America and (*2)Material Department, UCSB, United States of America*

WeD1-2   8:45 - 9:00

Sub-50-nm InGaAs MOSFET with n-InP source on Si substrate 141
*Atsushi Kato, Toru Kanazawa, Eiji Uehara, Yoshiharu Yonai, and Yasuyuki Miyamoto
Tokyo Institute of Technology, Japan

WeD1-3   9:00 - 9:15

Analysis on channel thickness fluctuation scattering in InGaAs-OI MOSFETs 143

*Sanghyeon Kim(*1), Masafumi Yokoyma(*1), Ryosho Nakane(*1), Osamu Ichikawa(*2), Takenori Osada(*2), Masahiko Hata(*2), Mitsuru Takenaka(*1), and Shinichi Takagi(*1)

*(*1)The University of Tokyo, Japan and (*2)Sumitomo Chemical Co. Ltd., Japan*

WeD1-4   9:15 - 9:30

Impact of Al$_2$O$_3$ ALD temperature on Al$_2$O$_3$/GaSb metal-oxide-semiconductor interface properties 145

*Masafumi Yokoyama(*1), Yuji Asakura(*1), Haruki Yokoyama(*2), Mitsuru Takenaka(*1), and Shinichi Takagi(*1)

*(*1)The University of Tokyo, Japan and (*2)NTT Photonics Laboratories, NTT Corporation, Japan*

WeD1-5   9:30 - 9:45

1/f -noise in Vertical InAs Nanowire Transistors 147

*Karl-Magnus Persson, Martin Berg, Erik Lind, and Lars-Erik Wernersson

Dept. of Electrical- and Information Technology, Lund University, Sweden

---

WeD2   **Integrated Lasers**

WeD2-1 (invited)   10:30 - 11:00

InP Based Photonic Integrated Circuits For DWDM Optical Communication 149

*Beck Mason, Michael Larson, Yuliya Akulova, and Srinath Kalluri

JDSU Transmission R&D, United States of America

WeD2-2   11:00 - 11:15

17-Gb/s Direct Modulation of Lambda-scale Embedded Active Region Photonic Crystal Lasers 151

*Koji Takeda(*1,*3), Tomonari Sato(*1,*3), Akihiko Shinya(*2,*3), Kengo Nozaki(*2,*3), Hideaki Taniyama(*2,*3), Koichi Hasebe(*1,*3), Takaaki Kakitsuka(*1,*3), Masaya Notomi(*2,*3), and Shinji Matsuo(*1,*3)

*(*1)NTT Photonics Labs., Japan, (*2)NTT Basic Res. Labs., Japan, and (*3)Nanophotonics Center, Japan*

WeD2-3   11:15 - 11:30

Room-temperature Continuous-wave Operation of Lateral Current Injection Membrane Laser 153

*Kyohei Doi(*1), Takahiko Shindo(*2), Mitsuaki Futami(*1), Jieun Lee(*1), Takuo Hiratani(*1), Daisuke Inoue(*1), Shu Yang(*1), Tomohiro Amemiya(*2), Nobuhiko Nishiyama(*1), and Shigehisa Arai(*1,*2)

*(*1)Department of Electrical and Electronic Engineering, Tokyo Institute of Technology, Japan and (*2)Quantum Nanoelectronics Research Center, Tokyo Institute of Technology, Japan*

WeD2-4   11:30 - 11:45

Mode Locked InAs/InP Quantum dash based DBR Laser monolithically integrated with a semiconductor optical amplifier 155

*Siddharth Joshi(*1), Nicolas Chimot(*1), Ricardo Rosales(*2), Sophie Barbet(*1), Alain Accard(*1), Abderrahim Ramdane(*2), and Francois Lelarge(*1)

*(*1)3-5 Lab, Marcoussis, France and (*2)Laboratoire de Photonique et de Nanostructures, CNRS, France*

Thursday, May 23, 2013

ThD1   **THz Detectors and Generators**

ThD1-1 (invited)   8:30 - 9:00

Asymmetric dual-grating gate InGaAs/InAlAs/InP HEMTs for ultrafast and ultrahigh sensitive terahertz detection 157

*Taiichi Otsuji(*1), Takayuki Watanabe(*1), Stephane Boubanga Tombet(*1), Tetsuya Suemitsu(*1), Dominique Coquillat(*2), Wojciech Knap(*2), Denis Fateev(*3), and Vyacheslav Popov(*3)

*(*1)Tohoku University, Japan, (*2)University of Montpellier and CNRS, France, and (*3)Kotelnikov Institute of Radio Engineering and Electronics (Saratov Branch), RAS, Russia*

ThD1-2   9:00 - 9:15

Improvement in Nonlinear Characteristics of Zero Bias GaAsSb-based Backward Diodes 159

*Tsuyoshi Takahashi(*1,*2), Masaru Sato(*1,*2), Yasuhiro Nakasha(*1,*2), and Naoki Hara(*1,*2)

*(*1)Fujitsu Laboratories Ltd., Japan and (*2)Fujitsu Limited, Japan*

ThD1-3   9:15 - 9:30

Characterization and Modeling of Zero Bias rf-Detection Diodes based on Triple Barrier Resonant Tunneling Structures 161

*Gregor Keller(*1), Anselme Tchegho(*1), Benjamin Muenstermann(*1), Werner Prost(*1), Franz-Josef Tegude(*1), and Michihiko Suhara(*2)

(*1)Center for Semiconductor Technology and Optoelectronics, University of Duisburg-Essen, Germany and (*2)Electrical and Electronic Engineering, Graduate School of Science and Engineering, Tokyo Metropolitan University, Japan

ThD1-4   9:30 - 9:45

Extremely-High Sensitive Terahertz Detector based on Dual-Grating Gate InP-HEMTs 163

*Yuki Kurita(*1), Guillaume Ducournau(*2), Kengo Kobayashi(*1), Yahya M. Meziani(*3), Vyacheslav V. Popov(*4), Wojciech Knap(*5), and Taiichi Otsuji(*1)

(*1)RIEC, Tohoku University, Japan, (*2)IEMN, France, (*3)Universidad de Salamanca, Spain, (*4)Kotelnikov Institute of Radio Engineering and Electronics RAS, Russia, and (*5)Univ. Montpellier 2, CNRS, France

ThD1-5   9:45 - 10:00

High Performance Modulation Doped AlGaAs/InGaAs Thermopiles (H-PILEs) for Uncooled IR FPA Utilizing Integrated HEMT-MEMS Technology 165

*Masayuki Abe(*1), Kian Siong Ang(*2), Rene Hofstetter(*2), Hong Wang(*2), and Geok Ing Ng(*2)

(*1)3D-bio Co., Ltd., Japan and (*2)Nanyang Technological University, Singapore

ThD1-6   10:00 - 10:15

Frequency Modulation in mm-Wave InGaAs MOSFET/RTD Wavelet Generators 167

Mikael Egard, Mats Arlelid, Lars Ohlson, Mattias Borg, Erik Lind, and *Lars-Erik Wernersson

Electrical and Information Technology, Lund University, Sweden, Sweden

ThD1-7   10:15 - 10:30

Ultrashort pulse generators using resonant tunneling diodes with improved power performance 169

*Dongpo Wu, Jie Pan, Katsutaro Mizumaki, Masayuki Mori, and Koichi Maezawa

Graduate School of Science and Engineering, University of Toyama, Japan

ThD2

High-Speed Circuits and Devices

ThD2-1 (invited)   11:00 - 11:30

Sub-50nm Indium Phosphide High Electron Mobility Transistor Technology for Terahertz Monolithic Microwave Integrated Circuits and Systems 171

*Stephen Sarkozy, Xiaobing Mei, Wayne Yoshida, Po-Hsin Lin, Ling-Shine Lee, Joe Zhou, Kevin Leong, Vesna Radisic, William Deal, and Richard Lai

Aerospace Systems, Northrop Grumman Corporation, United States of America

ThD2-2   11:30 - 11:45

35 nm mHEMT Technology for THz and ultra low noise applications 173

*Arnulf Leuther, Axel Tessmann, Michael Dammann, Hermann Massler, Michael Schlechtweg, and Oliver Ambacher

Fraunhofer IAF, Germany

ThD2-3   11:45 - 12:00

250-290 GHz Amplifier in 75-nm InP HEMT Technology Using Inverted Microstrip Transmission Line 175

*Hiroshi Matsumura, Shoichi Shiba, Masaru Sato, Tsuyoshi Takahashi, Toshihide Suzuki, Yasuhiro Nakasha, and Naoki Hara

Fujitsu Limited, Japan

ThD2-4   12:00 - 12:15

Comparative Study on Frequency Limits of Nanoscale HEMTs with Various Channel Materials 177

*Yutaro Nagai(*1), Shohei Nagai(*1), Jun Sato(*1), Shinsuke Hara(*1), Hiroki I. Fujishiro(*1), Akira Endoh(*2), Issei Watanabe(*2), and Akifumi Kasamatsu(*2)

*(*1)Tokyo University of Science, Japan and (*2)National institute of Information and Communication Technology, Japan*

ThD2-5   12:15 - 12:30

InP/InGaAs DHBT Technology Using SiN/SiO$_2$ Sidewall Spacers 179

*Norihide Kashio, Kenji Kurishima, Minoru Ida, and Hideaki Matsuzaki

NTT Photonics Laboratories, NTT Corporation, Japan

MoPLN-3 (Plenary)
10:30 - 11:10

Advances in III-V Semiconductor Photonics: Nanostructures and Integrated Chips

Osamu Wada

Center for Collaborative Research and Technology Development (CREATE)
Kobe University
1-1 Rokkodai, Nada, Kobe 657-8501, Japan
owada@kobe-u.ac.jp

Abstract—**This paper reviews the progress of III-V semiconductor technology for the application to photonic telecommunications and signal processing systems by focusing nanostructure materials and devices as well as monolithic/hybrid integration technologies for improving performance and functions. Also the role of combined use of nanostructure devices and integration technology is discussed for the development of future energy efficient photonic systems.**

Keywords—*III-V semiconductors, photonic devices, all-optical switches, nanostructures, quantum dots, optoelectronic integrated circuits (OEICs), photonic integrated ciruits (PICs), Si-photonics*

I. INTRODUCTION

III-V compound semiconductors are indispensable for real world photonic systems for optical communications, interconnects, sensing, etc. In parallel with the invention of low loss optical fibers III-V semiconductor light sources and photodetectors have been developed for the application first to optical telecommunication systems. Particularly GaAs- and InP-based material systems [1] have played a crucial role in realizing high-speed electronics and high-reliability photonic devices. Subsequent development of transmission systems up to the throughput of tens of Tb/s has been achieved by continuous improvement in performance, functionality, reliability and cost using these material systems. The bandwidth increase achieved by such development shows an increase rate exceeding. 150%/yr.

Many inventions and developments have contributed to this innovation. In this paper we focus on two aspects. One is novel material physics and technology based on nanostructure materials including quantum wells, wires and dots, and another is the integration technology to combine different types of devices in hybrid/monolithic form. On the other hand, the demand of lowering the energy consumption is a crucial requirement in every hierarchy of the network systems down to constituent devices.[2] This may not be achieved solely by incremental improvement of conventional materials and device technologies. We review in the following the progresses made in those focus areas and discuss how these will contribute to the future photonics technology.

II. NANOSTRUCTURE PHOTONIC DEVICES

Optical telecommunication systems have been extensively developed since its first installation in early 1980s using GaAs- and InP-based lattice-matched heterostructure devices. Development of high-quality, high-throughput growth techniques (e.g.,MOCVD and MBE) has provided highly controlled nanometer-size heterostructures with sufficient reliability. This has lead to the practical use of quantum nanostructures in 1990s, and many bulk lasers have been replaced by quantum well (QW) lasers due to their high performance, e.g., optical gain characteristics. Development of novel materials systems including strained layer heterostructure materials and nitride-based materials systems has widened the material choice for broadening the operation wavelength range and improving the device performance.

QWs have been applied to design variety of novel photonic functions not limited to sources and detectors, but modulators and optical switches. For example, superlattice and quantum well structures have been applied to quantum cascade lasers, quantum confined Stark effect (QCSE) modulators, superlattice avalanche photodiodes, spin-polarization all-optical switches, and intersubband (ISBT) transition all-optical switches.[2] Recent ultrafast ISBT switches have been shown to work at a bit rate of 160 Gb/s. [3]

Although quantum dot (QD) materials have been proposed for temperature-stable lasers already in 1982[4], their development has been retarded because most of the research efforts have been injected into QWs. QD materials have been applied to lasers and variety of photonic devices until now. Recent QD lasers have shown temperature-stable performance up to 220°C.[5] Polarization-insensitive semiconductor amplifiers (SOAs) made by using shape- and strain-controlled QDs have shown 40 Gb/s operation.[6] All-optical switches have been demonstrated by using QDs incorporated into vertical cavity structures, and energy reduction capability below 1 pJ/bit is being deployed.[7]

Recent advances of nanostructure fabrication have enabled the various nanometer-size laser structures such as vertical cavity surface emitting lasers (VCSELs)[8] and metal cavity lasers.[9] Lateral current injection 2D photonic crystal laser has demonstrated encouraging characteristics including 390 μA threshold current and 10 Gb/s modulation speed. [10]

III. INTEGRATION TECHNOLOGY

The idea of integration was proposed for improving performance, widening functions, increasing reliability and reducing cost, in the very early stage of the photonic device research.[11, 12] III-V semiconductor based integration has first been challenged in the form of optoelectronic integrated circuits (OEICs).[13, 14] The most successful examples of OEICs were photoreceiver chips for high-speed transmission and optical interconnection systems.[15, 16] The first practical photonic integrated circuits (PICs) were laser/absorption modulator integrated high-speed light sources.[17, 18] In parallel with the development of monolithic integration techniques, which sometimes suffer from the fabrication yield problem, hybrid integration technique has been used for fulfilling the concurrent requirement for integrating different devices. Recent development of transmitters and receivers has lead to the success of advanced transmission systems using various schemes such as coherent detection and multilevel modulation. Recent coherent (PM-QPSK) PIC chip has 500 Gb/s throughput.[19] Optical interconnect application has become an extremely important area of III-V photonics. VCSELs and arrayed photodiodes are often used for transmitter/receiver integration. Recent interconnect experiment has demonstrated the throughput over 3 Tb/s with the energy efficiency of the order of pJ/bit/ch. [20]

Heteroepitaxial growth techniques particularly for III-V on Si substrates have been extensively investigated, but the interface dislocation/defect density has not been reduced enough for laser application. Improved interface stabilization techniques have been invented in recent researches of GaPN-buffered GaAs epitaxy on Si [21] and InAs QD growth on Si [22], and basic devices and circuits have been demonstrated on Si substrates. Heterogeneous materials realized by either heteroepitaxy or wafer fusion technique provides unique device functions which cannot be realized by only III-V materials. For example, InGaAs/Si avalanche photodiodes with high gain-bandwidth product and Si waveguide integrated III-V lasers have been demonstrated using wafer bonding technique. [23]

IV. FUTURE PROSPECT

In the future photonic transmission and signal processing systems, III-V photonic devices and circuits must fulfill performance requirements at sufficiently low energy consumption. This cannot be achieved by incremental developments of existing material and device technologies. Nanostructure devices are crucial for energy reduction because of their small volume as well as quantum-effect induced stable operation characteristics, e.g., high temperature stability. Such nanostructure devices can be integrated with variety of electronic and photonic devices on common substrates to realize energy efficient, highly functional circuits for future systems. Recent demonstrations of lasers with nano-dimension active volumes and ultrafast photonic switches based on QDs are encouraging. Nano-device integrated chips will lead to the realization of photonic devices and systems for the low emission era.

REFERENCES

[1] O. Wada and H. Hasegawa ed., InP-Based Materials and Devices -Physics and Technology-, John Wiley & Sons, New York, 1999.

[2] O. Wada, "Progresses in Semiconductor-Based Ultrafast Signal Processing Devices for Green Photonics," IEEE J. Selected Topics Quantum Electron., vol. 17, 2011, pp.309-319

[3] R. Akimoto, S. Gozu, T. Mozume and H. Ishikawa, "Monolithically integrated all-optical gate switch using intersubband transition in InGaAs/AlAsSb coupled double quantum wells," Optics Express, vol. 19, 2011, pp. 13386-94

[4] Y. Arakawa and H. Sakaki, "Multidimensional quantum well laser and temperature dependence of the threshold current," Appl. Phys. Lett., vol. 40, 1982, pp. 939-941.

[5] T. Kageyama, K. Takada, K. Nishi, M. Yamaguchi, R. Mochida, Y. Maeda, H. Kondo, K. Takemasa, Y. Tanaka, T. Yamamoto, M. Sugawara and Y. Arakawa, "Long-wavelength quantum dot FP and DFB lasers for high temperature applications," SPIE Photonics West 2012, paper 8277-11.

[6] N. Yasuoka, H. Ebe, K. Kawaguchi,. M. Ekawa, S. Sekiguchi, O. Wada, M. Sugawara, and Y. Arakawa, "Polarization-insensitive Quantum Dot Semiconductor Optical Amplifiers Using Strain-controlled Columnar Quantum Dots," IEEE/OSA J. Lightwave Technology, vol. 30, 2012, pp. 68- 75.

[7] C. Y. Jin, O. Kojima, T. Kita, O. Wada, and M. Hopkinson, "Observation of Phase Shifts in a Vertical Cavity Quantum Dot Switch," Appl. Phys. Lett., vol. 98, 2011, pp. 231101-1~3.

[8] F. Koyama, "VCSELs: Their 30 Years History and New Challenges," Proc. SPIE, vol. 7135, pp. 0J-1~10.

[9] S. L. Chuang, C-. Y. Lu, A. Matsudaira, "Metal-cavity nanolasers," Tech. Dig. OFC/NF-OEC 2012, OW1G.2

[10] S. Matsuo, K. Takeda, T. Sato, M. Notomi, A. Shinya, K. Nozaki, "Electrically-pumped photonic crystal lasers for optical communications," Tech. Dig. ECOC 2012, paper Th.1.E.2.

[11] S. E. Miller, "Integrated Optics: An introduction," Bell System Technical J., vol. 48, 1969, 2059-2069.

[12] S. Somekh and A. Yariv, "Fiber-Optics communications," Proc. Intern. Telephony Conf. Los Angeles, CA, 1972, pp. 407-418.

[13] A. Yariv, "The beginning of integrated optoelectronic circuits," IEEE Trans. Electron Devices, vol. ED-31, 1984, pp. 1656-1661.

[14] O. Wada, T. Sakurai, and T. Nakagami, "Recent Progress in Optoelectronic Integrated Circuits (OEIC's)," IEEE J. Quantum Electronics, vol. QE-22, 1986, pp. 805-821.

[15] T. Iwama, T. Horimatsu, Y. Oikawa, K. Yamaguchi, M. Sasaki, T. Touge, M. Makiuchi, H. Hamaguchi, and O. Wada, "4 x 4 OEIC Switch Module Using GaAs Substrate," J. Lightwave Technology, vol. 6, 1988, pp. 772-778.

[16] J. D. Crow, "Optoelelctronic integrated circuits for high-speed computer networks," Tech. Dig. OFC'89, 1989, paper WJ3.

[17] H. Soda, H. Furutsu, K. Sato, N. Okazaki, Y. Yamazaki, H. Nishimoto and H. Ishikawa, "High-power and high-speed semi-inslating BH structure monolithic electro-absorption modulator/DFB laser light source," Electron. Lett., vol. 26. 1990, pp. 9-10

[18] T. L. Koch and U. Koren, "Semiconductor photonic integrated circuits," IEEE J. Quantum Electron., vol. 27, 1991, pp. 641-653.

[19] L. Coldren, M. Lu, L. Johansson, M. Rodwell, and J. Parker, "Single-Chip Integrated Transmitters and Receivers," Tech. Dig. ECOC2012, paper Tu.1.G.1.

[20] C. T. Schow, "Power-efficient transceivers for high-bandwidth, short-reach interconnects," Tech. Dig. OFC/NF-OEC 2012, 2012, paper OTh1E4.

[21] H. Yonezu, Y. Furukawa and A. Wakahara, "Convergence of III-V compound and Si technologies: photonic-electronic convergence," Conf. Proc. IPRM'2010, 2010, Kagawa, Japan, paper PL1.

[22] Q. Jiang, A. Lee, M. Tang, A. Seeds and H. Liuy, "Si-based long wavelength III-V quantum dot lasers," Tech. Dig. Compound Semiconductor Week (IPRM/ISCS) 2012, paper Tu-1D.4.

[23] M. J. R. Heck, H-. W. Chen, A. W. Fang, B. R. Koch, D. Liang, H. Park, M. N. Sysak and J. E. Bowers, "Hybrid Silicon Photonics for Optical Interconnects," IEEE J. Selected Topics in Quantum Electron., vol. 17, , 2011, pp. 333-346.

MoPLN-4 (Plenary)
11:10 - 11:50

III-V MOS Technology: From Planar to 3D and 4D

Peide D. Ye

School of Electrical and Computer Engineering and Birck Nanotechnology Center, Purdue University
West Lafayette, Indiana 47907, U.S.A.
E-mail: yep@purdue.edu

Abstract— Recently, III-V MOSFETs with high drain currents (I_{ds}>1mA/μm) and high transconductances (g_m>1mS/μm) have been achieved at sub-micron channel lengths (L_{ch}), thanks to the better understanding and significant improvement in high-k/III-V interfaces. However, to realize a III-V FET at beyond 14nm technology node, one major challenge is how to effectively control the short channel effects (SCE). Due to the higher permittivity and lower bandgap of the channel materials, III-V MOSFETs are more susceptible to SCE than its Si counterpart. Therefore, the introduction of 3-dimensonal (3D) structures to the fabrication of deep sub-100nm III-V FETs is necessary. In this talk, we will review the materials and device aspects of III-V MOS technology developed very recently. We will also report some of new progress by demonstration of 20-80 nm channel length III-V gate-all-around nanowire MOSFETs with EOT=1.2nm and lowest SS=63 mV/dec. The total drain current per pitch can be further enhanced by introducing 4D structures.

Keywords—InGaAs, gate-all-around (GAA), nanowire, MOSFET

I. INTRODUCTION

Despite intense research in the 1980s, the lack of high-quality native oxides on III-V semiconductors has stymied every effort to develop practical III-V MOSFETs. However, this situation has changed recently. The requirement for non-native, high-*k* dielectric oxides for Si CMOSFETs began in the late 1990s. [1] In particular, the development of ALD high-*k* dielectrics for Si MOSFETs, has since flourished. In 2007, Intel successfully integrated a Hf-based ALD high-*k* dielectric and metal gate process into its 45 nm node technology. This achievement has been heralded as one of the biggest technical leaps in Si CMOS development, after the introduction of poly-silicon gate in the 1960s. [2] The success of ALD high-*k* dielectrics on Si has raised the prospect of extending the approach to III-V semiconductors since the first studies in 2003. [3] Meanwhile, III-V MOSFETs have received renewed attention for continued scaling of CMOS technology beyond the 14 nm node. III-V MOSFETs are now intensively investigated by many academic groups as well as industrial giants. [4] In this talk, we will address the issues and progress on III-V MOS technology from materials, structures and device aspects.

II. MATERIALS

High-k/III-V interface is the key for III-V MOS technology development. Lacking a suitable gate insulator, practical III-V MOSFETs remain all but a dream for more than four decades. Historically, GaAs is the most studied III-V material for MOS application. Oxidation of GaAs generally produces a mixture of Ga_2O_3, As_2O_3, As_2O_5 and elemental As. The mixture tends to have poor thermal and chemical stability, incompatibility with standard IC processes, and even pins the Fermi level of GaAs. Due to the limitation of volume manufacture using *in-situ* MBE approach [5,6], ALD process on III-V is intensively studied and now well accepted by the device community in particular, on In-rich InGaAs. Interface trap density can be reliably reduced down to the middle of 10^{11}/cm^2-eV to 10^{12}/cm^2.

Here, we would like to use the new work on atomic layer epitaxial oxide on GaAs (111)A as one example showing that high-k/III-V still has lots of room to improve even at the most difficult GaAs surface. We present a new way of passivating the interface trap states by growing an epitaxial layer of high-k dielectric oxide, $La_{2-x}Y_xO_3$, on GaAs(111)A. High-quality epitaxial $La_{2-x}Y_xO_3$ thin films are achieved by an *ex-situ* ALD process, and GaAs MOS capacitors made from this epitaxial structure show very good interface quality with small frequency dispersion and low interface trap densities. In particular, the La_2O_3/GaAs interface, which has a lattice mismatch of only 0.04%, shows very low D_{it} in the GaAs bandgap of below 3×10^{11}/cm^{2}eV near the conduction band edge. The La_2O_3/GaAs capacitors also show the lowest frequency dispersion of any dielectric on GaAs. This is the first achievement of such low trap densities for oxides on GaAs.[6] Both GaAs NMOSFETs and PMOSFETs with good on-state and off-state performance are demonstrated. The work is in close collaborations with Prof. Roy Gordon's group at Harvard.

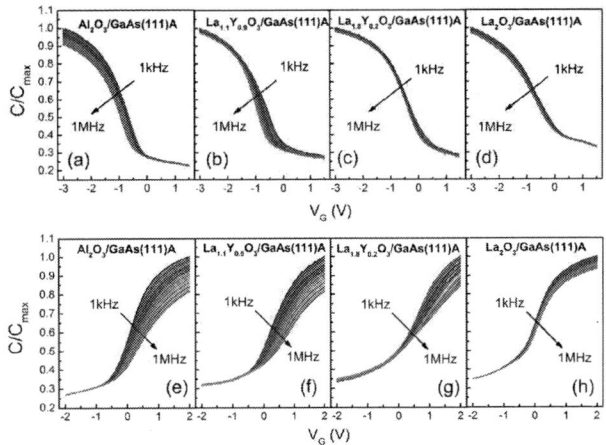

Figure 1. *C-V* characteristics of p-type and n-type GaAs MOS capacitors with stacks of (a, e) Ni/8nm Al_2O_3/GaAs(111)A, (b, f) Ni/6.5 nm Al_2O_3/7.5 nm $La_{1.1}Y_{0.9}O_3$/GaAs(111)A, (c, g) Ni/6.5 nm Al_2O_3/7.5 nm $La_{1.8}Y_{0.2}O_3$/GaAs(111)A, and (d, h) Ni/6.5 nm Al_2O_3/9 nm La_2O_3/GaAs(111)A, respectively. The figure is from Ref. [7].

III. STRUCTURES

Recently, III-V MOSFETs with high drain currents (I_{ds}>1mA/μm) and high transconductances (g_m>1mS/μm) have been achieved at sub-micron channel lengths, [8-11] thanks to the better understanding and significant improvement in high-k/III-V interfaces. However, to realize a III-V FET at beyond 14nm technology node, one major challenge is how to effectively control the SCE. Due to the higher permittivity and lower bandgap of the channel materials, III-V MOSFETs are more susceptible to SCE than its Si counterpart. Therefore, the introduction of 3D structures to the fabrication of deep sub-100nm III-V FETs is necessary. Fig.2 shows the evolution of III-V MOSFET development from planar devices to 3D/4D structures including III-V FinFET, GAA FETs and vertically stacked GAA FETs.

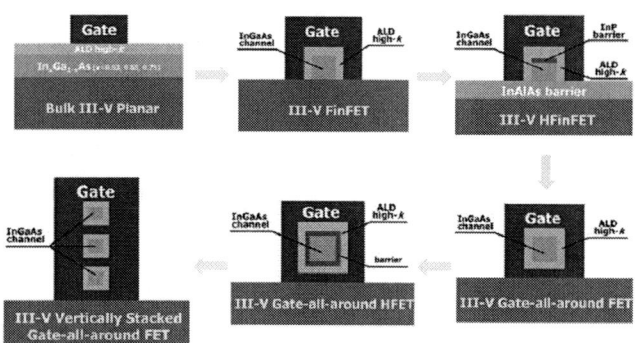

Figure 2. Evolution of III-V MOSFETs in the past years from planar MOSFETs to 3D/4D transistors.

IV. DEVICES

Here, we use some of the results from InGaAs GAA nanowire MOSFETs as one of examples showing the progress in device aspects. IEDM 2012, we experimentally demonstrate InGaAs GAA nanowire MOSFETs with an EOT down to 1.2nm by the successful integration of complex oxide dielectric LaAlO$_3$ (k~16). The reduction of EOT has allowed the demonstration of the first 20nm L_{ch} InGaAs MOSFETs with g_m of 1.65mS/μm at V_{ds}=0.5V and negligible SCE. A systematic scaling metrics study with L_{ch} between 20-80nm and nanowire size-dependent transport study with nanowire width (W_{NW}) between 20-35nm has also been carried out for three different gate stacks, demonstrating near-ideal SS of 63mV/dec and DIBL of 7mV/V as shown in Figure 3. [12] It is shown that the integration of 4nm LaAlO$_3$ with ultra-thin 0.5nm Al$_2$O$_3$ interfacial layer allow reduction of EOT=1.2nm with optimized D$_{it}$, offering excellent scalability, near-ballistic transport, and high g_m at low supply voltage.

ACKNOWLEDGMENT

The author would like to thank the support in the past years from National Science Foundation, SRC FRCP MSD Center, SRC GRC and CSR programs, NIST, and Air Force Office for Scientific Research. The author also would like to thank Roy G. Gordon, R.M. Wallace, M. S. Lundstrom, D. A. Antoniadis, J. A. del Alamo for their direct contributions and valuable discussions. In particular, the author would like to thank his postdoc and Ph.D. students Yi Xuan, Yanqing Wu, Min Xu, Jiangjiang Gu, Ling Dong and many others for their excellent work.

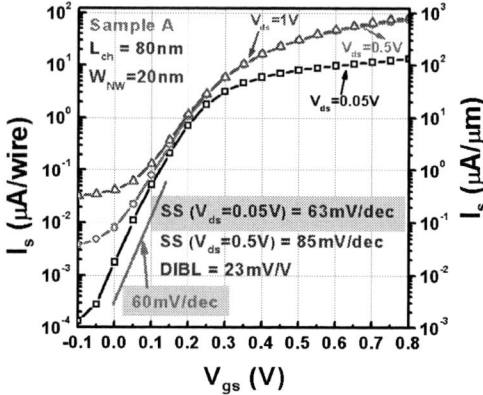

Figure 3. Transfer characteristics of L$_{ch}$=80nm InGaAs GAA MOSFET (Sample A, W$_{NW}$=20nm) with SS=63mV/dec at V$_{ds}$=0.05V. Three measured devices show SS=63mV/dec. Ref. [12].

REFERENCES

[1] G.D.Wilk, R.M.Wallace, and J.M.Anthony, *J. Appl. Phys.* **89**, 5243 (2001)
[2] New York Times, January 27, 2007.
[3] P.D.Ye, G.D. Wilk, J. Kwo, B. Yang, H.-J.L. Gossmann, M. Frei, S.N.G. Chu, J.P. Mannaerts, M. Sergent, M. Hong, K.K. Ng, J. Bude, *IEEE Electron Device Lett.* **24**, 209 (2003).
[4] M. Radosavljevic, B. Chu-Kung, S. Corcoran, G. Dewey, M. K. Hudait, J. M. Fastenau, J. Kavalieros, W. K. Liu, D. Lubyshev M. Metz, K. Millard, N. Mukherjee, W. Rachmady, U. Shah, and R. Chau, *IEEE International Electron Devices Meeting*, 2009, pp. 809-812.
[5] M. Hong, J. Kwo, A. R. Kortan, J. P. Mannaerts, A. M. Sergent, *Science*, vol. 283, 1897-1900, 1999.
[6] R. J. W. Hill, D. A. J. Moran, L. Xu, Z. Haiping, D. Macintyre, S. Thoms. A. Asenov, P. Zurcher, K. Rajagopalan, J. Abrokwah, R. Droopad, M. Passlack and L.G. Thayne, *IEEE Electron Device Letters*, vol. 28, pp. 1080-1082, 2007.
[7] X. W. Wang, L. Dong, J. Y. Zhang, Y. Q. Liu, P. D. Ye, and R. G. Gordon, *Nano Letters*, DOI: 10.1021/nl303669w, 2012.
[8]Y. Xuan, Y. Q. Wu and P. D. Ye, *IEEE Electron Device Letters*, vol. 29, pp. 294-296, 2008
[9] Y. Yonai, T. Kanazawa, S. Ikeda and Y. Miyamoto, *IEEE International Electron Devices Meeting*, 2011, pp. 307-310.
[10] M. Egard, L. Ohlsson, B.M. Borg, F. Lenrick, R. Wallenberg, L.-E. Wernersson, and E. Lind, *IEEE International Electron Devices Meeting*, 2011, pp. 13.2.1-13.2.3.
[11] D.-H. Kim, T.-W. Kim, R.J.W. Hill, C.D. Young, C.Y. Kang, C. Hobbs, P. Kirsch, J.A. del Alamo, and R. Jammy, IEEE Electron Device Letters, vol.34,no.2, 196-198, 2013.
[12] J. J. Gu, X. W. Wang, H. Wu, J. Shao, A. T. Neal, M. J. Manfra, R. G. Gordon, and P. D. Ye, in 2012 IEEE International Electron Devices Meeting (IEDM), 2012,633.

MoC3-1 (Invited)
14:00 - 14:30

Reshaping the optical properties of quantum dots via strain and electric fields

Armando Rastelli[1,2], Rinaldo Trotta[1,2], Eugenio Zallo[2], Paola Atkinson[2], Oliver G. Schmidt[2]

[1]Institute of Semiconductor and Solid State Physics, Johannes Kepler University of Linz, Austria
[2]Insitute for Integrative Nanosciences, IFW Dresden, Germany
armando.rastelli@jku.at

Abstract— **We introduce a new class of quantum dot-based devices, in which the semiconductor structures are integrated on top of piezoelectric actuators. This combination allows us on one hand to study in detail the effects produced by variable strains (up to about 0.2%) on the excitonic emission of single quantum dots and on the other to manipulate their electronic- and optical properties to achieve specific requirements for their use in quantum optics experiments and possibly future devices for quantum communication.**

Keywords—quantum dots, strain-tunable devices

I. INTRODUCTION

Optically active semiconductor quantum dots (QDs) can be made as nanoinclusions of a low energy bandgap material in a matrix with larger energy bandgap. One of the simplest ways to obtain QDs with excellent structural, electronic, and optical properties is represented by self-assembly of 3D nanoislands during lattice-mismatched heteroepitaxial growth. In this, so-called Stranski-Krastanow (SK) growth mode, elastic stress is one of the main driving forces leading to the formation and evolution of QDs. The most prominent example is represented by InGaAs QDs in GaAs matrix. A large number of experiments have demonstrated that these QDs are excellent quantum emitters which can be used as sources of triggered single photons, indistinguishable photons and polarization entangled photon pairs.

One open problem is that QDs are affected by dot-to-dot fluctuations. This makes it difficult to obtain QDs with electronic and optical properties which meet (sometimes very stringent) requirements for their use in advanced quantum optics experiments, especially involving independent sources. Post-growth techniques are therefore required to fine-tune the optical properties of QDs. Vertical electric fields (applied along the growth direction) represent the most powerful "tuning knob" to date. By using the so called "giant Stark effect" (GSE) both the emission energy and the excitonic fine-structure splitting (FSS) can be widely tuned [1, 2]. The latter is important for generation of entangled photon pairs. However electric fields alone allow the FSS of only selected QDs to be reduced to low enough values required for entanglement generation. Electric control can also be used to electrically pump single QDs. This feature, which takes profit of mature semiconductor technology, is one of the major advantages of epitaxial QDs compared to other solid state emitters. On the other hand the structure designs for electrical pumping and for exploiting the GSE are not compatible, making it clear that additional and independent "tuning knobs" will be indispensable to satisfy the increasing needs imposed by advanced quantum optics experiments, especially involving more than a single QD.

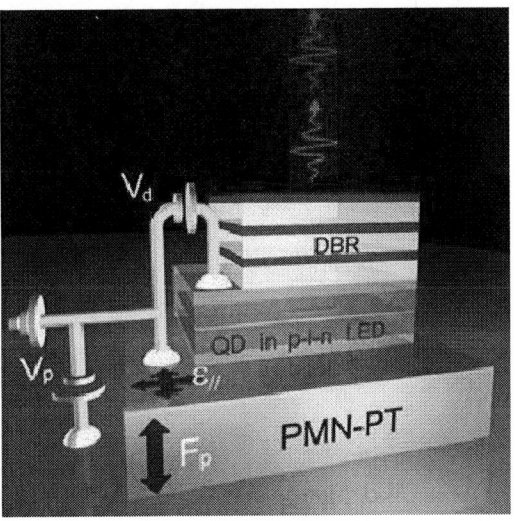

A diode membrane with embedded quantum dots integrated on top of a lead magnesium niobate –lead titanate (PMN-PT) piezoelectric substrate. The devices features electric and elastic control.

II. RESULTS

In this contribution we introduce a new class of QD-based devices, in which the semiconductor structures are integrated on top of piezoelectric actuators. This combination allows us on one hand to study in detail the effects produced by variable strains (up to about 0.2%) on the excitonic emission of single QDs and on the other to add a powerful "tuning knob" to QDs. We first discuss the effects of nominal biaxial strain on the emission of single QDs embedded in optical microcavities [3]. Afterwards we show that strain does not only affect the emission energy of QDs but also the relative binding energies of excitonic species confined in QDs [4] and the FSS of neutral excitons[5].

Finally we show integration of diode membranes on the piezoelectric actuators. This allows us to combine electric control and piezoelectric-induced strain on the same QD (see Figure), to create powerful tunable devices. They allow

978-1-4673-6130-9/13 $31.00 © 2013 IEEE

erasing the FSS of any QD (using the GSE in reverse bias combined with strain) [6] or producing a tunable and frequency-stabilized QD resonant-cavity light-emitting-diode (LED) [7].

REFERENCES

[1] R.B. Patel et al. Nature Photon. 4, 632 (2010)

[2] A.J. Bennet et al Nature Phys. Nature Physics 6, 947 (2010)

[3] T. Zander, A. Herklotz, S. Kiravittaya, M. Benyoucef, F. Ding, P. Atkinson, S. Kumar, J. D. Plumhof, K. Dörr, AR, O. G. Schmidt, Optics Express 17, 22452 (2009)

[4] F. Ding et al Phys. Rev. Lett. 104, 067405 (2010)

[5] J. D. Plumhof, V. Krapek, F. Ding, K. D. Jöns, R. Hafenbrak, P. Klenovsky, A. Herklotz, K. Dörr, P. Michler, AR, O. G. Schmidt, Phys. Rev. B (R) 83, 121302 (2011)

[6] R. Trotta, E. Zallo, C. Ortix, P. Atkinson, J.D. Plumhof, J. van den Brink, A. Rastelli, O.G. Schmidt, Universal Recovery of the Energy-Level Degeneracy of Bright Excitons in InGaAs Quantum Dots without a Structure Symmetry, Phys. Rev. Lett. 109, 147401 (2012)

[7] R. Trotta, P. Atkinson, J. D. Plumhof, E. Zallo, R.O. Rezaev, S. Kumar, S. Baunack, R. Schröter, A. Rastelli, O. G. Schmidt, Nanomembrane Quantum-Light-Emitting Diodes Integrated onto Piezoelectric Actuators, Adv. Mater, 24, 2668 (2012)

MoC3-2 (Oral)
14:30 - 14:45

Suppression of multi-photon emission in 1.5-μm quantum-dot single-photon source

Toshiyuki Miyazawa[1], Kazuya Takemoto[2], Yoshiki Sakuma[3], Haizhi Song[2], Motomu Takatsu[2], Tsuyoshi Yamamoto[1], Yasuhiko Arakawa[2, 4]

[1] Fujitsu Laboratories Ltd.
10-1, Morinosato-Wakamiya, Atsugi-shi, Kanagawa, 243-0197, Japan
miyazawa.toshi@jp.fujitsu.com
[2] Institute for Nano Quantum Information Electronics, The University of Tokyo
4-6-1, Komaba, Meguro-ku, Tokyo, 153-8505, Japan
[3] National Institute for Materials Science
1-1, Namiki, Tsukuba-shi, Ibaraki, 305-0044, Japan
[4] Institute of Industrial Science, The University of Tokyo
4-6-1, Komaba, Meguro-ku, Tokyo, 153-8505, Japan.

Abstract—Photon-correlation function $g^{(2)}(\tau)$ of a single 'double-capped' InAs/InP quantum dot (QD) was obtained with high accuracy using a superconducting single-photon detector (SSPD). By using resonant excitation to the e_1h_0 state in the single InAs/InP QD, $g^{(2)}(0)$ was decreased to the detection limit. Rabi oscillation of the PL intensity of e_0h_0 state is also observed under the resonant excitation into the e_1h_0 state. These results suggest that the coherent exciton generation into e_1h_0 state is effective to improve the purity of the single-photon emission.

Keywords— quantum dot, single-photon source, quasi-resonant excitation, $g^{(2)}(0)$, Rabi oscillation

I. INTRODUCTION

1.55-μm band single-photon sources are suitable for long-distance quantum-key distribution systems as well as quantum-information networks. We have successfully transmitted secure keys over a 50 km single-mode fiber [1] using a single-photon source consisted of 'double-capped' InAs/InP QD. To achieve secure-key distribution to a further distance, $g^{(2)}(0)$ which reflects multi-photon emission, must be decreased to the level of 10^{-3} [2]. Recently, the $g^{(2)}(0)$ was improved by the resonant excitation of several higher energy states (i.e., quasi-resonant excitation [3, 4]), as reported in Ref. 5. However, due to high dark counts and after-pulse events of InGaAs single-photon detecting avalanche-photodiodes, the $g^{(2)}(0)$ remained high value of 0.017 after subtracting the background counts.

In this study, we used SSPDs to evaluate the $g^{(2)}(0)$ of the e_0h_0 state in the InAs/InP QD for various excitation energies. The low noise and after-pulse properties of SSPD allowed us to evaluate the $g^{(2)}(0)$ around the level of 10^{-3}. We also measured the power dependences of the PL intensity to investigate coherent properties of the photoemission from the exciton state.

II. EXPERIMENTS

We used self-assembled InAs QDs grown on InP (100) substrate using 'double-capping' method [6] as a single-photon source. To accumulate emitted photons from an InAs/InP QD into single-mode fiber efficiently, we used an optical structure called single-photon horn structure which includes InAs/InP QDs [5]. Note that all measurements mentioned below were performed at 10 K using a continuous flow helium cryostat.

We measured continuous wave (CW) PL and CW-PLE spectra to clarify the electronic structure of exciton states in the InAs/InP QD. We used a tunable CW laser as an excitation light source for these experiments. Wavelength of the CW laser was set to 1310 nm for CW-PL and changed from 1450 to 1565 nm for CW-PLE. The detection energy of the PLE spectrum was set to the emission energy of the neutral exciton.

After determination of exciton energies, we performed Hanbury Brown and Twiss (HBT) type [7] photon-correlation measurement with two SSPDs to obtain the g(2)(τ) as a function of time delay τ. In the measurement, wavelengths of excitation pulses τ_{exc} were set to 1310, 1519 and 1538 nm. A distributed-feedback laser was used as a light source of 1310-nm pulse to achieve excitation of the continuum states of the QD. In generating of 1519 and 1538 nm-pulses, wavelength tunable fiber-laser was used to realize the quasi-resonant excitation. To improve monochromaticities of these excitation pulses, pulse durations of these lasers were set to be ~30 ps by decreasing the spectral width through a 0.3-nm width band-pass filter. Repetition rates of these lasers were set to be 30 MHz.

We also measured the excitation power dependence of exciton emission intensities using pulsed-PL setup to clarify the coherent exciton generation [8, 9]. For the pulsed-PL measurements, we used above mentioned pulsed-lasers which were operated at wavelengths of 1310, 1519, and 1538 nm.

III. RESULTS AND DISCUSSIONS

Figure 1 shows CW-PL spectrum and CW-PLE spectrum of a single InAs/InP QD. From the existence of fine-structure splitting, we determined the PL peaks at 1578 and 1587 nm as a neutral exciton emission and a negatively charged exciton, respectively [10]. Prominent peaks at 1538 and 1519 nm in the PLE spectrum were identified as the e_1h_0 and the e_1h_1 state, respectively [10], where e_nh_m represents exciton state consisted of *n*-th excited electron, and *m*-th excited hole state, respectively.

978-1-4673-6130-9/13 $31.00 © 2013 IEEE

MoC3-2 (Oral)
14:30 - 14:45

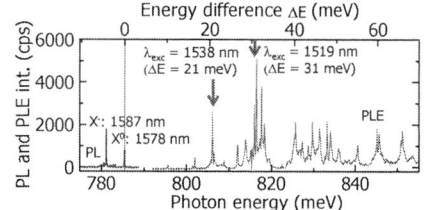

Figure 1. PL and PLE spectra of single InAs/InP QD.

We then measured photon-correlation function $g^{(2)}(\tau)$ using pulsed lasers and two SSPDs with HBT setup. Owing to the low dark and after-pulse counts of SSPD, we demonstrated antibunching behavior of the $g^{(2)}(\tau)$ without subtracting any background counts as shown in Figs. 2 (a)-(b). Note that the vertical axis indicates the log-plot of the $g^{(2)}(\tau)$. The measured $g^{(2)}(0)$ were estimated as 0.16, and 0.008 using 1310 nm-, and 1519 nm-excitation pulses, respectively. On the other hand, by using the 1538 nm-excitation pulse, there was no photon-correlation peak around $\tau = 0$. These experiments indicate that the $g^{(2)}(0)$ was remarkably suppressed by decreasing the excitation energy. The $g^{(2)}(0)$ obtained using 1538 nm-excitation pulse was lower than the detection limit ($g^{(2)}(0) = 0.003$) of the system.

Figures 3(a)-(c) show power dependences of the PL intensities of the ground state exaction. The power dependence of the PL intensities shows saturation behavior using 1310 nm-excitation pulse. On the other hand, the PL intensities show Rabi oscillations using 1519 and 1538 nm excitation pulses as shown in Fig 3 (b) and (c), respectively. These PL intensities oscillate sinusoidally with the square root of the excitation power. The first maximum of the PL intensities are clearly observed using 1519 and 1538 nm-excitation pulses even though damping behavior was observed in these Rabi oscillations. The oscillation remained up to the rotation angle of 3π which reflects a coherent control of the exciton populati on in the e_1h_0 state using 1538 nm-excitation pulse. This coherent control of the e_1h_0 exciton brings the determinisitic preparation of the e_0h_0 exciton and contributes to the suppression of $g^{(2)}(0)$.

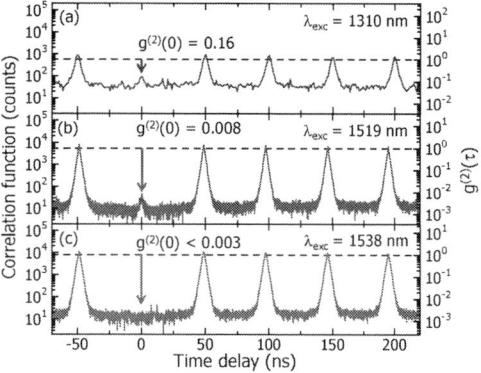

Figure 2. $g^{(2)}(\tau)$ of the photoemission from exction state. (a), (b), and (c) are obtained using pulses of 1310, 1519, and 1538 nm in wavelength, respectively.Dotted lines indicate the estimated $g^{(2)}(\tau) = 1$.

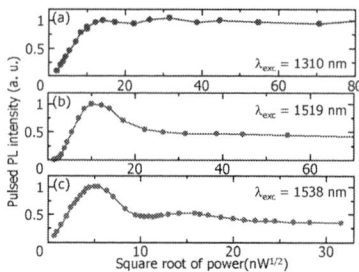

Figure 3. Power dependence of the PL intensity. (a), (b), and (c) are obtained using pulses of λ_{exc} = 1310, 1519, and 1538 nm in wavelength, respectively.

IV. SUMMARY

We successfully demonstrated suppression of the $g^{(2)}(0)$ to the detection limit (< 0.003) of the system with SSPDs, by tuning the excitation energy to the energy of the e_1h_0 state in the single 'double-capped' InAs/InP QD. Clear Rabi oscillation of the pulsed-PL intensity was also observed by resonant excitation into the e_1h_0 state. These results suggest that coherent exciton generation into the e_1h_0 state is effective to reduce the $g^{(2)}(0)$ of the e_0h_0 state photoemission.

Figure Labels: Use 8 point Times New Roman for Figure labels.

ACKNOWLEDGMENTS

This work was supported by the Project for Developing Innovation Systems of the Ministry of Education, Culture, Sports, Science and Technology (MEXT), Japan. We are grateful to Dr. Y. Nambu of NEC Corporation for technical support.

REFERENCES

[1] K. Takemoto, Y. Nambu, T. Miyazawa, K. Wakui, S. Hirose, T. Usuki, M. Takatsu, N. Yokoyama, K. Yoshino, A. Tomita, S. Yorozu, Y. Sakuma, and Y. Arakawa , Appl. Phys. Express **3** (2010) 092802.

[2] E. Waks, C. Santori, and Y. Yamamoto, Phys. Rev. A **66**, 042315 (2002).

[3] A. Malko, M. H. Baier, K. F. Karlsson, E. Pelucchi, D. Y. Oberli, and E. Kapon, Appl. Phys. Lett., **88**, 081905 (2006).

[4] H. Kumano, S. Kimura, M. Endo, H. Sasakura, S. Adachi, S. Muto and I. Suemune, J. Nanoelectron. Optoelectron., **1**, 39 (2006).

[5] K. Takemoto, S. Hirose, M. Takatsu, N. Yokoyama, Y. Sakuma, T. Usuki, T. Miyazawa, and Y. Arakawa, phys. stat. sol. (c) **5**, No. 9, 2699 (2008).

[6] Y. Sakuma, K. Takemoto, S. Hirose, T. Usuki, and N. Yokoyama, physica E, **26**, 81 (2005).

[7] R. Hanbury Brown and R. Q. Twiss, Nature **178**, 1447 (1956).

[8] C. Santori, D. Fattal, J. Vucˇkovic, G. S. Solomon, E. Waks, and Y. Yamamoto, Phys. Rev. B **69**, 205324 (2004).

[9] P. Ester, L. Lackmann, S. M. de Vasconcellos, M. C. Hübner, A. Zrenner, and M. Bichler, Appl. Phys. Lett. **91**, 111110 (2007).

[10] T. Miyazawa, K. Takemoto, T. Nakaoka, T. Saito, S. Hirose, Y. Sakuma, N. Yokoyama, and Y. Arakawa, Phys. Status Solidi c **8**, No. 2, 417 (2011).

MoC3-3 (Invited)
14:45 - 15:15

Wurtzite Gallium Phosphide has a Direct-Band Gap

S. Assali, I. Zardo, S. Plissard, , M. A. Verheijen, J. E. M. Haverkort and E. P. A. M. Bakkers

Department of Applied Physics, Eindhoven University of Technology, 5600 MB Eindhoven, The Netherlands

Gallium Phosphide (GaP) with the normal cubic crystal structure has an indirect band gap, which severely limits the emission efficiency. We report the fabrication of GaP nanowires with pure hexagonal crystal structure and demonstrate the direct nature of the band gap. We observe strong photoluminescence at a wavelength of 594nm with short lifetime, typical for a direct band gap. Furthermore, by incorporation of aluminum or arsenic in the GaP nanowires, the emitted wavelength can be tuned across an important range of the visible light spectrum (555-690nm). This approach of crystal structure engineering enables new pathways for tailoring materials properties enhancing functionality.

Keywords—Gallium Phosphide, Crystal structure, Direct Band Gap

I. INTRODUCTION

Widely used semiconductors like silicon (Si), germanium (Ge) and gallium phosphide (GaP) have an indirect band gap for the normal cubic (diamond or zinc blende) crystal structure, limiting their use for photonic devices. A unique feature of the vapor-liquid-solid (VLS) nanowire growth mechanism is that well-known semiconductors can be grown with different crystal structures. Changing the crystal structure can dramatically modify the opto-electronic properties of a material with a given chemical composition. It has been predicted that Ge and GaP with hexagonal (wurtzite) crystal structure have a direct band gap[1-4]. The calculated direct band gap for wurtzite GaP ranges between 2.10 - 2.18eV[2-4], and therefore is a promising candidate for light emission in the green-yellow region of the visible spectrum. Similarly, the band gap of Aluminum Phosphide (AlP) is predicted to change from indirect to direct when the structure is converted from cubic to hexagonal with gap energy of 2.97eV[3]. Therefore, by forming ternary $Al_xGa_{1-x}P$ and $GaAs_yP_{1-y}$ compounds, the emission wavelength can theoretically be tuned over a wide range. However, the change from indirect to a direct band gap by changing from zinc blende to wurtzite has not yet been demonstrated experimentally for any material at ambient pressure conditions.

II. RESULTS

For this study, GaP nanowires are grown at relatively high temperatures of >700°C and at a low V/III ratio on (111) oriented zinc blende GaP substrates[5,6] using the VLS mechanism and patterned gold islands as catalysts. Radial growth, which leads to tapered nanowires, has been suppressed

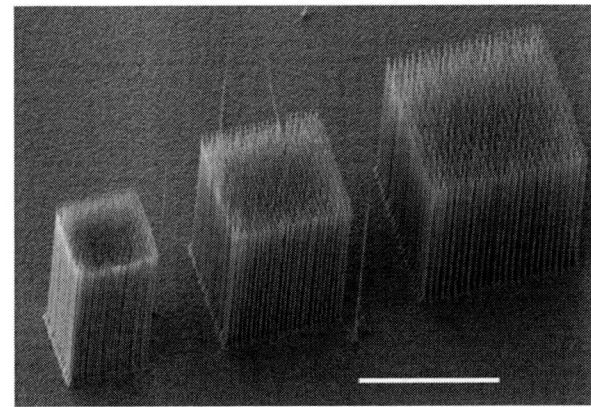

Figure 1. Scanning Electron Microscopy (SEM image of an array of wurtzite Gallium Phosphide nanowires. The position of the wires is predefined by positioning of the catalyst particles by using electron beam lithography. The scale bar corresponds to 10 micrometers.

by using HCl during growth[16]. In order to control the nanowire position two lithography techniques are used; electron beam lithography to fabricate small arrays with varying pitch and diameter, and nanoimprint to pattern large-scale areas with a constant pitch and diameter. The scanning electron microscopy (SEM) image in Fig. 1 shows uniform arrays of 20 μm long non-tapered nanowires with almost 100% yield defined by electron beam lithography. The different fields have a different wire-to-wire spacing.

In order to assess the crystal structure of the nanowires, high-resolution transmission electron microscopy (HRTEM) studies are performed. The TEM image in Fig. 2a and the corresponding Fast-Fourier Transform (Fig. 2b) demonstrate the wurtzite crystal structure of the GaP nanowires. Less than 1 stacking fault per micrometer is typically observed for the tens of wires studied. An example of a whole wire, which has been imaged from top to bottom by TEM shows just a few stacking faults across the whole length of the nanowire. A wider band gap ternary AlGaP shell is grown around the GaP core to suppress possible undesirable surface effects, as will be discussed below. Shells are grown under different conditions compared to axial nanowire growth, and the thickness is controlled by the shell growth time and the composition by the Al/Ga gas input ratio. The core/shell structure is visible from the High Angle Annular Dark Field (HAADF) TEM image in Fig. 2c and the shell composition and thickness for the different

MoC3-3 (Invited)
14:45 - 15:15

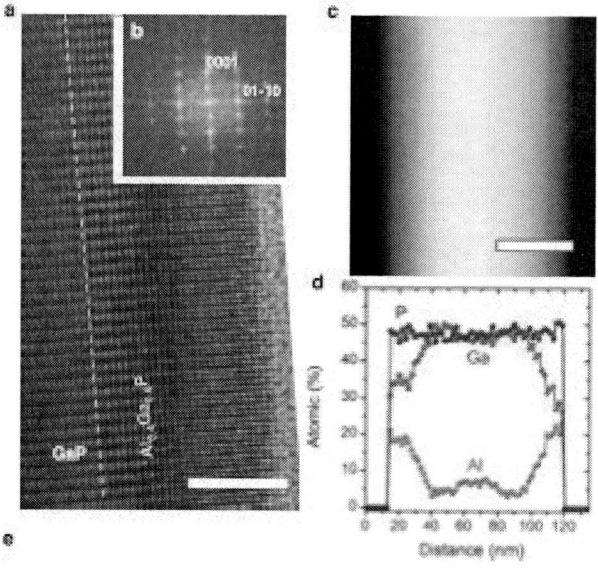

Figure 2. Transmission Electron Microscopy (TEM) of a wurtzite Gallium Phosphide / Alumium gallium phosphide core/shell nanowire. A) high resolution image, B) the inset shows the fast Fourier transform, c) high-angle annular dark field image of the core shell wire, d) energy dipersive X-ray line scans shows the chemical composition of the wire.

samples have been determined from Energy-Dispersive X-ray (EDX) line scans as shown in Fig. 2d. This particular wire has a 10 nm thick $Al_{0.4}Ga_{0.6}P$ shell. Important to mention is that AlGaP is nearly lattice matched to GaP, and as a result no defects are generated in the shell.

To verify the direct nature of the band gap of WZ GaP, the wires are studied with low-temperature micro-photoluminescence (PL). For this e-beam defined nanowire array samples are used with a wire-to-wire spacing of 1.0 micrometer. The emission intensity and radiative lifetime of the WZ nanowires are compared with a ZB (001) bulk GaP sample as a reference.

Spectra are obtained under identical experimental conditions such that we can quantitatively compare the PL intensity of the WZ wires and that of the ZB substrate. The integrated PL emission of the main ZB peak at 2.317 eV is compared to that of the main WZ wire peak at 2.09 eV for different wire diameters. The WZ PL intensity increases with wire diameter, due to increased volume, and levels off at the largest diameters. Important is that the integrated PL emission of WZ wires (100 nm diameter) is 65 times higher than that of

the ZB GaP(001) substrate. This represents a lower limit of the enhancement factor of the WZ PL intensity, since this value is not corrected for the probed materials volume. Note that the wires cover less than 1 % of the surface area, and moreover, the in and out coupling of light from the microscope objective is much better for a planar substrate compared to vertical nanowires. The phonon assisted band-to-band emission of a GaP(111) substrate, observed at 2.28 eV at higher excitation power, is below the detection limit under these measurement conditions. The high emission intensity, the excitonic nature of the emission together with the very good agreement between the experimental and predicted WZ GaP band gap values, strongly suggest the presence of direct band gap transitions in the WZ GaP nanowires.

To further substantiate the direct nature of the band gap, we perform time-resolved PL measurements on WZ $GaP/Al_{0.4}Ga_{0.6}P$ core/shell nanowires as well as on ZB bulk GaP. A long lifetime of 254 ± 3 ns is obtained for the A-line of the ZB bulk GaP, which is in the typical range for an indirect band gap transition. In strong contrast, a lifetime of 0.78 ± 0.01 ns is observed for the WZ core/shell nanowires, which compares favorably to lifetimes reported for direct band gap semiconductor like GaAs. The striking difference in lifetimes marks the transition from indirect to direct band gap material by transforming the crystal structure from ZB to WZ.

ACKNOWLEDGMENT

The authors thank D. Kriegner, G. Bauer, A. Meijerink, A. Belabbes, F. Bechstedt for helpful discussions. Y. Cui for the nanoimprint substrate preparation, A. J. Standing, T. Hoang and T.T.T. Vu for the technical support in the optical setup, and N. Akopian for comments to the manuscript. We thank the Dutch Organization for Scientific Research (NWO-VICI 700.10.441) for funding.

REFERENCES

[1] J.D. Joannopoulos *et al.* Electronic properties of complex crystalline and amorphous phases of Ge and Si. Density of States and Band Structures, *Phys. Rev. B* **7**, 2644-2657 (1973).

[2] Chin-Yu Yeh *et al.* Relationships between the band gaps of the zinc-blende and wurtzite modifications of semiconductors, *Phys. Rev. B* **50**, 2715-2718 (1994).

[3] A. De & Craig E. Pryor. Predicted band structures of III-V semiconductors in the wurtzite phase. *Phys. Rev. B* **81**, 155210 (2010).

[4] A. Belabbes, C. Panse, J. Furthmuller and F. Bechstedt. Electronic bands of III-V semiconductor polytypes and their alignment. *Phys. Rev. B* **86**, 075208 (2012).

[5] R. E. Algra *et al.* Crystal structure transfer in core/shell nanowires. *Nano Lett.* **11**, 1690–1694 (2011).

[6] M. T. Borgstrom *et al.* Synergetic nanowire growth. *Nature Nano.* **2**, 541-544 (2007).

978-1-4673-6130-9/13 $31.00 © 2013 IEEE

MoC3-4 (Oral)
15:15 - 15:30

Self-aligned quantum-dot growth
for single-photon sources

U. W. Pohl, A. Strittmatter, J.-H. Schulze, D. Quandt, T. D. Germann,
W. Unrau, T. Heindel, O. Hitzemann, D. Bimberg, S. Reitzenstein

Technische Universität Berlin
Institut für Festkörperphysik, Hardenbergstrasse 36,
10623 Berlin, Germany
pohl@physik.tu-berlin.de

Abstract—The buried oxide current-aperture in a *pin* diode-structure is used to create a strain field for the self-aligned nucleation of site-controlled single quantum dots. A single-photon source fabricated applying this approach shows spectrally very narrow emission lines (FWHM ≤ 25 μeV) and spectrally pure single-photon emission with a second-order autocorrelation $g^{(2)}(0) = 0.05$.

Keywords—Site-controlled quantum dots; single-photon sources

I. INTRODUCTION

Site-controlled growth of quantum dots (QDs) recently received much advertence for single-photon applications. QDs grown self-organized in the Stranski-Krastanow mode show very narrow emission linewidth and can be easily integrated with established semiconductor technology. A single-photon source based on a single active QD requires a spatial alignment of the QD with a surrounding cavity for achieving an efficient light extraction. Various approaches have been developed for controlling the site of the QD; they basically induce QD nucleation by a local variation of either the surface morphology or the surface strain. A drawback of commonly applied methods is the close proximity of the nanostructured surface to the QD nucleation site; defects created by patterning lead to line broadening due to spectral diffusion.

II. STRAIN-INDUCED CONTROL OF QUANTUM-DOT NUCLEATION

In this study we employ the long-range modulation of surface strain induced by a large buried stressor [1], allowing to embed the active QD in an as-grown low-defect environment. A two-step procedure is applied for metalorganic vapor-phase epitaxy of site-controlled QDs implemented in a pin diode structure with a current-confining oxide aperture. To define a stressor for the modulation of the surface strain (and simultaneously the current aperture), first a sandwich of $Al_{0.9}Ga_{0.1}As/AlAs/Al_{0.9}Ga_{0.1}As$ layers with 120 nm total thickness is grown and buried beneath 50 nm of GaAs. Then, circular mesas are etched using conventional photolithography, and the AlAs layer of the sandwich is partially oxidized to AlO_x via the mesa sidewalls, leaving a non-oxidized center. The selective AlAs oxidation is accompanied by a volume contraction. As a result, the GaAs layer on top of the sandwich is compressively strained above the oxidized part and tensely

strained in the center [1]. In a second epitaxy step another 50 nm of GaAs is grown on the topmost strained GaAs layer of the mesa, followed by a thin InGaAs layer for QD formation and a GaAs cap layer.

QDs tend to nucleate at locations of maximum tensile lateral strain of the GaAs surface. A systematic investigation of the dependence of QD nucleation site on geometrical quantities like GaAs thickness above the aperture and aperture diameter shows that such maximum actually occurs above the *edge* of the aperture. Fig. 1 shows the alignment of QDs above the edge of a large aperture with quadratic shape. For smaller apertures the locations of maximum lateral strain merge to a single maximum, yielding the ability to nucleate only single quantum dots above the center of the aperture.

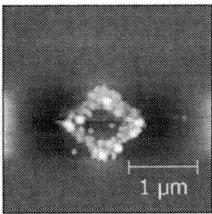

Figure 1. Atomic force micrograph of uncapped $In_{0.65}Ga_{0.35}As/GaAs$ quantum dots nucleated above a large quadratic aperture.

III. QUANTUM-DOT EMISSION

In device structures fabricated for electrical injection, an *n*-doped AlGaAs/GaAs bottom distributed-Bragg-reflector (DBR) and a *p*-type top contact layer are added to the structure described above as illustrated in Fig. 2a. To obtain apertures with a diameter suitable for the nucleation of only single QDs, arrays of mesas with diameters varied from 22.0 μm to 24.8 μm in 200 nm steps are fabricated and all subjected to the same oxidation procedure. While the smallest mesas yield closed apertures without positioned QD nucleation, site-controlled nucleation of single QDs was found on medium-sized mesas. Since the AlO_x current aperture simultaneously forms the stressor for defining the preferential location for QD nucleation, the described fabrication process leads to a self-aligned site control for optimum current injection. Fig. 2b shows the electroluminescence of such QD at 1 μA injection current imaged at low temperature.

978-1-4673-6130-9/13 $31.00 © 2013 IEEE

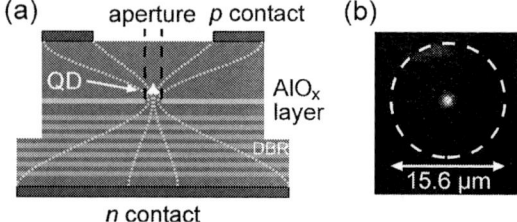

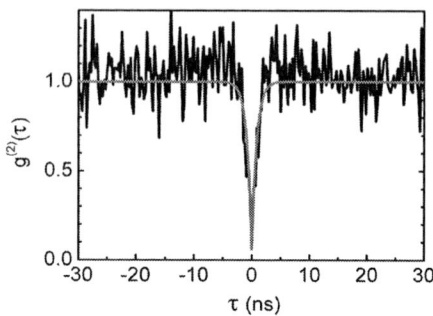

Figure 2. (a) Schematic of the device structure indicating the location of the quantum dot above the AlO$_x$ oxide aperture. Dotted lines indicate the current path. (b) CCD image of the device electroluminescence. The dashed circle marks the inner diameter of the circular p contact.

The emission spectrum shows several lines with narrow linewidths (FWHM) below 50 µeV; lowest observed values are limited by the resolution of the setup (25 µeV). Polarization-resolved measurements reported in [2] show that the lines originate from excitonic few-particle complexes of the same QD as indicated by the labels in Fig. 3. The lines originating from the exciton (X) and the biexciton (XX) yield an anti-binding of the latter with –0.72 meV binding energy and a fine-structure splitting of 84 µeV. If the injection current is reduced, eventually only a single emission line from the exciton is observed.

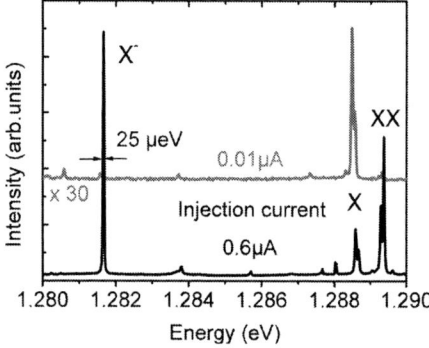

Figure 3. Electroluminescence spectrum of the device depicted in Fig. 2a.

Figure 4. Measured (black line) and fitted (grey line) photon-autocorrelation function of the negatively charged trion X$^-$.

IV. SINGLE-PHOTON SOURCE

The negatively charged trion X$^-$ does not have a dark-state configuration like the exciton and is therefore particularly interesting for an efficient single-photon source. The intense emission line assigned to X$^-$ in Fig. 3 is thus selected for a photon-correlation measurement using a Hanbury-Brown-Twiss interferometer. The corresponding autocorrelation function $g^{(2)}(\tau)$ recorded at 0.32 µA injection current is depicted in Fig. 4. The raw data recorded with 4 ps channel width are binned to 256 ps per data point, a value well below the timing resolution of the avalanche photo-diodes used in the interferometer (800 ps); the grey line is a fit to the raw data. The measurement shows a pronounced anti-bunching with $g^{(2)}(0) = 0.05$, proving a close to pure single-photon emission.

ACKNOWLEDGMENT

We gratefully acknowledge strain modelling by A. Schliwa and financial support by DFG within CRC 787.

REFERENCES

[1] A. Strittmatter, A. Schliwa, J. H. Schulze, T. D. Germann, A. Dreismann, O. Hitzemann, E. Stock, I. Ostapenko, S. Rodt, W. Unrau, U. W. Pohl, V. Haisler, A. Hoffmann, D. Bimberg, and V. Haisler,"Lateral positioning of InGaAs quantum dots using a buried stressor," Appl. Phys. Lett. 100, 093111 (2012).

[2] W. Unrau, D. Quandt, J.-H. Schulze, T. Heindel, T.D. Germann, O. Hitzemann, A. Strittmatter, S. Reitzenstein, U.W. Pohl, and D. Bimberg, "Electrically driven single photon source based on a site-controlled quantum dot with self-aligned current injection," Appl. Phys. Lett. 101, 211119 (2012).

MoC3-5 (Oral)
15:30 - 15:45

Single-Photon Emission in Telecommunication Band from an InAs Quantum Dot in a Pillar Structure

X. Liu[1], N. Kobayashi[1], K. Akahane[2], M. Sasaki[2], H. Kumano[1], and I. Suemune[1]

[1]Research Institute for Electronic Science,
Hokkaido University,
Sapporo 001-0021, Japan
E-mail: liuxm@es.hokudai.ac.jp
[2]National Institute of Information and Communications Technology,
Koganei, Tokyo 184-8795, Japan

Abstract— **We report on the experimental demonstration of single-photon source based on a single InAs quantum dot emitting in the telecommunication band. The low-density quantum dots grown by epitaxial method are embedded in an as-etched pillar structure. Photoluminescence spectrum of a single InAs quantum dot exhibits intense and narrow emission lines. Photon antibunching behavior is clearly observed using superconducting single-photon detectors with high sensitivity, which indicates the single-photon emission.**

Keywords—InAs; InP; telecommunication wavelength; Photon antibunching; single-photon source

I. INTRODUCTION

Single-photon emitters such as molecules, atoms, color centers in diamond and semiconductor quantum dots (QDs) have attracted much attention due to their potential applications such as quantum communications, quantum cryptography and optical quantum information processing [1-8]. In particular, applications for a high-rate and long-distance quantum key distribution (QKD) require the efficient generation of single photons in the telecommunication wavelength range (1.3 μm-1.55 μm) with an optical fiber connection. Among these systems, single QDs are very promising candidates utilizing their discrete level, which offer high emission rates, narrow spectral line width and wide tunability of emission wavelengths. Only a few experiments have demonstrated single-photon emission based on single QDs and their applications in the telecommunication band [5-8].

Single-photon emission from InAs QDs has been reported with those prepared with metal-organic vapor-phase epitaxy. Molecular-beam epitaxy (MBE) of InAs dots results in elongated quantum dashes on InP(001) substrates and circular InAs QDs on InP(311)B substrates. This leads to high QD density on the order of 10^{11} cm^{-2} and inter-dot coupling, which prevent single-photon emission. In this study, we prepare low-density InAs QDs on InP(311)B substrates with MBE. Conventional wetting layer is absent in the present QD sample. The single-photon emission from an InAs QD embedded in an as-etched pillar structure is investigated. We demonstrate that the emission line in the telecommunication band exhibits a clear antibunching dip in the second-order correlation measurements with the help of superconducting single-photon detectors (SSPDs) with high sensitivity.

II. EXPERIMENT

A. Sample preparation

The low-density InAs QDs of $\sim 10^{10}$ cm^{-2} were grown by use of MBE on InP(311)B substrates. A 300-nm-thick lattice-matched In$_{0.53}$Ga$_{0.25}$Al$_{0.22}$As buffer layer was first grown at 470 °C after thermal cleaning of the substrate. For the growth of the 5-monolayer (ML) InAs QD layer the growth temperature was increased to 530 °C. Subsequently, the QDs were capped with a 150-nm-thick In$_{0.53}$Ga$_{0.25}$Al$_{0.22}$As barrier at 470 °C. Photoluminescence (PL) of the as grown, unprocessed sample is shown in Fig. 1(a). It can be seen that the sample exhibits the QD emission in the range of 1.2-1.6 μm. No emission of wetting layer was observed. For the fabrication of well-controlled pillar structure with high yield, electron-beam lithography and sequential etching were performed. Fig. 1(b) shows the SEM image of a fabricated pillar structure with a typical height of ~ 800 nm.

B. Experimental setup

The sample placed inside a cryostat was excited using continuous wave (CW) at 633 nm. A near infrared microscope objective with a numerical aperture of 0.42 was applied for a free space excitation. The PL spectrum was measured by directing the emission into a 50 cm double grating spectrometer equipped with a liquid-nitrogen-cooled InGaAs

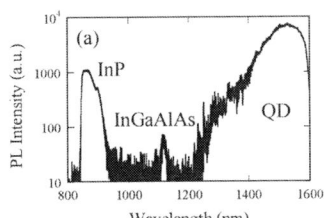 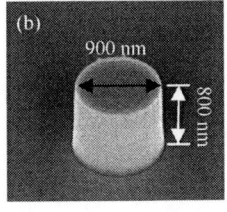

Figure 1. (a) PL spectrum of 5 ML InAs QD sample at 4 K, together with the emission from the InP substrate and the In$_{0.53}$Ga$_{0.25}$Al$_{0.22}$As barrier. (b) Side-view SEM image of an as-etched pillar structure.

This work was partially supported by the Grand-in-Aid for Scientific Research (S), No. 24226007, Nanotechnology Platform by the Ministry of Education, Culture, Sports, Science and Technology, and SCOPE (Strategic Information and Communications R&D Promotion Programme) from the Ministry of Internal Affairs and Communications.

978-1-4673-6130-9/13 $31.00 © 2013 IEEE

MoC3-5 (Oral)
15:30 - 15:45

photodiode array detector. For the second-order correlation measurements, the emission was coupled into a single mode fiber with a microscope objective. To spectrally select a single QD emission, a tunable BPF with a full width at half maximum (FWHM) of 0.5 nm was used. The coupled emission photons through the tunable BPF were divided by a fiber beam splitter with a coupling ratio of 50:50 and then directed into two SSPDs. The generated events at the SSPDs were sent to a time-to-amplitude converter (TAC) as the start and stop signals. In order to compensate the electric delay of the SSPD system and obtain the information at the negative time delay, an optical delay of ~125 ns was inserted into one arm of beam splitter.

C. Results

Fig. 2(a) shows the PL spectrum of an InAs QD embedded in a pillar structure observed in the 1359-1373 nm wavelength range at 4 K. It can be seen that the PL spectrum shows a neutral exciton line (X^0), a neutral biexciton line (XX^0) and a negatively charged exciton line (X^-) that peak at 1362.8 nm, 1365.6 nm and 1370.8 nm, respectively. These emission lines from a single QD are identified by the power and polarization dependent measurements. As expected, the X^0 and XX^0 lines exhibit reversed polarization with identical splitting. And the integrated intensities of the X^0 and XX^0 lines show almost linear and quadratic dependence on the excitation power as shown in Fig. 2(b). The FWHM of the emission X^- is very narrow and estimated to be 46 µeV.

Single-photon emission of the dominant emission X^- is demonstrated by measuring the second-order correlation function $g^{(2)}(\tau)$ with a Hanbury-Brown-Twiss (HBT) setup [9]. The excitation power is 0.95 µW, which is close to the saturation power. The single count rate at each SSPD is around 7 KHz. Fig. 3 shows the normalized coincidence with a time bin of 244 ps and an integration time of ~5 hours. It can be seen that the measured data exhibit a clear antibunching dip at the zero time delay. The solid curve is the least-square fit to the data with an expression:

$$g^{(2)}(\tau) = 1 - [1 - g^{(2)}(0)]\exp(-|\tau|/T) \quad (1)$$

where τ is the delay time, T is the sum of the emission lifetime and the inverse pumping rate, and $g^{(2)}(0)=0$ is an indication of perfect single-photon state. The fitting result shows T = 0.95 ns and $g^{(2)}(0) = 0.41 < 0.5$, indicating that the emission X^- we observed is non-classical light and the InAs QD acts as a single-photon source.

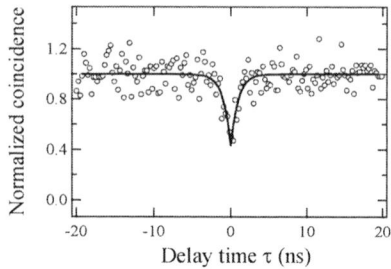

Figure 3. Normalized coincidence of a second-order correlation measurement at an excitation power of 0.95 µW. The solid curve is the least-square fit of (1) to the data.

III. CONCLUSION

Using an as-etched pillar structure incorporating an InAs/InP QD, we have demonstrated the single-photon emission at ~1.37 µm. In this work, we collected photons from the as-etched pillar structure, which resulted in relatively low extraction efficiency. In order to obtain high extraction efficiency of the single-photon source for practical applications, we will embed our as-etched pillar structure into metal film as the metal-embedded pillar structure including QDs can provide more emission of photons due to the photon reflection at the metal surface. The research to this direction is under way.

REFERENCES

[1] M. Nothaft, S. Hohla, F. Jelezko, N. Fruhauf, J. Pflaum, and J. Wrachtrup, "Electrically driven photon antibunching from a single molecule at room temperature," Nat. Commun., vol. 3:628, pp. 1–6, January 2012.

[2] B. Darquié, M. P. A. Jones, J. Dingjan, J. Beugnon, S. Bergamini, Y. Sortais, G. Messin, A. Browaeys, and P. Grangier, "Controlled Single-Photon Emission from a Single Trapped Two-Level Atom," Science, vol. 309, pp. 454–456, July 2005.

[3] A. Beveratos, R. Brouri, T. Gacoin, A. Villing, J.-Ph. Poizat, and P. Grangier, "Single Photon Quantum Cryptography," Phys. Rev. Lett., vol. 89, pp. 187901-1–187901-4, October 2002.

[4] P. Michler, A. Kiraz, C. Becher, W. V. Schoenfeld, P. M. Petroff, L. Zhang, E. Hu, and A. Imamoglu, "A Quantum Dot Single-Photon Turnstile Device," Science, vol. 290, pp. 2282–2285, December 2000.

[5] P. M. Intallura, M. B. Ward, O. Z. Karimov, Z. L. Yuan, P. See, A. J. Shields, P. Atkinson, and D. A. Ritchie, "Quantum key distribution using a triggered quantum dot source emitting near 1.3 µm," Appl. Phys. Lett., vol. 91, pp. 161103-1–161103-3, November 2007.

[6] K. Takemoto, M. Takatsu, S. Hirose, N. Yokoyama, Y. Sakuma, T. Usuki, T. Miyazawa, and Y. Arakawa, "An optical horn structure for single-photon source using quantum dots at telecommunication wavelength," J. Appl. Phys., vol. 101, pp. 081720-1–081720-5, April 2007.

[7] K. Takemoto, Y. Nambu, T. Miyazawa, K. Wakui, S. Hirose, T. Usuki, M. Takatsu, N. Yokoyama, K. Yoshino, A. Tomita, S. Yorozu, Y. Sakuma, and Y. Arakawa, "Transmission Experiment of Quantum Keys over 50 km Using High-Performance Quantum-Dot Single-Photon Source at 1.5 µm Wavelength," Appl. Phys. Express, vol. 3, pp. 092802-1–092802-3, September 2010.

[8] M. D. Birowosuto, H. Sumikura, S. Matsuo, H. Taniyama, P. J. van Veldhoven, R. Noetzel, and M. Notomi, "Fast Purcell-enhanced single photon source in 1,550-nm telecom band from a resonant quantum dot-cavity coupling," Sci. Rep., vol. 2, pp. 1– 5, March 2012.

[9] R. Hanbury Brown and R. Q. Twiss, "Correlation between Photons in two Coherent Beams of Light," Nature, vol. 177, pp. 27–29, January 1956.

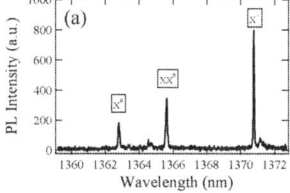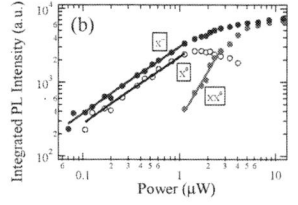

Figure 2. (a) PL spectrum of an InAs QD at 4 K. The emission lines X^0, XX^0 and X^- correspond to the neutral exciton, neutral biexciton and negatively charged exciton, respectively. (b) Integrated PL intensity as a function of the CW excitation power. The solid lines are guide to the eyes.

MoD3-1 (Invited)
14:00 - 14:30

IPRM2013, May 19 - 23, 2013, Kobe, Japan
The 25th International Conference on Indium Phosphide and Related Materials

Travelling Wave Mach-Zehnder Modulators

(Invited)

Kelvin Prosyk, Abderrahmane Ait-Ouali, Junfu Chen, Michael Hamacher, Detlef Hoffmann, Ronald Kaiser, Ron Millett, Alessio Pirastu, Marco Totolo, Karl-Otto Velthaus, Ian Woods

Cogo Optronics Canada Inc.
Ottawa, Canada

Abstract—**Travelling wave electrode Mach-Zehnder modulators in InP/InGaAsP are investigated for application to advanced digital optical fibre formats. Single-end 50-Ohm and novel differential 100-Ohm designs are described.**

Keywords—modulator; Mach-Zehnder; travelling wave; IQ; 100G; QPSK

I. INTRODUCTION

The first series push-pull travelling wave electrode (TWE) Mach-Zehnder modulator (MZM) to employ periodic loading of the main transmission line with the shunt capacitance of lumped waveguide electrodes was published in 1982, implemented in lithium niobate [1]. Taking advantage of the improved velocity match available in GaAs/AlGaAs, Walker realized significant improvements in 1991 [2]. The Heinrich-Hertz Institute transferred the concept into InP-based materials in 1998, gaining further advantages in chip size and drive voltage [3]. Later publications of TWE modulators in InP followed, showing steadily increasing bandwidth [5]-[6].

We have studied the TWE electrode in InP:Fe/InGaAsP with an objective to simultaneously optimize all critical performance parameters in a manner suitable for manufacturing small-footprint, low-power, and high bit-rate transmitters for emerging modulation formats [6]-[7]. Crucial to this is the development of a full C-band in-phase quadrature modulator (IQM) for 100 Gb/s dual polarization quadrature phase shift keying (DP-QPSK). In this presentation, the progress to date will be summarized and new steps towards further reduction in power consumption are outlined.

II. MODULATOR DESIGNS

A. 43 Gb/s VSR MZM

The development of a commercially optimized MZM began with a high bit rate very short reach (VSR) application that uses non-return to zero (NRZ) on-off keying (OOK) format. The performance that must be simultaneously achieved includes: 50Ω single end drive; Vpi = 2.5V; 40 GHz bandwidth; extinction ratio (ER) > 20 dB; insertion loss (IL) < 8 dB to a lens system with a 20° circular far field; and zero chirp. In addition to these specifications, there are the usual requirements of great enough margin to specification to ensure high yield, and a guaranteed operating life span of 20 years.

To meet these demands, an adiabatically tapered spot size converter (SSC) and deeply etched waveguides were used with

a TWE, as shown in Fig. 1. A selective etch followed by enhanced selective area growth was used to create a thin, highly doped overclad in the TWE, and a thick, undoped overclad at the SSC. As a consequence, the modulated signal is transferred efficiently to the multiquantum well (MQW) core with a minimum of RF loss, while the optical mode has room to expand vertically with a minimum of doping loss. The deep etch allows for compact routing of light, low capacitance, and negligible radiation mode coupling to the output. Although wavelength tunability is not a main objective for VSR applications, an IL < 6.0 dB, ER > 25 dB and Vpi = 2.5V across a 40nm C-band were demonstrated in a 43 Gb/s, chirp free device [6].

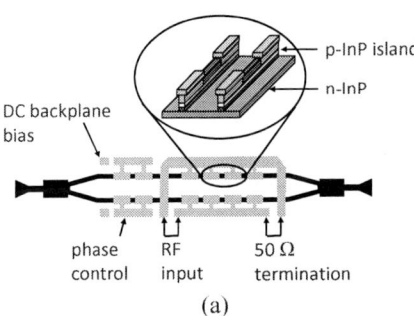

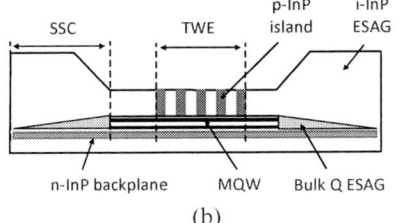

Figure 1. (a) Overview of 43Gb/s MZM; (b) side view schematic

B. 100 Gb/s DP-QPSK IQM

Having verified the performance of a single MZM, the fabrication of an IQM is straightforward [7]. An example of an IQM with RF inputs routed to one side is shown in Fig. 2(a). One-side routing is suitable for 100 Gb/s packages where the optical fibre and RF feedthrough are located at 90 degrees to each other, such as the Optical Interworking Forum (OIF) Gen-1 standard, shown in Fig. 2(b). Since the OIF package was

978-1-4673-6130-9/13 $31.00 © 2013 IEEE

MoD3-1 (Invited)
14:00 - 14:30

designed for large lithium niobate modulators, the InP prototype demonstrator was mostly empty space inside.

The package was tested with a Tektronix optical modulation analyzer at 100Gb/s. The constellation diagram and reconstructed eyes are shown in Fig. 3 at 1550nm with both polarizations driven simultaneously, using a pseudo-random bit sequence (PRBS) length of 2^{31}-1. The average error vector magnitude was 11%, primarily limited by package-induced ripple in the small signal electro-optic frequency response.

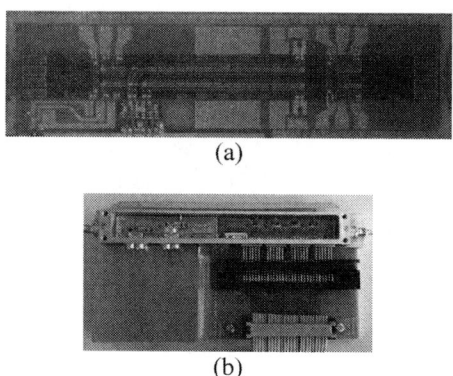

(a)

(b)

Figure 2. (a) IQM chip mounted on a test carrier; (b) two IQMs in DP configuration in a Gen 1 OIF prototype demonstration package.

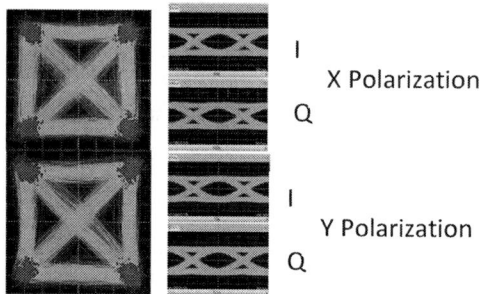

I
X Polarization
Q

I
Y Polarization
Q

Figure 3. Constellation diagrams and reconstructed eyes for both polarizations driven simultaneously at PRBS length of 2^{31}-1.

C. 100Ω Differential IQM

Driver power consumption can be reduced by approximately 50% if the single end driver is replaced by a differential driver, and the 50Ω travelling wave electrode increased to 100Ω while maintaining the Vpi at 2.5V across the electrodes (i.e., 1.25V differential). A new TWE configuration was explored based on a conduction-backed coplanar stripline with lateral ground traces (CB-GSSG). The extra grounding serves to isolate each MZM from the surrounding environment and from each other. The chip was mounted on a prototype test carrier containing a large gold-coated surface to act as a lower ground plane. Also mounted on the ground plane, on either side of the chip, were AlN blocks the same height as the chip containing a CB-GSSG transmission line launcher, termination network, and via'd upper ground surface. A schematic of the

chip is shown in Fig. 4(a). Bandwidth was > 25GHz, and inter-MZ crosstalk was < -37 dB as shown in Fig. 4(b)-(c). The optical path was similar to the single-end IQM of section B.

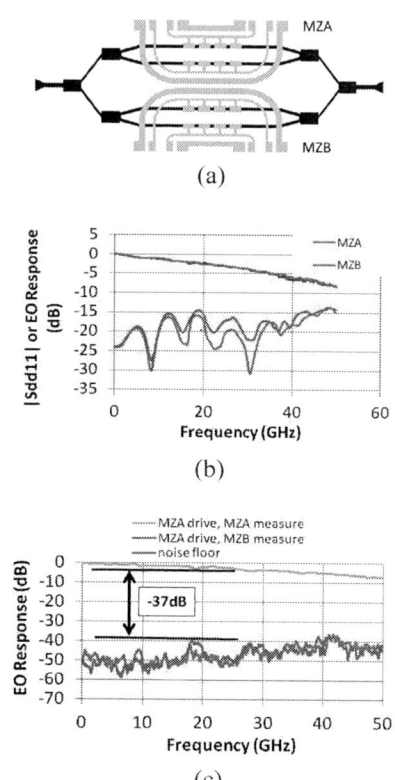

(a)

(b)

(c)

Figure 4. (a) 100Ω CB-GSSG chip; (b) RF performance; (c) crosstalk..

REFERENCES

[1] U. Langmann, D. Hoffmann, "Capacitively loaded transmission line for subnanosecond stepped Δβ operation of an integrated optical directional coupler switch", *Proc. IEEE MTT-S Int. Microwave Symp.*, pp. 110-112, 1982.

[2] R.G. Walker: "High-speed III-V semiconductor intensity modulators", IEEE J. Quantum Electron., vol. *QE-21*, pp, 654-667, 1991.

[3] L. Morl, C. Bornholdt, D. Hoffmann, K. Matzen,G.G. Mekonnen, F.W. Reier, "A travelling wave electrode Mach-Zehnder 40 Gb/s demultiplexer based on strain compensated GaInAa/AlInAs tunnelling barrier MQW structure", Proceedings l0th Int. Conf. on InP and Related Materials,Tsukuba, Ibaraki, Japan, pp. 403-406, 1998.

[4] S. Akiyama, et al., "40 Gb/s InP-based Mach-Zehnder Modulator with a driving voltage of 3 Vpp", Proceedings 16th Int. Conf. on InP and Related Materials, Kagoshima, Japan, 2004.

[5] H. N. Klein, H. Chen, D. Hoffmann, S. Staroske, A. G. Steffan, K.-O. Velthaus, "1.55μm Mach-Zehnder Modulators on InP for optical 40/80 Gbit/s transmission networks", Proceedings 18th Int. Conf. on InP and Related Materials, Kagoshima, Japan, 2006.

[6] K. Prosyk, A. Ait-Ouali, C. Bornholdt, M. Gruner, M. Hamacher, D. Hoffmann, R. Kaiser, R. Millett, K.-O. Velthaus, I. Woods, "High performance 40GHz InP Mach-Zehnder modulator", OFC/NFOEC Technical Digest, OW4F.7, 2012.

[7] K. Prosyk, T. Brast, M. Gruner, M. Hamacher, D. Hoffmann, R. Millett, K.-O. Velthaus., "Tunable InP-based optical IQ modulator for 160 Gb/s", ECOC Postdeadline Papers, Th.13.A.5, 2011.

MoD3-2 (Oral)
14:30 - 14:45

IPRM2013, May 19 - 23, 2013, Kobe, Japan
The 25th International Conference on Indium Phosphide and Related Materials

Novel planar structure single-RF drive MZ optical modulator on InP(110) substrate

Yoshihiro Ogiso, Masakazu Arai, Eiichi Yamada, Hiromasa Tanobe, Yasuo Shibata, Masaki Kohtoku

NTT Photonics Laboratories, NTT corporation
Atsugi, Kanagawa, Japan
e-mail: ogiso.yoshihiro@lab.ntt.co.jp

Abstract—**A MZ optical modulator (MZM) with a novel planar structure was proposed. A single-RF drive push-pull MZM with a simple coplanar waveguide (CPW) was fabricated by utilizing a (110)-oriented InP substrate in terms of the crystallographic orientation dependence of the electro-optic Pockels effect. The device exhibits a 3 dB-EO bandwidth of 30 GHz and 40 Gb/s NRZ high speed modulation.**

Keywords—component; MZ Modulator; InP(110)

I. INTRODUCTION

Advanced modulation such as a high-order in-phase quadrature modulation (IQM) plays an important role in huge-capacity transmission systems [1]. InP-based Mach-Zehnder optical modulators (MZMs) have been studied intensively as an alternative to LiNbO₃-based modulators because of the small chip size and potential for inexpensive fabrication. A compact module size is essential as regards realizing compact transceivers. Although, the InP-based MZM chip is very small, the module size tends larger than expected. One of factor preventing the realization of a compact module is the space needed for the RF interface. The number of RF connectors required for a single-RF electrode MZM is half that needed for a conventional dual-RF electrode MZM. Therefore, single-RF electrode is preferable as regards reducing the module size [2]. However, the structure and fabrication process of a single-RF MZM on an InP substrate are complicated.

In this paper, a novel planar structure single-RF InP-based MZM is proposed and demonstrated. The application of a lateral electrical field enables simple push-pull operation with a single-RF electrode. A planarized buried waveguide without irregular regrowth was realized by overcoming the difficulties involved in for epitaxial growth on an InP(110) substrate [3]. The simple planar MZM structure with a simple coplanar (CPW) makes it possible to achieve high-speed operation. The device exhibits 3 dB-EO bandwidth of 30 GHz and 40 Gb/s high-speed modulation.

II. DEVICE STRUCTURE

We proposed the novel InP-based MZM shown in Fig. 1. The feature of the structure is a single-RF drive operation with a simple buried waveguide and electrode. MZ arm waveguide stripes were formed along the [1 $\overline{1}$ 0] direction on a (110)-oriented substrate. And the electrodes were arranged so that the electrical field could be applied in a lateral direction. In

this configuration, when a voltage is applied to the center signal electrode, the direction of the electrical field in each MZ arm reverses as shown in Fig. 1. Since the change in the sign of the refractive index due to the Pockels effect depends on the direction of the electrical filed, a single-RF drive push-pull operation is achieved. To apply the electrical field effectively, we employed a semi-insulating (SI) core layer buried in an n-type cladding layer. In this structure, the optical and electrical losses are expected to be smaller than with the conventional p-i-n type InP-based MZM, because it does not employ a p-type cladding layer.

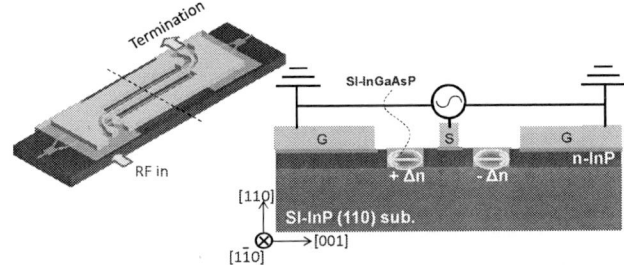

Fig. 1 Schematic and cross-section diagrams of the proposed InP-based MZM

III. DEVICE FABRICATION

A. Crystal growth on InP(110) substrate

The (110) substrate is usually known to be a difficult facet for epitaxial growth due to the difficult nucleation. A (110) InP substrate with a vicinal surface is widely used to obtain a smooth epitaxial layer. In this work, we used a 3° off (110) towards [111] B substrate and obtained an InP/ InGaAsP epitaxial layer with good surface morphology. The entire layer structure was grown using horizontal low-pressure metal-organic vapor-phase epitaxy (MOVPE). The growth temperature was 570 °C. Asymmetrical irregular regrowth is easily induced on a (110)-oriented substrate because of the in-plane anisotropy as shown in Fig. 2 (a). The asymmetric irregular regrowth was successfully suppressed and a planer regrowth surface was obtained as shown in Fig. 2 (b) by utilizing vertical and symmetrical dry etching and a sidewall wet etching.

B. MZM device fabrication

978-1-4673-6130-9/13 $31.00 © 2013 IEEE

Figure 3 shows a cross-sectional SEM image and a microscope image of an InP(110) MZM in which a SI bulk core waveguide was introduced to act as a current blocking layer. The MZM was composed of two waveguide structures where a deep-etched mesa waveguide and a buried n-SI-n symmetric structure were introduced in the I/O passive waveguide regions and an electro-optic control region, respectively. An Fe-doped SI-InGaAsP (electron trap density: 6.4×10^{16} cm^{-3}) mesa waveguide on a SI-InP(110) substrate was buried in Si-doped n-InP (1×10^{18} cm^{-3}) cladding layers and Si-doped n-InGaAsP (5×10^{18} cm^{-3}) contact layers. The buried n-SI-n structure was 0.65 µm thick. A 10 µm thick Au traveling-wave electrode was designed to match the velocity difference between a lightwave and a microwave. To reduce the driving voltage, the electrode length and the entire chip size were set 6.0 mm and 0.35 mm x 9.0 mm, respectively, as shown in Fig. 3. These sizes are relatively long compared with a conventional InP-based MZM utilizing the quantum-confined Stark effect.

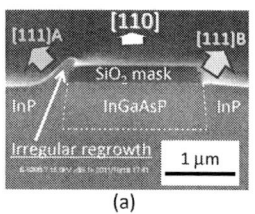 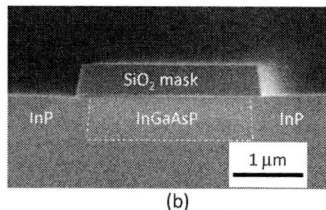

(a)   (b)

Fig. 2   Cross-sectional SEM images of buried InP/InGaAsP structure with (b) and without (a) symmetrical etching techniques

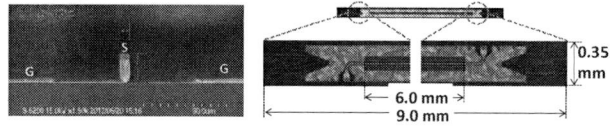

Fig. 3   Cross-sectional SEM image (left) and microscope image (right) of the InP(110) MZM

IV. HIGH FREQUENCY CHARACTERISTICS

First, we measured the insertion loss. The SI-core waveguide propagation loss of below 1.5 dB/cm, which is nearly equal to that of an undoped intrinsic core waveguide. Then, we measured the high frequency characteristics. Figure 4 shows the small-signal electrical and electro-optical S-parameters of the MZM. The input polarization state was the TM mode. A 6-dB electrical-electrical (E/E) S21 bandwidth of 30 GHz and a 3-dB electro-optical (E/O) bandwidth of 30 GHz were achieved. The results indicated that the MZM satisfied the lightwave - microwave velocity matching condition. In addition, the electrical reflection (S11) was below -20 dB, which is sufficiently small for practical use. We also investigated the dynamic performance. A 40-Gb/s non-return-to-zero (NRZ) signal with a pseudo random binary sequence of 2^{31}-1 was applied to a single-RF electrode. The operating wavelength, input optical power and driving voltage were 1550 nm, +13 dBm, and 5.5 V$_{pp}$, respectively. Figure 5 shows the corresponding eye diagram and a clear eye opening was obtained. Further optimization of waveguide dimensions and Si, Fe-doping densities can reduce the driving voltage.

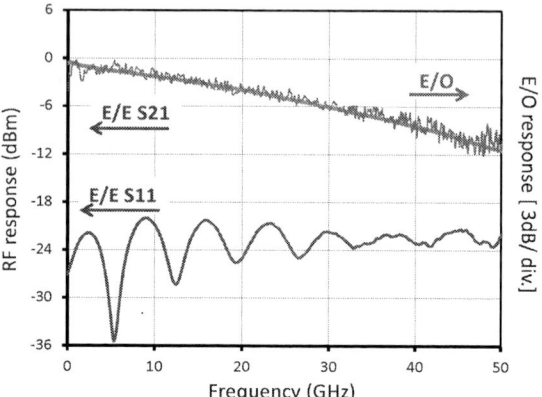

Fig. 4   Electrical and electro-optical S-parameter

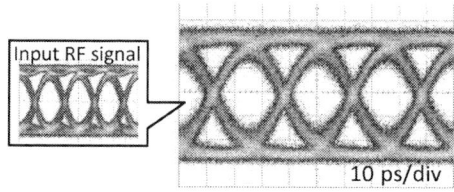

Fig. 5   40 Gb/s NRZ modulation eye diagram

V. CONCLUSION

A novel planar structure MZM was proposed. A single-RF drive push-pull MZM with a simple CPW was fabricated by utilizing a (110)-oriented InP substrate in terms of the crystallographic orientation dependence of the electro-optic Pockels effect. The device exhibits a 3 dB-EO bandwidth of 30 GHz and 40 Gb/s NRZ high speed modulation. We believe the (110)-oriented planar MZM to be suitable for high-speed and high-order IQM including nested parallel MZMs.

REFERENCES

[1] A. Sano, T. Kobayashi, K. Ishihara, H. Masuda, S. Yamamoto, K. Mori, E. Yamazaki, E. Yoshida, Y. Miyamoto, T. Yamada, and H. Yamazaki, "240-Gb/s polarization-multiplexed 64-QAM modulation and blind detection using PLC-LN hybrid integrated modulator and digital coherent receiver", European Conference on Optical Communication (ECOC) 2009, PDP PD2.4, 2009

[2] K.-O. Velthaus, M. Hamacher, M. Gruner, T. Brast, R. Kaiser, K. Prosyk, I. Woods, D. Hoffmann, and M. Schell, "High performance InP-based Mach-Zehnder modulator for 10 to 100 Gb/s optical fiber transmission systems", Conf. InP and Related Materials (IPRM 2011), Th-9.2.1, May 2011

[3] R. Bhat, M. A. Koza, D. M. Hwang, M. J. S. P. Brasil, R. E. Nahory, and K. Oe, "OMCVD growth of InP, InGaAs, and InGaAsP on (110) InP substrates", Journal of Crystal Growth, vol. 124, pp. 311-317, 1992

MoD3-3 (Oral)
14:45 - 15:00

Flat-top Optical Frequency Comb Block Generation using InP-based Mach-Zehnder Modulator

Takeaki Saikai, Takahiro Yamamoto, and Hiroshi Yasaka
Research Institute of Electrical Communication RIEC
Tohoku University
Kanagawa, Japan

Eiichi Yamada
NTT Photonics Laboratories
NTT Corporation
Kanagawa, Japan

Abstract—A nine-channel optical frequency comb block with a low intensity deviation is generated successfully by adopting an asymmetric push-pull drive method for a symmetric dual drive Mach-Zehnder modulator. The flat-top optical frequency comb block with an intensity deviation of less than 1 dB is realized for the first time by using a compact and low-drive voltage InP-based Mach-Zehnder modulator.

Keywords—Optical modulator, Mach-Zehnder modulator, Light source, Optical frequency comb.

I. INTRODUCTION

A multi-channel optical transmitter is a key component for introducing the channel block operation function into future photonic network systems to increase their capacity and functionality. The transmitter can be realized by combining a multi-wavelength light source and a multi-wavelength WDM modulator. An eight-channel WDM modulator was demonstrated by monolithically integrating two arrayed-waveguide grating (AWG) filters, an eight-channel semiconductor optical amplifier (SOA) array and an eight-channel electro-absorption (EA) modulator array on an InP substrate [1]. As regards the multi-wavelength light source, optical frequency comb block generators have been demonstrated by utilizing several technologies. Of these, the sideband generation method with a dual drive Mach-Zehnder (MZ) modulator [2-3] is promising because it is simple and suitable for generating an optical frequency comb with a small number of channels. By adopting an InP-based MZ modulator to an optical frequency comb block generator, compact and low drive voltage optical transmitter will be realized, because the InP-based MZ modulator is very compact and has very low half-wavelength voltage V_π of less than 3 V. The InP modulator has an advantage that it can be integrated monolithically with the WDM modulator and has a possibility for realizing an one chip multi-channel optical transmitter. We generated a nine-channel optical frequency comb block and tuned its wavelength from 1525 to 1565 nm by using a compact and low-drive voltage InP-based MZ modulator [3]. However, the intensity deviation of the nine-channel optical frequency comb block remained high (around 5 dB). It is not sufficient to use it as a multi-wavelength light source for multi-channel optical transmitter.

To reduce the intensity deviation, we propose and demonstrate an asymmetric push-pull drive method for an MZ modulator. By introducing the method, we reduce the intensity deviation and confirm the generation of a flat-top nine-channel optical frequency comb block with an intensity deviation of less than 1 dB.

II. NUMERICAL ANALYSIS

First we calculate the output spectrum of the optical frequency comb block from a dual drive MZ modulator under an asymmetric push-pull drive condition. In this condition, the phase changes in the two phase control waveguides of the MZ modulator ϕ_1 and ϕ_2 are expressed as follows.

$$
\begin{cases}
\phi_1 = \dfrac{\pi}{2}A \cdot \sin \omega t + \dfrac{\pi}{2}B \\
\phi_2 = -x \cdot \dfrac{\pi}{2}A \cdot \sin \omega t - \dfrac{\pi}{2}B
\end{cases}
\tag{1}
$$

Here A is the modulation amplitude and B is the DC bias voltage applied to the waveguides, which are normalized by the half-wavelength voltage V_π. x denotes the modulation amplitude ratio of the two RF modulation signals applied to the two phase control waveguides and has a value of 0 to 1. ω denotes the modulation angular frequency. Under this condition, the output electric field of the MZ modulator is expressed as follows.

$$
\begin{aligned}
E_{out} \Big/ \left(\frac{E_0}{2} e^{j\omega_0 t} \right) &= \left[e^{j\left(\frac{\pi}{2}A\sin \omega t + \frac{\pi}{2}B \right)} + e^{-j\left(x \cdot \frac{\pi}{2}A\sin \omega t + \frac{\pi}{2}B \right)} \right] \\
&= e^{j\frac{\pi}{2}B} \left[J_0\left(\frac{\pi}{2}A \right) + \sum_{n=1}^{\infty} J_n\left(\frac{\pi}{2}A \right) e^{jn\omega t} + \sum_{n=1}^{\infty} (-1)^n J_n\left(\frac{\pi}{2}A \right) e^{-jn\omega t} \right] \\
&+ e^{-j\frac{\pi}{2}B} \left[J_0\left(x \cdot \frac{\pi}{2}A \right) + \sum_{n=1}^{\infty} (-1)^n J_n\left(x \cdot \frac{\pi}{2}A \right) e^{jn\omega t} + \sum_{n=1}^{\infty} J_n\left(x \cdot \frac{\pi}{2}A \right) e^{-jn\omega t} \right]
\end{aligned}
\tag{2}
$$

Where $J_n(m)$ is an nth order Bessel function.

Here the intensity deviation of the nine-channel optical frequency comb block is calculated by using equation (2), because the intensities of the nine central sideband components are important for extracting the eight-channel optical frequency comb block by using an eight-channel WDM modulator. Fig. 1 shows the calculated intensity deviation of the nine-channel optical frequency comb block as a function of the modulation amplitude ratio x under the optimum values of A and B.

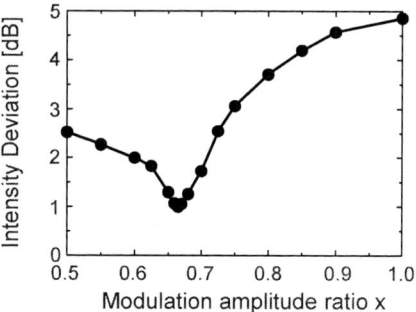

Figure 1. Calculated intensity deviation of a nine-channel optical frequency comb block as a function of modulation amplitude ratio x under optimum of modulation amplitude A and DC bias voltage B values.

The intensity deviation of the nine-channel optical frequency comb block is around 5 dB when the ratio $x = 1$ (symmetric operation condition shown in Ref. 3). It decreases gradually and has a minimum value of 0.99 dB when x is 0.665. Fig. 2 shows the intensity deviation of the nine-channel optical frequency comb block at $x = 0.665$ as a function of the modulation amplitude A and DC bias voltage B. It indicates that a minimum intensity deviation of 0.99 dB is obtained when the values of A and B are 3.22 and 0.41, respectively.

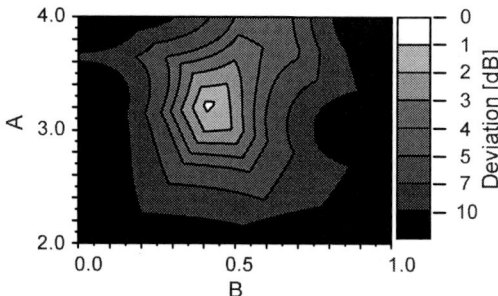

Figure 2. Calculated intensity deviation as a functions of modulation amplitude A and DC bias voltage B when $x = 0.665$.

III. EXPERIMENT

An optical frequency comb block generation experiment is carried out using an InP-based MZ modulator module [4]. The experimental setup is shown in Fig. 3. The 3-dB bandwidth of the MZ modulator module is 12.5 GHz. The modulation frequency $f_m = \omega/2\pi$ is set at 12.5 GHz. The relative phase of the RF modulation signals is precisely controlled by using a phase shifter. The maximum amplitude of the RF modulation signal is set at 6.6 V_{p-p}. The amplitude to the one phase control waveguide is reduced by using a variable gain RF amplifier. A single mode laser diode with a wavelength of 1535 nm is used as a seed light source. Fig. 4 shows the modulator output spectrum under optimum conditions. A flat-top optical frequency comb block can be obtained. And the measured intensity deviation of the nine-channel comb block is 0.77 dB. This indicates that a nine-channel flat-top optical frequency comb block can be generated by using a compact and low drive voltage InP-based MZ modulator when the asymmetric push-pull drive method is employed.

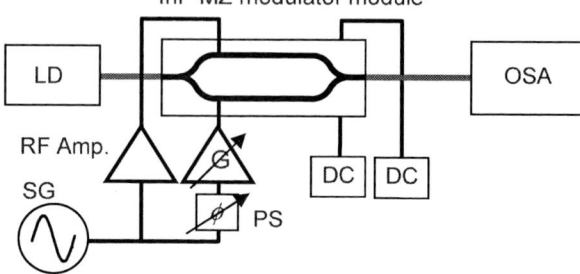

Figure 3. Experimental setup. LD: Input seed light source, RF Amp.: RF amplifier, SG: Modulation signal generator, PS: Phase shifter, DC: DC bias voltage source, OSA: Optical spectrum analyzer.

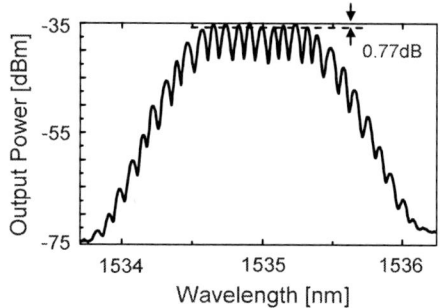

Figure 4. Measured output spectrum of InP-based MZ modulator.

IV. CONCLUSION

The generation of a flat-top optical frequency comb block is demonstrated numerically and experimentally by applying an asymmetric push-pull drive scheme to an MZ modulator. By using an InP-based MZ modulator we confirm the generation of an optical frequency comb block with a low intensity deviation of less than 1 dB. This indicates that semiconductor MZ modulators can be employed for a compact and low drive voltage optical frequency comb block generator.

REFERENCES

[1] Y. Suzaki et al., "Monolithically Integrated Eight-Channel WDM Modulators with Cyclic AWGs for Multi-Channel-Block Operation over C-Band," IEE Electronics Letters, vol. 41, No. 9, pp. 551-552, 2005.

[2] T. Sakamoto et al, "Widely wavelength-tunable ultra-flat frequency comb generation using conventional dual-drive Mach-Zehnder modulator," Electron. Lett., vol. 43, No. 19, pp. 1039-1040, 2007.

[3] T. Yamamoto et al., "Optical Frequency Comb Block Generation by using Semiconductor Mach-Zehnder Modulator," IEEE Photonics Technol. Lett., vol. 25, No. 1, pp. 40-42, 2013.

[4] K. Tsuzuki et al., "1.3-V_{pp} push-pull drive InP Mach-Zehnder modulator module for 40 Gbit/s operation," Proc. 31st Europ. Conf. Opt. Commun. (ECOC'05), Glasgow, Scotland, Th2.6.3, pp. 905-906, 2005.

MoD3-4 (Oral)
15:00 - 15:15

IPRM2013, May 19 - 23, 2013, Kobe, Japan
The 25th International Conference on Indium Phosphide and Related Materials

A 1×4 MMI-Integrated High-Power Waveguide Photodetector

Efthymios Rouvalis, Philipp Müller, Dirk Trommer, Jens Stephan,
Andreas G. Steffan, and Günter Unterbörsch

u²t Photonics AG
Reuchlinstraße 10/11,
10553 Berlin, Germany

Abstract—We report on the design, packaging and experimental characterization of a waveguide photodetector monolithically integrated with a 1×4 MMI coupler and a vertically-tapered spot-size converter. The total device consists of an array of 4 photodiodes connected in parallel in order to increase the saturation current. A 3-dB bandwidth of up to 30 GHz was measured. High output power was achieved with approximately 10 dBm at 10 GHz and 7.5 dBm at 20 GHz. From 1-dB compression measurements at 10 GHz and 20 GHz, a saturation photocurrent as high as 40 mA was found. At 10 GHz and 20 GHz a third order intercept point (OIP3) of over 20 dBm was measured for photocurrent levels up to 35 mA.

Keywords—microwave photonics, p-i-n photodiode, multi-mode interference (MMI) coupler, third-order intermodulation distortion (IMD3), millimeter-wave source,1-dB compression.

I. INTRODUCTION

High-power and high-linearity InP-based photodiodes (PDs) are of great importance in analog photonic links and have been used in a variety of microwave photonic applications such as radio-over-fiber links, phased array radars, etc. So far, a variety of different PDs have been proposed that have shown high output power and high linearity [1], [2]. The majority of these devices have been top- or back- vertically-illuminated photodiodes (VPDs) with a p-i-n or a uni-traveling-carrier (UTC) epitaxy. Despite high output power levels that have already been demonstrated, most of these devices have a rather low bandwidth that is typically RC-limited. The motivation for high bandwidth InP-based PDs led to the development of waveguide photodiodes (WGPDs) with 3-dB bandwidths exceeding 100 GHz [3], [4]. WGPDs are also attractive since they are suitable for monolithic integration with other passive and active structures [5].

However, most WGPDs that have been reported so far do not have the high power capability of VPDs. Power saturation and thermal failure strongly depend on the width of the active area which is typically 2-4 μm. This width needs to be kept small in order to increase the 3-dB bandwidth, resulting in high photocurrent density. In this paper, we demonstrate a 4×PD array integrated with a 1×4 Multi-Mode Interference (MMI) coupler and a spot size converter. The device proposed here effectively increases the saturation current by a factor of 4, compared to a single WGPD. A high 3-dB bandwidth of up to 30 GHz was measured. An output power of 10 dBm at 10 GHz was obtained together with a high saturation current of about 40 mA and high linearity.

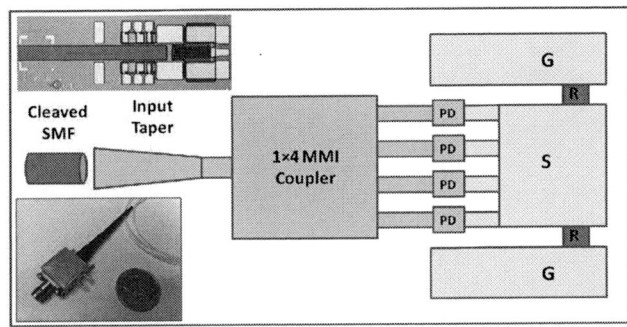

Fig. 1. Simplified schematic of the 1×4 MMI-integrated waveguide high-power photodiode chip. A vertical taper waveguide was included for efficient spot-size conversion. The combined RF output is fed into a G-S-G coplanar waveguide. A cleaved SMF was used at the optical input of the chip. Lateral tapering shown here is exclusively for illustration purposes. Top inset: photo of the chip, bottom inset: photo of the packaged module.

II. DEVICE CONCEPT AND DESIGN

In most typical VPDs the saturation current depends strongly on the area of the detector that is illuminated, assuming uniform illumination. In WGPDs the width of the active area plays a much more important role than the length. However, by increasing the width of the PD, the total active area increases and this results in a lower total 3-dB bandwidth. This can be overcome by distributing the optical signal, and thus the photocurrent density, over a number of photodiodes that are connected in parallel. In our case, this is realized by splitting the optical signal via a 1×4 MMI coupler and feeding an array of four identical photodiodes that are electrically connected in parallel, as shown schematically in Fig. 1. A vertical input taper waveguide was included to improve fiber-to-chip-coupling. A 100 Ω resistor was also integrated to reduce the ripple in the RF response and also the effective load resistor. The required DC bias was applied to all 4 PD-sections through the same DC port.

The total p-n junction capacitance is 4 times larger than that one of an individual PD. However the parasitic capacitance from the contact pads should remain the same [6]. By taking into account the parallel combination of the 100 Ω internal resistor and the 50 Ω termination together with the increased capacitance, then the total RC limited bandwidth is expected to be reduced approximately by a factor of 4 resulting in values of about 25 GHz.

978-1-4673-6130-9/13 $31.00 © 2013 IEEE

MoD3-4 (Oral)
15:00 - 15:15

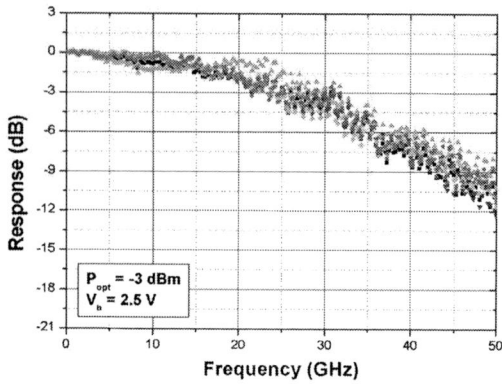

Fig. 2. Relative frequency response of 6 different 1×4 MMI-integrated fiber-coupled modules measured using an LCA at an optical input power of -3 dBm and a reverse bias voltage of 2.5 V.

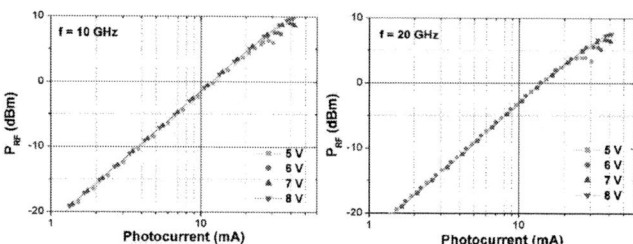

Fig. 3. Saturation measurements at 10 GHz (left) and 20 GHz (right) for reverse bias ranging from 5 V to 8 V as a function of the measured DC photocurrent. The maximum power measured was 9.7 dBm at 10 GHz and 7.5 dBm at 20 GHz.

III. EXPERIMENTAL RESULTS

The PDs were mounted in modules for fiber-to-chip coupling with cleaved SMF and an RF interface that operates at frequencies over 50 GHz. Average dark current values were 0.01 μA and 0.38 μA at a reverse bias voltage of 2 V and 8 V respectively. After fiber coupling, the devices were characterized in terms of responsivity. The maximum responsivity was 0.4 A/W at 1.55 μm.

As a next step, six packaged modules were characterized in terms of 3-dB bandwidth using a calibrated Lightwave Component Analyzer (LCA) up to 50 GHz. The calibration was performed on a broadband short, open and a 50 Ω load. For these measurements the polarization was optimized externally using a polarization controller. The applied reverse bias was only 2.5 V and the optical input power was fixed to -3 dBm. The results are shown in Fig. 2. A 3-dB bandwidth of over 25 GHz was found for all devices with some of them exhibiting values in excess of 30 GHz.

An optical heterodyne system using two tunable laser sources was built to measure the RF output power from the photodiodes. An EDFA and a VOA were used in order to amplify and control the power level of the optical signal in the PD. The power of the generated heterodyne RF signal was measured using a broadband calibrated RF power meter. The results for 10 GHz and 20 GHz at a reverse bias voltage ranging from 5 V to 8 V are shown in Fig. 3. The same

packaged module at a photocurrent up to 40 mA delivered 9.7 dBm and 7.5 dBm at 10 GHz and 20 GHz respectively. It was also found that the saturation DC photocurrent was indeed 4 times larger than that of the individual PD. It is noteworthy that the maximum delivered RF power was limited by the optical power that was available at the output of the EDFA that was used to amplify the heterodyne signal. In addition, no temperature controller was needed to obtain this performance.

The linearity of the PD was characterized using a typical two tone experimental setup [7]. The Third-Order Intercept Point (OIP3) was taken as a figure of merit. The measurement was performed at 10 and 20 GHz and a reverse bias of 8 V. Two tones were taken at a frequency offset of 10 MHz (i.e. 10 and 10.01 GHz, 20 and 20.01 GHz respectively) and the applied RF power to the modulators was 11 dBm. The two optical signals were combined, amplified using an EDFA and then attenuated through a VOA. The generated fundamental and 3rd order intermodulation distortion (IMD3) signals were measured using an electrical spectrum analyzer. The experimental results for 10 GHz and 20 GHz are shown in Fig. 4. The PD demonstrated a high linearity with an OIP3 between 20 and 30 dBm for photocurrents up to 35 mA and over 15 dBm even up to almost 40 mA.

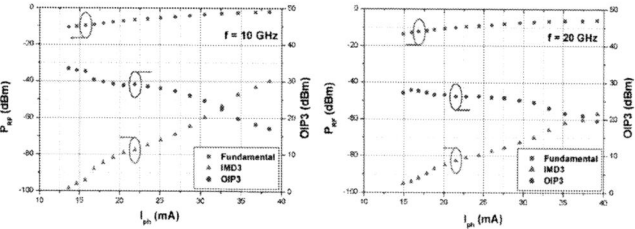

Fig. 4. Fundamental tone (green squares) and IMD3 (red triangles) power (left y-axis), and OIP3 (blue dots, right y-axis) as a function of DC photocurrent at a frequency of 10 GHz (left) and 20 GHz (right) and reverse bias of 8 V.

REFERENCES

[1] M. Chtioui et al., "High-Power High-Linearity Uni-Traveling-Carrier Photodiodes for Analog Photonic Links," *IEEE Photonics Technology Letters*, vol. 20, no. 3, pp. 202-204, 2008.

[2] Z. Li et al., "High-power high-linearity flip-chip bonded modified uni-traveling carrier photodiode," *Optics Express*, vol. 19, no. 26, pp. B385-B390, 2011.

[3] A. Umbach et al., "Photoreceivers for 100 Gbit/s Applications," *in Proc. of IEEE/LEOS Summer Topical Meetings 2007*, TuE4.2.

[4] A. Beling et al., "High-Power InP-based Waveguide Photodiodes and Photodiode Arrays Heterogeneously Integrated on SOI," *in Proc. of International Conference on Indium Phospide and Related Materials (IPRM) 2012.*

[5] E. Rouvalis et al., "High-speed photodiodes for InP-based photonic integrated circuits," *Optics Express*, vol. 20, no. 8, pp. 9172-9177, 2012.

[6] D. Trommer, A. Umbach, and G. Unterbörsch, "InGaAs Photodetector with Integrated Biasing Network for mm-Wave Applications," *in Proc. of International Conference on Indium Phospide and Related Materials (IPRM) 1998*, TuP-52, pp. 276-279.

[7] H. Jiang et al., "The Frequency Behavior of the Third-Order Intercept Point in a Waveguide Photodiode," *IEEE Photonics Technology Letters*, vol. 12, no. 5, pp. 540-542, 2000.

978-1-4673-6130-9/13 $31.00 © 2013 IEEE

MoD3-5 (Oral)
15:15 - 15:30

Study of lowering onset gain for a high-speed InGaAs/InAlAs avalanche photodiode

Masahiro Nada, Yoshifumi Muramoto, Haruki Yokoyama, Tadao Ishibashi and Hideaki Matsuzaki

NTT Photonics Laboratories
NTT Corporation
Atsugi City, Kanagawa, Japan
nada.masahiro@lab.ntt.co.jp

Abstract—**We investigate a novel InAlAs avalanche photodiode with double p-field control layers, which effectively eliminates the potential barrier between InGaAs absorption and InAlAs avalanche layers. The fabricated APD exhibits a low onset gain required for future optical fiber communications systems. A large 3-dB bandwidth of 22 GHz is maintained with a low gain of 2.8 for an APD with 100-nm avalanche-layer, and a GB product of 235 GHz is also achieved.**

Keywords—*Avalanche photodiode, field control layer, InAlAs.*

I. Introduction

High-speed avalanche photodiodes (APDs) are key devices for future optical fiber communications systems with a serial baud rate of beyond 10 Gbit/s. For example, 100-Gbit/s Ethernet systems employ four lanes of 25 Gbit/s. To extend the transmission distance maintaining its capacity, APDs with a gain-bandwidth (GB) product of higher than 200 GHz will be required. For those applications, unique InAlAs/InGaAs APD structure has been proposed, and demonstrated to have high responsivity with a GB product of 235 GHz [1, 2]. One- and four-lane receiver optical subassembly (ROSA) modules using such APDs have also been fabricated, and over-40-km transmission with the serial baud rate of 25.8 Gbit/s has been demonstrated [2, 3].

The most important figure of merit of an APD is multiplied responsivity, which determines the minimum sensitivity of a receiver. High-power tolerance is also an important parameter, because in the case of the application of APD for a receiver, the maximum optical power input to the APD tends to become higher for higher bit rate operations. In this case, the large photocurrent caused by a high-gain property of the APD can lead to failure of a trans-impedance amplifier receiving the APD's output. In order to meet such requirements, it is essential that the APDs be operated at a low gain. However, for the state-of-the-art APD with a high GB product of over 200 GHz, InAlAs is used as an avalanche layer, which has large potential barrier for electrons against the InGaAs absorption layer. The carrier-trapping or -blocking by such large potential barrier around the interface between InGaAs absorption and InAlAs avalanche layers makes the onset gain inherently large [1, 4].

In this study, we investigate lowering the onset gain for an InAlAs/InGaAs high-speed APD, and obtain a low gain operation along with a high-speed operation. For this purpose,

we employed an APD design with double p-field control layers. We confirm APD operation with an onset gain of 2.8 for 3-dB bandwidth (f_{3dB}) of 22 GHz.

II. Design of APD

Fig. 1 shows a schematic cross section of the fabricated APD based on the inverted p-down configuration [5]. The epitaxial structure was grown on a semi-insulating InP substrate using the MOCVD method. It consists of p-contact, p-InGaAs absorption, undoped-InGaAs absorption, InGaAs 2nd p-field control, InAlGaAs gap graded, InAlAs 1st p-field control, 100-nm InAlAs avalanche, n-field control, edge-field buffer, and n-contact layers. To obtain both a large f_{3dB} and high responsivity, we employed a 1-μm of hybrid (p-doped and undoped) InGaAs absorption layer, which we call the "maximized induced-current (MIC) design" [6].

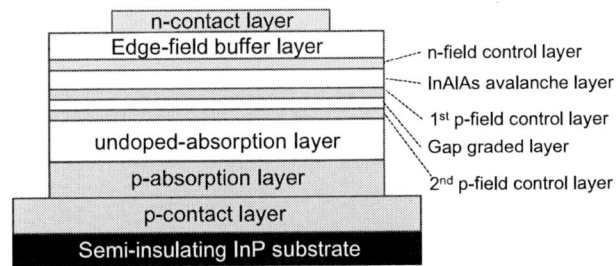

Figure 1. Schematic cross section of fabricated APD.

The calculated band diagram of the proposed APD around the InGaAs absorption and InAlAs avalanche layers at a bias voltage near the onset condition without illumination is shown in Fig. 2. In order to explain a role of 2nd p-field control layer, that of an APD without 2nd p-field control layer is also shown. Note that the electric fields at avalanche layer are the same for both APDs, i.e., the gains are the same. Since there is a potential barrier for electrons of as large as 0.5 eV between the InGaAs absorption and InAlAs avalanche layers for the APD, photo-injected electrons can be trapped at the gap-graded region. In order to effectively eliminate the potential barrier, higher bias voltage to bend the conduction band from the InGaAs absorption to InAlAs avalanche layers is needed, results in that the APD needs the larger gain at avalanche layer. This is the reason why the APD without 2nd p-field control layer exhibits large onset gain. In contrast, the APD with the

978-1-4673-6130-9/13 $31.00 © 2013 IEEE

2nd p-field control layer has no potential barrier between the absorption and avalanche layers at onset condition. This is because the electric field at gap grading layer is already strong when the 2nd p-field control layer is depleted and the electric field in the InGaAs absorption layer starts to rise. The injected electrons can pass through the gap-grading region under the lower electric field condition for the InAlAs avalanche layer. Thus, our proposed APD exhibits lower onset gain compared to the APD without 2nd p-field control layer.

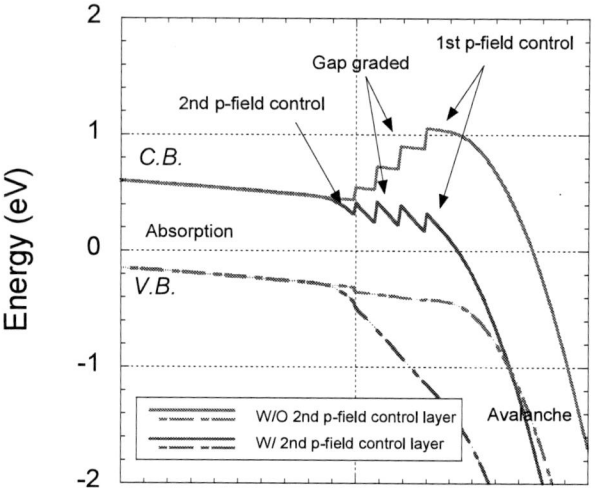

Figure 2. Band diagram around the 2nd p-field control layer.

lower electric filed in the InAlAs avalanche layer. Its onset gain is successfully reduced to 2.8, while the f_{3dB} and GB-product are maintained as high as 22 GHz and 235 GHz, respectively. The proposed APD is promising for ensuring the high-power tolerance required in high-speed optical fiber communications systems with the serial baud rate of beyond 10 Gbit/s .

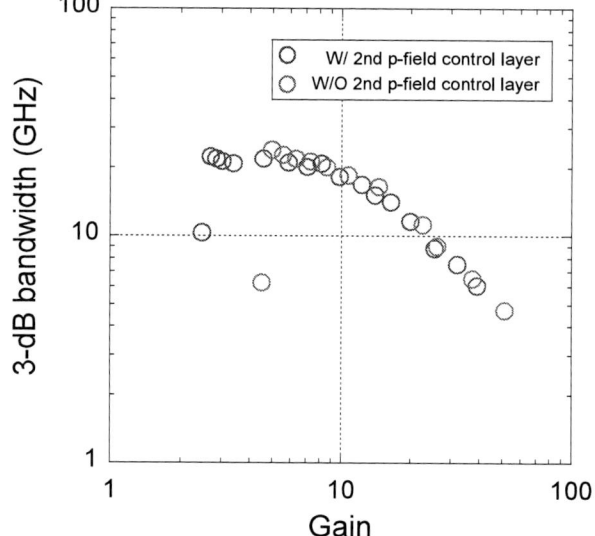

Figure 3. Gain-bandwidth characteristics of fabricated APD.

III. RESULT AND DISCUSSION

To confirm the advantage of the double p-field control layers for a low gain and a high-speed operation, relationships between the f_{3dB} and gain are investigated for the APDs with and without the 2nd p-field control layer as shown in Fig. 3. The f_{3dB}s of both APDs markedly rise at a certain onset gain, and reach to higher than 20 GHz, which is enough to 25-Gbit/s operations. The APD with the 2nd p-field control layer exhibits a low onset gain of 2.8 with the maximum f_{3dB} of 22 GHz. On the other hand, the APD without the 2nd p-field control layer shows a relatively high onset gain of as large as 4.8 to ensure the f_{3dB} of over 20 GHz. This pronounced effect in lowering the gain for a high-speed operation is due to the 2nd p-field control layer. When the gain is increased further, the f_{3dB}s of the both APDs are gradually decreased, and show the same behavior against the gain. The evaluated GB product of our proposed APD is 235 GHz, which allows 25-Gbit/s operation while maintaining high gain.

IV. SUMMARY

We proposed a novel APD structure employing the double p-field control layers, and investigated its onset behavior. The potential barrier for electrons between InGaAs absorption and InAlAs avalanche layers is effectively eliminated by its design, and thus the proposed APD can operate at a high-speed with a

ACKNOWLEDGMENT

The authors thank K. Hachisuka, S. Chiba, I. Kotaka, and S. Ando for device fabrication, and T. Akeyoshi and S. Suzuki for their continuous encouragement.

REFERENCES

[1] M. Nada, Y. Muramoto, H. Yokoyama, T. Ishibashi, and S. Kodama, 'InAlAs APD with high multiplied responsivity-bandwidth product (MR-bandwidth product) of 168 A/W·GHz for 25 Gbit/s high-speed operations', Electron. Lett., vol. 48, p. 397, 2012.

[2] M. Nada, Y. Muramoto, H. Yokoyama, T. Ishibashi, and S. Kodama, 'High-sensitivity 25 Gbit/s avalanche photodiode receiver optical sub-assembly for 40 km transmission', Electron. Lett., vol. 48, p. 777, 2012.

[3] T. Yoshimatsu, M. Nada, M. Oguma, H. Yokoyama, T. Ohno, Y. Doi, I. Ogawa, and E. Yoshida, 'Compact and High-Sensitivity 100-Gb/s (4 × 25 Gb/s) APD-ROSA with a LAN-WDM PLC Demultiplexer', ECOC2012, Th.3.B.

[4] M. Lahrichi, G. Glastre, E. Derouin, D. Carpentier, N. Lagay, J. Decobert, and M. Achouche, 240-GHz Gain-Bandwidth Product Back-Side Illuminated AlInAs Avalanche Photodiodes," IEEE Photon. Tech. Lett., vol. 22, p. 1373, 2010.

[5] M. Nada, Y. Muramoto, H. Yokoyama, T. Ishibashi, and S. Kodama,: 'Inverted InAlAs/InGaAs avalanche photodiode with low–high–low electric field profile', Jpn. J. Appl. Phys., vol. 51, p. 1373, 2012.

[6] Y. Muramoto, and T. Ishibashi, 'InP/InGaAs pin photodiode structure maximising bandwidth and efficiency', Electron. Lett., vol.39, p. 174, 2003.

MoD3-6 (Oral)
15:30 - 15:45

IPRM2013, May 19 - 23, 2013, Kobe, Japan
The 25th International Conference on Indium Phosphide and Related Materials

Monolithic Integration of InP-Based Waveguide Photodiodes with MIM Capacitors for Compact Coherent Receiver

Ryuji MASUYAMA, Hideki YAGI, Naoko INOUE, Yutaka ONISHI, Tomokazu KATSUYAMA,
Takehiko KIKUCHI, Yoshihiro YONEDA and Hajime SHOJI

Transmission Devices R & D Laboratories, Sumitomo Electric Industries, LTD.
1,Taya-cho, Sakae-ku, Yokohama, 244-8588, Japan
Phone: +81-45-853-7318 Fax: +81-45-852-2913
E-mail: masuyama-ryuuji@sei.co.jp

Abstract— We have demonstrated monolithic integration of InP-based waveguide photodiodes (WGPDs) with metal-insulator-metal (MIM) capacitors using the 3-inch diameter wafer process. The uniformity of MIM capacitance and the dielectric breakdown voltage of the capacitor obtained within +/-2 % and over 100 V, respectively. The dark current of WGPDs was less than 3 nA at a reverse voltage of 1.6 V, owing to the InP passivation structure formed on WGPD surface. The fabricated integrated chip size was 2.0 mm x 5.1 mm. These results indicate that integrated WGPDs can provide both smaller and easier assembly for the compact coherent receiver.

Keywords— *waveguide photodiode, capacitor, coherent receiver, monolithic integration*

I. INTRODUCTION

To realize a compact coherent receiver, monolithic integration of a 90° hybrid and waveguide photodiodes (WGPDs) is very effective [1]-[3]. In optical receiver modules, capacitors are usually assembled with photodiodes (PDs) and pre-amplifiers to get good and stable characteristics. If it is possible to integrate the passive components in the optical modules on an identical substrate, the size and assembly cost of optical modules can be significantly reduced. And also, the integrated chips with the large dimensions need to be fabricated by a larger diameter wafer process to reduce the fabrication cost. In this paper, we have demonstrated monolithic integration of InP-based WGPDs with metal-insulator-metal (MIM) capacitors for the compact coherent receiver using the 3-inch diameter wafer process.

II. DEVICE STRUCTURE AND FABRICATION PROCESS

Fig. 1 shows (a) the photomicrograph of the InP-based integrated PD chip and the wire connection with pre-amplifiers, and (b) SEM photograph of WGPD with the MIM capacitor. The chip size was 2.0 mm x 5.1 mm. The InP-based integrated PD chip consists of two parts, a GaInAsP core waveguide part which functions as a 90° hybrid and an InP/GaInAs WGPD part which is arranged on four terminals of the 90° hybrid. These waveguides are monolithically formed by the butt-joint regrowth process, i-line stepper optical

lithography and reactive ion etchting techniques. The four WGPDs function as photodiode pairs which consist of inner-CHs and outer-CHs. The four output pads on the rear side of integrated PD chips and the four input pads on the front side of pre-amplifiler chips are designed to face each other to minimize the wire length for better frequency response.

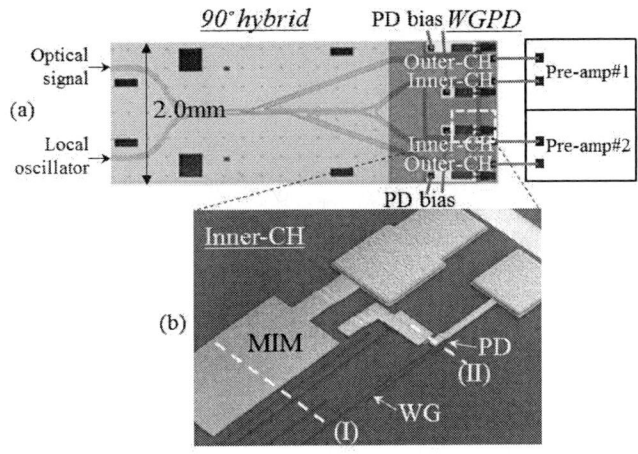

Figure 1. (a) photomicrograph of the InP-based integrated PD chip and the wire connection with pre-amplifiers, and (b) SEM photograph of WGPD with the MIM capacitor.

Fig. 2 shows the schematic cross-sectional view at the positions of broken lines (I) and (II) in Fig. 1 (b). The lower capacitor electrode is connected with the n-ohmic electrode of WGPD. The upper capacitor electrode is connected with the ground electrode through the via hole which formed in the semi-insulating substrate. This design can reduce parasitic indactance as low as possible, because it does not need a wire connection between the integrated capacitor and the pre-amplifier. Furthermore, we have applied an air bridge structure for the p-ohmic electrode to reduce parasitic capacitance of WGPD.

978-1-4673-6130-9/13 $31.00 © 2013 IEEE

MoD3-6 (Oral)
15:30 - 15:45

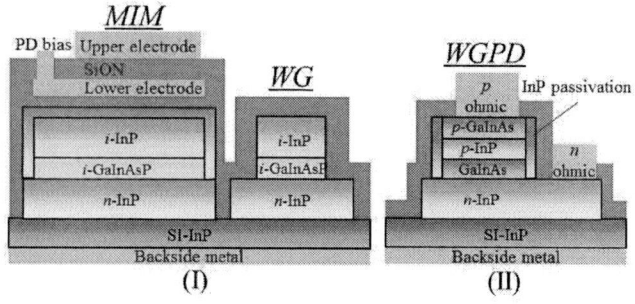

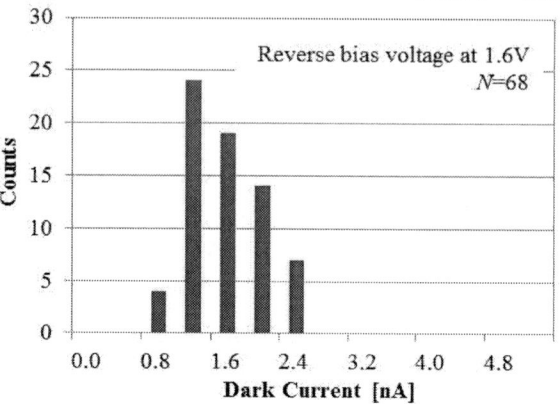

Figure 2. Schematic cross-sectional view at the positions of broken lines (I) and (II) in Fig. 1 (b).

Figure 4. Histogram of dark currents for WGPD on a 3-inch diameter semi-insulating substrate.

WGPDs have the InP passivation structure on their surfaces [4]-[5]. The InP passivation layer was re-grown by organometallic vapor phase epitaxy to reduce a dark current. The insulating films of capacitors were silicon oxynitride (SiON) deposited by chemical vapor deposition. The ohmic electrodes and the lower electrode of the MIM structure were formed by a lift-off technique. The upper electrode of the MIM structure was formed by gold plating. The MIM capacitors were formed in a GaInAsP core waveguide part where the flat surface of the epitaxially grown layer leads to the uniform performance. The area of the each MIM capacitor was 31,200 μm^2. The unit-area capacitance was 0.16 fF/μm^2.

WGPDs with capacitors exhibited a high responsivity of 0.14 A/W at a wavelength of 1.55 μm and wide bandwith of 22 GHz at a reverse voltage as low as 1.6 V. The bandwidth is corresponding to the bandwidths limited by the RC time and the carrier transit time, and no degradation of the frequency response due to the integrated capacitor was observed. We confirmed the dynamic characteristics enough for the coherent receiver oprerating up to 32 Gbaud.

III. PERFORMANCE

Fig. 3 shows the histogram of capacitances for the MIM capacitor on a 3-inch diameter semi-insulating substrate. The measured capacitances had a good uniformity, and it was within +/-2 % for the target of 5.0 pF. The dielectric breakdown voltage of the capacitors was over 100 V.

IV. SUMMARY

We have demonstrated monolithic integration of InP-based WGPDs with MIM capacitors using the 3-inch diameter wafer process. The uniformity of the MIM capacitance and the dielectric breakdown voltage of the capacitor obtained within +/-2 % and over 100 V, respectively. The fabricated integrated chip size was 2.0 mm x 5.1 mm. These results indicate that integrated WGPDs can provide both smaller and easier assembly for the compact coherent receiver.

REFERENCES

[1] V. Houtsma, N. G. Weimann, T. Hu, R. Kopf, A. Tate, J. Frackoviak, R. Reyes, Y. K. Chen, L. Zhang, C. R. Doerr and D. T. Neilson, "Manufacturable Monolithically Integrated InP Dual-Port Coherent Receiver for 100G PDM-QPSK Applications," OFC 2011, paper OML2, 2011.

[2] P. Runge, S. Schubert, A. Seeger, K. Janiak, J. Stephan, D. Trommer, A. Matissl, "Monolithic InP Receiver Chip with a 90° Hybrid and 56GHz Balanced Photodiodes," ECOC2012, paper Mo.2.E.3, 2012.

[3] H. Yagi, N. Inoue, Y. Onishi, R. Masuyama, T. Katsuyama, T. Kikuchi, Y. Yoneda and H. Shoji, "High-efficient InP-based balanced photodiodes integrated with 90° hybrid MMI for compact 100 Gb/s coherent receiver," to be presented at OFC2013.

[4] R. Yamabi, Y. Tsuji, K. Hiratsuka and H. Yano, "Fabrication of mesa-type InGaAs pin PDs with InP passivation structure on 4-inch diameter InP substrate," Proc. 16th Int. Cnf. Indium Phosphide and Related Materials, pp.245 (2004).

[5] R. Yamabi, T. Kagiyama, Y. Yoneda, S. Sawada and H. Yano, "Mesa-type InGaAs pin PDs with InP-passivation structure monolithically integrated with resistors and capacitors with large capacitance," Proc. 19th Int. Cnf. Indium Phosphide and Related Materials, pp.87 (2007).

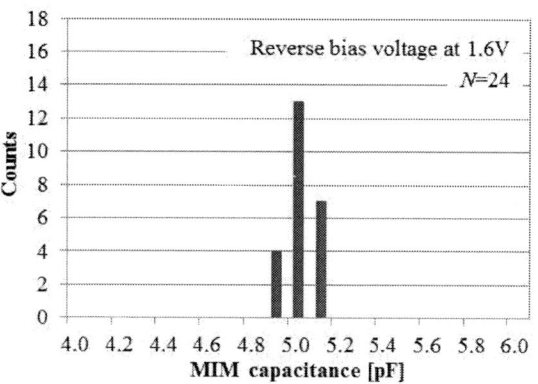

Figure 3. Histogram of capacitances for the MIM capacitor on a 3-inch diameter semi-insulating substrate.

Fig. 4 shows the histogram of dark currents for WGPD on a 3-inch diameter semi-insulating substrate. The dark current of WGPDs was less than 3 nA at a reverse voltage of 1.6 V, owing to the InP passivation structure formed on the WGPD surface.

978-1-4673-6130-9/13 $31.00 © 2013 IEEE

MoD3-7 (Oral)
15:45 - 16:00

Monolithic InP Receiver Chip with a 90° Hybrid and a Variable Optical Attenuator for 100GBit/s Colourless WDM Detection

P. Runge[1], S. Schubert[1], A. Seeger[1], T. Gärtner[1], K. Janiak[1], J. Stephan[2], D. Trommer[2] and M. Lønstrup Nielsen[2]

[1] Fraunhofer Heinrich-Hertz-Institute, Einsteinufer 37, 10587 Berlin, Germany
[2] u²t Photonics AG, Reuchlinstr. 10/11, 10553 Berlin, Germany
patrick.runge@hhi.fraunhofer.de

Abstract—**We demonstrate a monolithically integrated quadrature coherent receiver photonic integrated circuit (PIC) on an InP substrate with a 90° optical hybrid, a variable optical attenuator (VOA) and four pin-photodetectors. With an attenuation of more than 20dB the VOA enables the usage of the receiver PIC for colourless WDM detection.**

Keywords—*coherent detection; monolithic InP receiver; 90° hybrid; variable optical attenuator*

I. INTRODUCTION

In today's 100G long haul WDM optical fiber transmission links 112Gbits/s dual-polarisation quadrature phase-shift keying (DP-QPSK) is the commonly used modulation format. Typical receiver frontends of these transmission links are intradyne coherent receiver with a 90° hybrid mixing the received data signal (S) and the local oscillator signal (LO). As a result of this mixing process the phase modulated signal is converted into an amplitude modulation being detectable by the photodiodes.

For separating the different WDM channels, wavelength selective components such as AWGs are used on chip-level [1]. Resent considerations revealed that the wavelength selective components become obsolete if the wavelength of the local oscillator is tunable and the power ratio between S and LO can be adjusted [2].

According to the colourless WDM detection concept, Fig.1 illustrates a corresponding receiver chip by adding a variable optical attenuator (VOA) to the signal path. Lately, such a device has been presented on a silica-based photonic integrated circuit (PIC) with hybrid integrated photodiodes [3], where the VOA has been realised with a tunable Mach-Zehnder structure. Another possibility is the integration of a semiconductor optical amplifier in the signal path [4] being driven in the operation point of an electro-absorption modulator. But the latter realisation concept is not necessarily transparent to all modulation formats and baud rates because of the carrier relaxation times in the active material.

In this contribution, we demonstrate a monolithic InP based receiver chip with a 90° hybrid, a VOA and four pin-photodiodes. The fabricated devices are supposed to operate as a colourless WDM detector in either the C-band or the L-band.

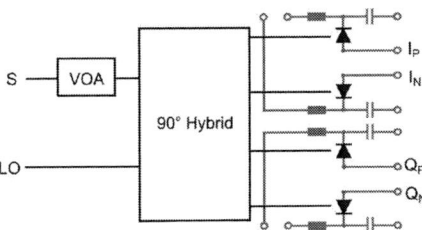

Figure 1: Conceptual scheme of the receiver PIC for colorless WDM

II. DESIGN AND FABRICATION

Receiver PICs according to Fig.1 have been fabricated. The receiver PICs include two full fibre spot-size converters (one for S and one for LO), an optical attenuator in the signal path, an optical 90° hybrid, and four pin-photodiodes.

A DC reverse voltage is applied to the array of photodiodes using an integrated bias network consisting of integrated capacitors and resistors (Fig.1).

A chip prototype has been fabricated on a 3-inch InP wafer. Semi-insulating InGaAsP/InP waveguide layers and the doped detector layers are grown by single-step MOVPE. The structuring of the photodiodes and the waveguides is done by etching. The waveguide integrated pin photodiodes comprise an InGaAs absorption layer and heterostructure contact layers in order to allow high responsivities and high bandwidths. The fabricated photodiodes have a 3dB bandwidth of 30GHz on chip level. The optical facet of the chips is AR coated to reduce insertion loss and to avoid back reflection into the local oscillator.

III. MEASUREMENT RESULTS

The focus of the presented results is on the performance of the VOA of the receiver PIC since the performances of the 90° hybrid and the photodiodes are already well known from previous publications [5, 6].

The attenuation of the signal on the receiver PIC in dependence of the applied bias current is shown in Fig.2. An attenuation of more than 20dB can be obtained. Unfortunately, the maximum possible attenuation could not be obtained from these measurements because the step size of the VOA current has been set to large. Moreover, a fixed measurement range was used for the detection of the PD current causing a fluctuation for low PD current which are in the region of the lower range limit.

The attenuation for a sweep of the bias current in dependence of the wavelength is presented in Fig.3. An attenuation of more than 20dB is expected over a spectrum of at least 40nm. A small drift of the attenuation characteristic for different wavelengths can be observed. The effect can be put into perspective if the slight drift in Fig.2 is taken into account. Theoretically, in Fig.2 the minima of the PD currents should be for all four PDs at the same VOA current. It is assumed that the effect is caused by the imperfect measurement setup.

Furthermore, a polarization dependent loss (PDL) of the VOA of smaller than 0.4dB could be measured over the entire C-band and L-band (Fig.4). The small PDL indicates that the contacts on the waveguides of the VOA do not interact with the propagating signal.

IV. CONCLUSION

We proposed and realised a monolithic InP coherent receiver chips with optical 90° hybrids, VOAs and photodiodes. The VOAs show excellent performance regarding the attenuation and the PDL. Packaging the receiver PIC including a free space optic with a low loss polarisation beam splitter [5], the performance of the device is extended to dual-polarisation modulation formats. Also known from [6] devices for either C-band or L-band are possible.

With an attenuation of more than 20dB over an entire band the performance of a packaged receiver chips should satisfy the demands for colourless WDM detection.

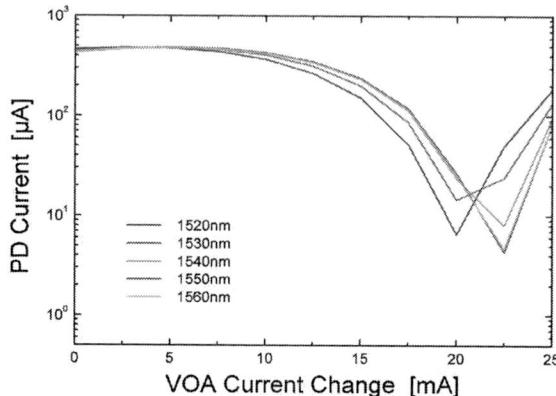

Figure 3: Wavelength dependence of the variable optical attenuator

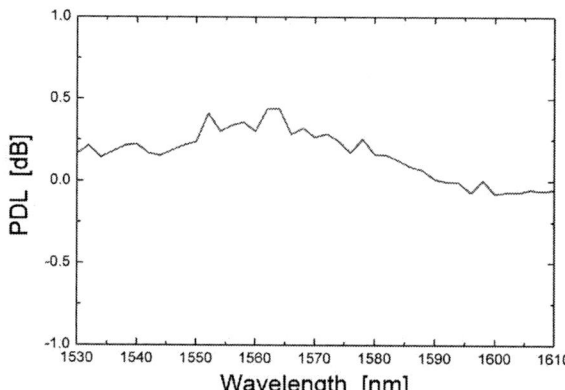

Figure 4: Polarisation dependent loss of the variable optical attenuator

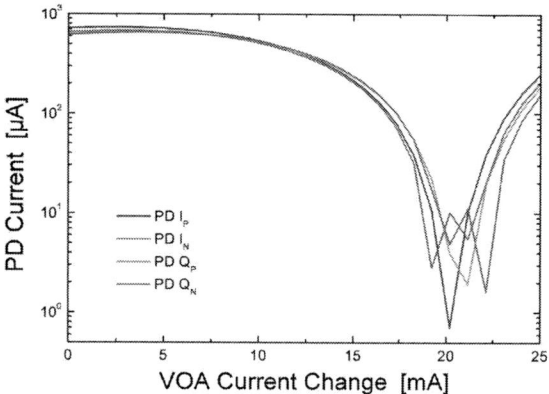

Figure 2: Attenuation characteristics of the receiver PIC

REFERENCES

[1] R. Nagarajan, J. Rahn, M. Kato, J. Pleumeekers, D. Lambert, V. Lal, H-S. Tsai, A. Nilsson, A. Dentai, M. Kuntz, R. Malendevich, J. Tang, J. Zhang, T. Butrie, M. Raburn, B. Little, W. Chen, G. Goldfarb, V. Dominic, B. Taylor, M. Reffle, F. Kish, and D. Welch, "10 Channel, 45.6Gb/s per Channel, Polarization-Multiplexed DQPSK, InP Receiver Photonic Integrated Circuit," J. Lightw. Technol., vol 29, pp. 386-395, February 2011.

[2] B. Zhang, C. Malouin, and T. J. Schmidt, "Towards full band colorless reception with coherent balanced receivers," Opt. Express, vol. 20, pp. 10339–10352, April 2012.

[3] S. Tsunashima, F. Nakajima, Y. Nasu, R. Kasahara, Y. Nakanishi, T. Saida, T. Yamada, K. Sano, T. Hashimoto, H. Fukuyama, H. Nosaka, and K. Murata, "Silica-based, compact and variable-optical attenuator integrated coherent receiver with stable optoelectronic coupling system," Opt. Express, vol. 20, pp. 27174–27179, November 2012.

[4] K. N. Nguyen, P. J. Skahan, J. M. Garcia, E. Lively, H. N. Poulsen, D. M. Baney, and D. J. Blumenthal, "Monolithically integrated dual-quadrature receiver on InP with 30 nm tunable local oscillator," Opt. Express, vol. 20, pp. B716–B721, December 2011.

[5] A. Beling, N. Ebel, A. Matiss, G. Unterbörsch, M. Nölle, J. K. Fischer, J. Hilt, L. Molle, C. Schubert, F. Verluise, and L. Fulop, "Fully-Integrated Polarization-Diversity Coherent Receiver Module for 100G DP-QPSK," Proc. OFC'11, OML5, 2011.

[6] P. Runge, S. Schubert, A. Seeger, K. Janiak, J. Stephan, D. Trommer, P. Domburg, and M. Lønstrup Nielsen, "Monolithic InP receiver chip with a 90° hybrid and 56GHz balanced photodiodes," Opt. Express, vol. 20, pp. B250–B255, December 2012.

Highly non-linear phenomena and coherent effects in 1500 nm QD lasers and amplifiers

Gadi Eisenstein, Amir Capua, Ouri karni

Electrical Engineering Dept. Technion – Israel Institute of Technology
Haifa 32000 Israel
J. P. Reithmaier
Institute of Nanostructure Technologies and Analytics, Kassel University
Kassel 34132 Germany
gad@ee.technion.ac.il

Abstract—**We present dynamical properties of quantum dot lasers and amplifiers operating under extreme nonlinear conditions. Several unique phenomena including coherent quantum interactions are demonstrated**

Keywords Quantum dots, semiconductor nonlinearities, coherent light matter interaction

I. INTRODUCTION

The dynamical properties of quantum dot (QD) and quantum dash (Qdash) lasers and amplifiers have been mainly studied previously in the context of their use in telecom applications [1, 2]. Device characteristics such as threshold, minimum temperature sensitivity and wide band modulation in lasers as well as gain, saturation and noise properties of amplifiers have been demonstrated often and are rather well understood.

QD gain media can also serve for more fundamental investigations which shed light on basic dynamical properties while enabling to demonstrate new phenomena. We report here several such phenomena taking place in QD and QDash gain media. Operating a 1550 nm QDash amplifier under very large optical and electrical excitations leads to an instantaneous gain response, observed using a unique multi wavelength pump probe measurement system [3]. Using a high resolution X-FROG set up [4], we observed quantum coherent light-matter interactions in an electrically driven laser amplifier operating at room temperature. To this end, we demonstrated clear Rabi oscillations and self-induced transparency and proved that quantum mechanical principles can be clearly implemented in practical devices.

II. InAs/InP QD AND QDASH GAIN MEDIA

The devices we study are based on MBE grown InAs/InP QD and QDash material. The QDashes are well developed and have been used successfully for several years. They offer amplifiers with a broad gain bandwidth and a fast response [1] as well as direct laser modulation at rates up to 10 Gbit/s [5]. Newly developed InAs/InP QDs [6] offer significant improvements in terms of record modal gain (above 10 cm^{-1} and digital modulation at up to 20 Gbit/s [7]. The optical gain material was studied using Atom Probe Tomography and the QDs were found to be somewhat elongated with an elongation ratio of up to 5:1, small enough to ensure three dimensional carrier confinement.

III. RESULTS

The results we present hereon relate to two unique phenomena. The first is two-photon induced responses in which nonlinear absorption initiates an instantaneous gain response and also laser oscillations. The second describes coherent light matter interactions using an ultra-short optical pulse and an X-FROG measurement.

A. Two photon induced gain

A multi wavelength pump probe measurement system in which the pump is a 100 fs pulse, the probe is a tunable CW signal and detection involves nonlinear sampling of the output CW probe by a replica of the pump pulse was used to study QD and QDash amplifiers. Under large optical and electrical excitations, the obtained responses are as shown in Fig. 1.

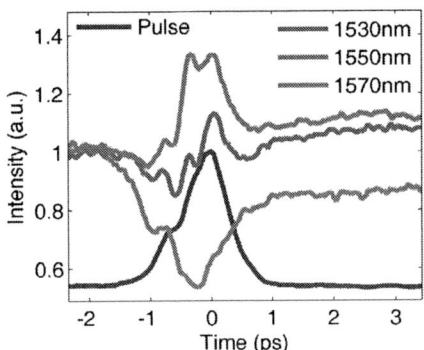

Fig. 1 Multi wavelength pump probe responses

The black trace in Fig. 1 describes the autocorrelation of the output perturbation pulse. It defines the time frame where t=0. The pump pulse is centered at 1550 nm and probing at the same wavelength (1550 nm) yields the green trace which shows classical saturation at t=0 followed by a recovery to a level lower than that prior to the pulse arrival. In contrast, the 1570 nm probe exhibits an almost instantaneous gain increase at t=0 followed by a steady state transmission level higher than

1. The gain increase is due to carriers generated by two photon absorption (TPA) which relax from high energy levels to all the ground states and avail the gain increase. The long term high transmission is a known signature of TPA. A similar response is observed at 1530 nm except that the response starts later and is weaker. This results from the fact that the pulse has some energy at 1530 nm and hence conventional one photon processes compete with the TPA at early times before the nonlinear response takes over.

A second TPA induced effect is shown in Fig. 2. A QD laser is driven below threshold emitting spontaneous emission.

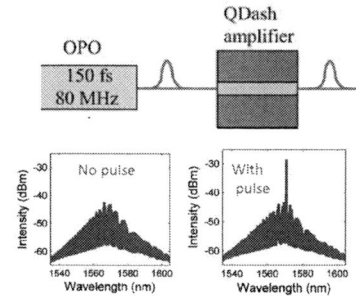

Fig. 2 Two photon induced laser oscillations

Injection of an energetic pulse induces long term extra gain which is sufficient to reach threshold so that the laser emits now a narrow line.

B. *Quantum coherent light matter interactions*

Direct observation of coherent quantum light matter interactions requires that the electronic wave function persists longer than the observation time. At room temperature, the semiconductor de-phasing time is 1-2 ps and therefore, all previous measurements were performed at cryogenic temperatures. Rather than increasing the de-phasing time (by cooling), it should also be possible to observe coherent interactions by shortening the measurement time. This is the approach we chose where using an ultra-sensitive X-FROG system which enables to record the complete complex electric field (amplitude and phase) with a single fs resolution, we could indeed observe directly Rabi oscillations and self-induced transparency in an electrically driven room temperature QD laser amplifier.

Fig. 3 shows time dependent amplitude and instantaneous frequency traces measured at the output of a QD laser amplifier into which a short, 100 fs pulse was injected at various intensities. As the pulse intensity grows, the amplitude profile tends to break up while the instantaneous frequency exhibits oscillations. The left figures are for a pulse centered at 1560

nm while the right figures are for 1530 nm. An FDTD simulation of the system at hand reveals that all the slight wiggles in the phase response are real and represent a portion of the coherent effect observed here.

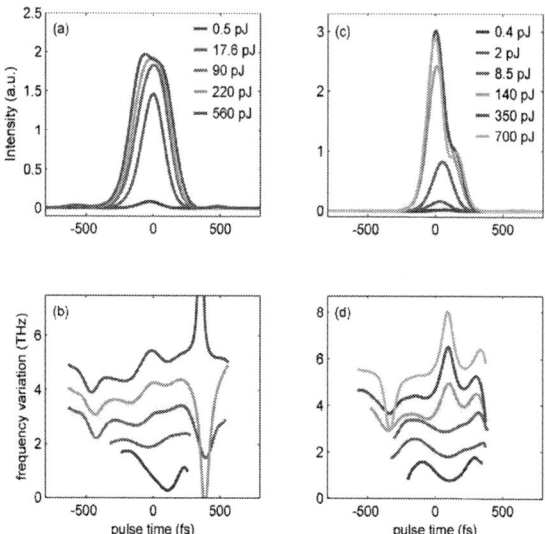

Fig. 3 Rabi oscillations in a QD optical amplifier operating at RT. Left figures 1560 nm right figures 1530 nm

IV. CONCLUSIONS

To conclude, we have described several unique properties of 1550 nm QD and QDash gain media including highly nonlinear phenomena (a TPA induced instantaneous gain response and lasing) as well as quantum coherent interactions which bring quantum mechanical principles to the realm of practical electrically driven semiconductor devices operating at room temperature.

REFERENCES

[1] J.P. Reithmaier, G. Eisenstein et al., J. of Physics D, **38**, 2088, 2005

[2] D.Bimberg et al. IEEE J Sel. Topics in QE **3**, 196, 1997

[3] A. Capua, G. Eisnetein, J.P. Reithmaier, Appl. Physics Lett., 131108, 2010

[4] A. Capua, G. Eisnetein, J.P. Reithmaier, Optics Express **20**, 347, 2012

[5] B. Dagens, IEEE PTL, **20**, 903, 2008

[6] C. Gilfert, J. Reithmaier et al., Appl. Physics Letters, **98**, 201102, 2011.

[7] J.P. Reithmaier, G. Eisenstein, ISLC 2012 PD. 3

MoD4-1 (Invited)
16:30 - 17:00

Selective area MOVPE of InGaAsP and InGaN systems as process analytical and design tools for OEICs

Yukihiro Shimogaki[1], Masakazu Sugiyama[2], and Yoshiaki Nakano[3]

[1] Department of Materials Engineering, The University of Tokyo, Tokyo, Japan
[2] Institute of Engineering Innovation, The University of Tokyo, Tokyo, Japan
[3] Research Center for Advanced Science and Technology, The University of Tokyo, Tokyo, Japan
e-mail shimo@dpe.mm.t.u-tokyo.ac.jp

Abstract—**Numerical simulation on growth rate non-uniformity of selective area growth (SAG) in metal-organic vapor phase epitaxy (MOVPE) is an effective method to examine the surface reaction kinetics, which is normally hindered by mass transport rate of precursors. SAG is also an effective tool to fabricate opto-electronic integrated circuits (OEICs) to reduce the process steps. In the present talk, the kinetic analyses on InGaAsP and InGaN MOVPE processes using SAG will be presented.**

Keywords—*Selective Area Growth, InGaAsP, InGaN, OEICs*

I. INTRODUCTION

Metal-organic vapor phase epitaxy (MOVPE) is a useful process to grow III-V and III-Nitride compound semiconductor epitaxial thin films, which will be used as optical devices and high-performance electronic devices and solar cells [1-3]. As the mass production of these devices are required, it is important to design large scale reactor and its optimum process conditions. The reaction mechanism involved in the above mentioned MOVPE processes are complicated and difficult to understand. Especially, the surface reaction mechanism and its reaction rate constant are hard to theoretically estimate using quantum chemical calculations. Experimental approach to examine these surface chemistries is also difficult, because the rate of surface reaction is normally hindered by the mass transport rate of the reacting species. We have developed numerical analysis on the growth rate non-uniformity of the SAG to extract the surface kinetics [4-8]. The obtained kinetic information can be used to design a suitable mask to fabricate OEICs. The SAG-MOVPE process can be developed to fabricate monolithically integrated multi-color LEDs based on InGaN/GaN material system [9, 10].

II. BASIC CONCEPT OF SAG AND ITS APPLICATIONS

A. Mechanism of SAG and its Applications

When SiO₂ thin films are formed on III-V or III-N substrate and patterned using lithographic procedures as masks for SAG, epitaxial growth occurs only on a bare substrate as shown in Fig. 1. As the sticking probability, η, of incoming growth species (which is unknown but may be a gas-phase reaction products such as CH₃Ga (MMGa)) is zero, the growth species are not consumed on the mask and they will be accumulated on

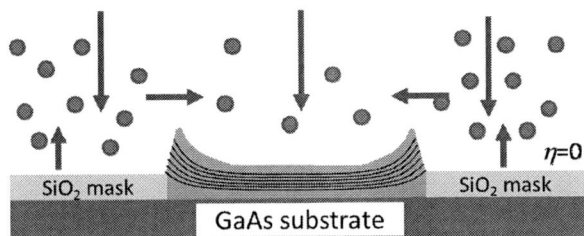

Figure 1. Mechanism of SAG.

the vicinity of the mask. Then the growth species will diffuse into the epitaxial growth region and enhance the growth rate. This growth rate enhancement may be proportional to the mask size, which will realize the local control of the growth rate using SAG techniques. For example, the well width of MQWs can be modified by the size and shape of the mask, which will cause the change of effective band gap energy of the material. This technique was successfully applied to the fabrication of four channel DFB laser array for CWDM systems [11].

B. Numerical Analysis on SAG to Extract Surface Kinetics

As shown in Fig. 1, the lateral diffusion of the growth species from the mask side to the growth area is the main cause of the growth rate enhancement (GRE). The lateral diffusion occurs both via the surface and the gas-phase. The diffusion length of the surface diffusion is normally within a micrometer, while that of gas-phase will be more than 10 μm. When we design the growth area (width of the un-masked area in Fig. 1) more than 10 μm, we can neglect the effect of surface diffusion. Then we can estimate the GRE by solving the diffusion equation shown in Fig. 2. C is the gas-phase concentration of the growth species. D is the gas-phase diffusion coefficient and k_s is the linear surface reaction rate constant of the growth species, respectively. The value of D/k_s, which is a kind of diffusion length, is the only parameter that controls the GRE profile. The GRE profile can be experimentally obtained using surface profiler and can be adjusted to the numerical calculation by choosing appropriate D/k_s value. D can be estimated from Chapman-Enskog equation or can be obtained from macro-scale growth rate profile in the MOVPE reactor [12]. Thus, k_s can be experimentally obtained even if the film growth rate is limited by the diffusional mass transport of the growth species.

MoD4-1 (Invited)
16:30 - 17:00

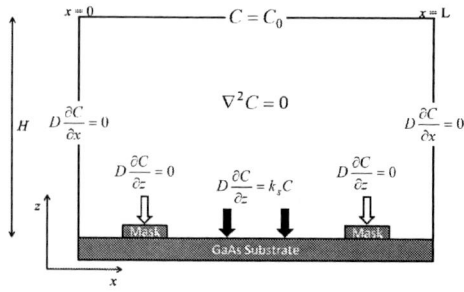

Figure. 2 Basic equation and boundary conditions for numerical analysis on SAG GRE profile.

The kinetic analysis on the surface chemistry of CVD processes using step coverage analysis in a trench structure had been developed by the authors [13, 14]. The advantage of using SAG as an analytical tool to investigate the surface kinetics is its capability to examine the crystal orientation effects as will be discussed below.

III. KINETIC ANALYSIS ON MOVPE SURFACE CHEMISTRY

Fig. 3 shows the temperature dependency of the k_s for GaAs MOVPE obtained from SAG analysis. GaAs (100) substrates with different offset angle were used.

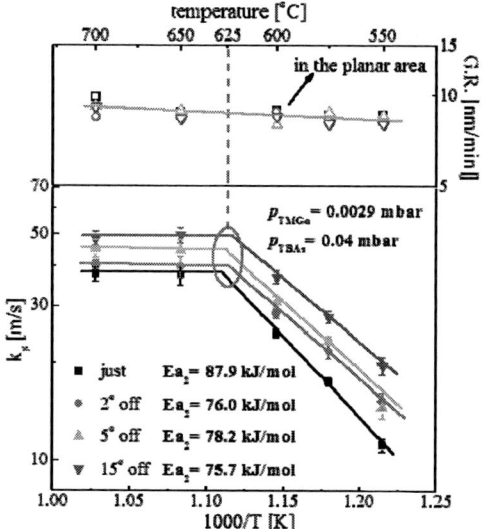

Figure. 3 Temperature dependency of GaAs growth rate and k_s value.

The growth rate of GaAs did not show temperature and substrate offset angle dependencies, because the growth rate is limited by the diffusional mass transport of the growth species. The k_s value obtained from SAG analysis, however, showed temperature and offset angle dependencies. The absolute value of k_s is higher for higher offset angle and the activation energy, Ea, is bit higher for exact (100) surface (offset angle = 0°). The higher offset angle results in a surface with higher step density and the growth can easily proceed by step flow mode, while on the exact surface, 2-dimensional nucleation is required for the film growth and Ea becomes higher than step flow growth. The activation energy was changed over 625°C due to the surface reconstruction during GaAs growth.

IV. MASK AND PROCESS DESIGN FOR SAG

Fig. 4 shows the PL peak wavelength distribution of the InGaAsP-MQWs with different mask width. The measured wavelength could be exactly estimated by numerical analysis based on the obtained kinetics, which suggests the possibility of designing the optimum mask size and shape, and process conditions for OEICs. The fabrication of InGaN/GaN based LEDs using SAG techniques will be also presented in the conference.

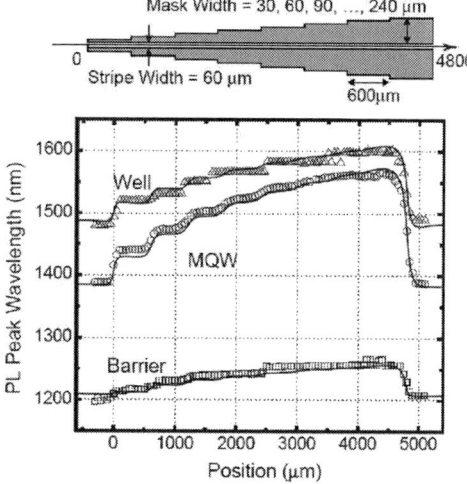

Figure. 4 PL wavelength distribution of InGaAsP MQWs by SAG.

ACKNOWLEDGMENT

The authors would like to acknowledge Dr. Haizheng Song and Dr. Tomonari Shioda for their great contributions.

REFERENCES

[1] J. J. Coleman, R.M. Lammert, M. L. Osowski, and A. M. Jones, IEEE J. Sel. Top. Quantum Electron, 3, 874-884 (1997).

[2] S. Pellegrino and L. Tarricone, Mater. Chem. Phys., 66, 189-196 (2000).

[3] G.B. Stringfellow, Mater. Sci. Eng. B, 87, 97-116 (2001).

[4] H.J. Oh, M. Sugiyama, Y. Nakano, and Y. Shimogaki, Jpn. J. Appl. Phys., 42, 6284-6291 (2003).

[5] H.J. Oh, M. Sugiyama, Y. Nakano, and Y. Shimogaki, J. Crystal Growth, 261, 419-426 (2004).

[6] Y. Wang, H. Song, M. Sugiyama, Y. Nakano, and Y. Shimogaki, Jpn. J. Appl. Phys., 47, 7788-7792 (2008).

[7] T. Shioda, Y. Tomita, M. Sugiyama, Y. Shimogaki, and Y. Nakano, Jpn. J. Appl. Phys., 46, L1045-1047 (2007).

[8] T. Shioda, M. Sugiyama, Y. Shimogaki, and Y. Nakano, Appl. Phys. Exp., 1, 071102 (2008).

[9] T. Shioda, M. Sugiyama, Y. Shimogaki, and Y. Nakano, J. Crystal Growth, 311, 2809-2812 (2009).

[10] T. Shioda, Y. Tomita, M. Sugiyama, Y. Shimogaki, and Y. Nakano, IEEE J. Selected Topics in Quantum Electonics, 15, 1053-1065 (2009).

[11] J. Darja, M. J. Chan, M. Sugiyama, and Y. Nakano, IEICE Electronics Express, 3, 522-528 (2006).

[12] I.T. Im, H.J. Oh, M. Sugiyama, Y. Nakano, and Y. Shimogaki, J. Crystal Growth, 261, 214-224 (2004).

[13] H. Komiyama, Y. Shimogaki, and Y. Egashira, Chemical Engineering Science, 54, 1941-1957 (1999).

[14] L.S. Hong, Y. Shimogaki, and H. Komiyama, Thin Solid Films, 365, 176-188 (2000).

MoD4-2 (Oral)
17:00 - 17:15

MOVPE growth of InAs/InP QDs on directly-bonded InP/Si substrate

Keiichi Matsumoto, Xinxin Zhang, Yoshonori Kanaya and Kazuhiko Shimomura

Department of Engineering and Applied Sciences, Sophia University,
7-1 Kioi-cho, Chiyoda-ku, 102-8554
Tokyo, Japan
e-mail: kshimom@sophia.ac.jp

Abstract— InP/Si substrate has been fabricated by employing wet-etching and wafer direct bonding technique. The surface of the InP/Si substrate was very smooth and no strain was observed. On top of the substrate, InAs/InP quantum dots (QDs) have been monolithically grown using metal organic vapor phase epitaxy (MOVPE). According to photo-luminescence (PL) measurement, almost the same intensity, peak wavelength and full width half of maximum (FWHM) have been observed compared to QDs on InP substrate.

Keywords—Directly-bonded InP/Si substrate, InAs/InP, Quantum dots, stranski-krastanov growth mode, double-cap procedure.

Recently, there has been great interest in integrating InP and other high mobility materials on Si substrates to make high performance complementary metal-oxide semiconductor (CMOS) devices. In our last work, we have developed InP/Si substrate by employing wafer direct bonding technique and metal organic vapor phase epitaxy (MOVPE) growth of GaInAs/InP multiple quantum well (MQW) structure has been demonstrated [1.2]. In this report, we show the realization of InAs/InP quantum dots (QDs) structure on the directly-bonded InP/Si substrate.

The preparation of the InP/Si substrate and growth of InAs/InP QDs structure are explained as follows. The substrates used in this work were mirror-polished InP substrate and Si substrate. First, GaInAs / InP template layer / GaInAs / InP / InP substrate structure was grown by low pressure MOVPE. After that, both as-grown InP and Si were cleaned with ultra sonic cleaning in order to avoid particles between substrates. Then the as-grown sample was dipped in HCl:H$_2$O solution to etch-off the InP substrate. After that, both GaInAs / InP / GaInAs structure and Si substrate were cleaned with a H$_2$SO$_4$:H$_2$O$_2$:H$_2$O solution to make their surfaces hydrophilic and obtain InP template, followed by de-ionized water rinse. Then surfaces were contacted in de-ionized water. After dried with N$_2$ gas flow in atmosphere, InP/Si substrate was realized. After that, the substrate was loaded into a furnace and heated in a N$_2$ atmosphere at 450 degree-C for 60 min. Thorough this process, InP/Si substrate was firmly fused.

Figure 1(a) and (b) show the scanning electron microscopy (SEM) image of directly-bonded InP/Si substrate. From fig. 1(a), both substrates were firmly bonded using direct bonding technique. The surface was mirror-like and, as shown in this fig. 1(b), very smooth.

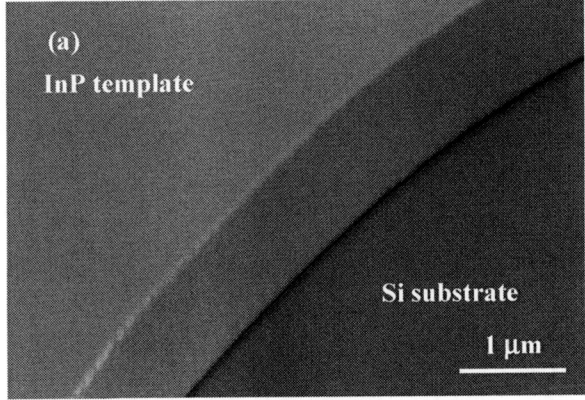

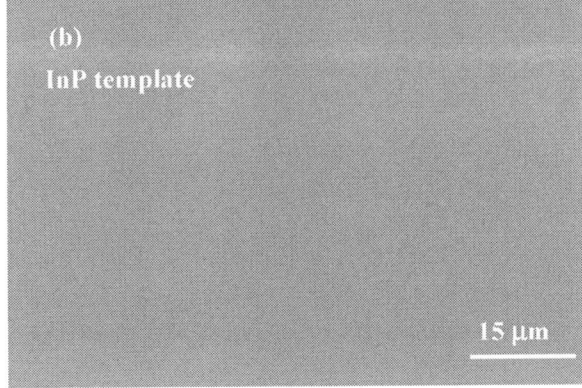

Fig. 1. SEM photograph of **(a)** margin part **(b)** center part of directly bonded InP/Si substrate.

According to contact mode atomic force microscopy (AFM), the root-mean-square (RMS) surface roughness of InP template on Si substrate was under 2.5 nm and almost the same value compared to InP substrate. The lattice strain distribution was measured by X-ray diffraction (XRD) system and results are shown in Fig. 2. As shown in this figure, no strain was generated during direct bonding process. Then, on top of the substrate, InAs QDs were grown by Stranski-Krastanov (S-K) grown mode using MOVPE growth technique and double cap procedure [3]. The growth temperature and pressure of QDs were 540 degree-C and 15 torr, and other layers were 640

978-1-4673-6130-9/13 $31.00 © 2013 IEEE

degree-C and 100 Torr. The precursors used in this growth were TBA, TBP, TMI, and TEG. Fig. 3 shows the schematic layer structure of the fabricated 3 stacked QDs, which is applicable for broadband LED [4]. As can be seen in Table I, the height of QDs of each layer was 2.5 nm, 4.5 nm and 5.5 nm from the substrate, and all the thickness of InP second cap layer was 22 nm. The Ga composition of $Ga_xIn_{1-x}As$ buffer layer under the QDs was 0.47, 0.38, and 0.38 from the substrate, and the thickness was 1.5 nm.

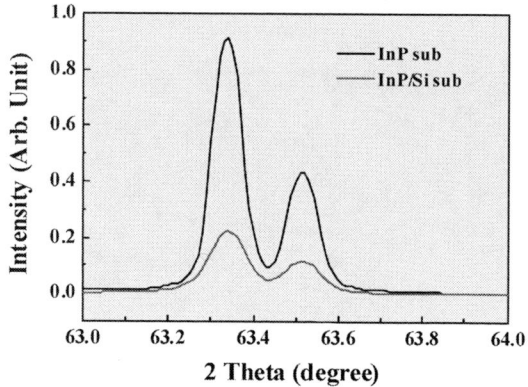

Fig. 2. X-ray diffraction curve of the directly-bonded InP/Si substrate.

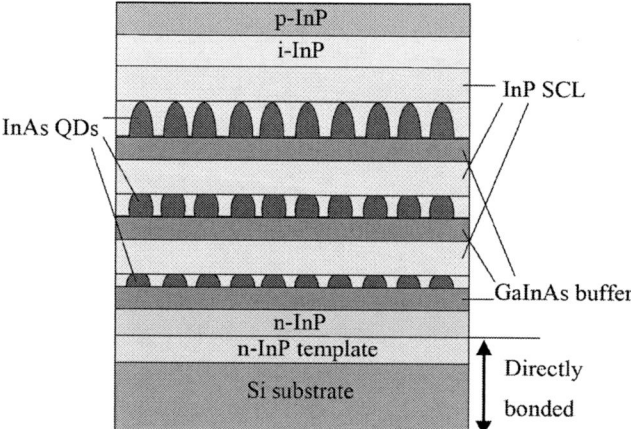

Fig. 3. Schematic layer structure of 3 stacked InAs QDs.

TABLE I. Thickness of FCL and Ga composition of buffer layer.

FCL	$Ga_xIn_{1-x}As$
5.5 nm	x=0.38
4.5 nm	x=0.38
2.5 nm	x=0.47

Fig. 4 shows the photo-luminescence (PL) spectrum for InAs QDs grown on InP/Si substrate and InP substrate. As presented in Table II, the peak wavelength of the PL spectrum from InP/Si substrate was 1650 nm and almost the same with that from InP substrate. Furthermore, the peak intensity was also comparable to InP substrate and degraded only 9 %. The full width half of maximum (FWHM) of PL spectrum from InP/Si and InP substrate were 330 nm and 420 nm, which is 78.5 % as against InP sample. These results suggest the three QDs layers grown on InP/Si substrate were effectively luminescenced.

These crystal growth techniques using InP/Si substrate are expected to achieve a good quality and integration of several functional optical devices on Si substrate.

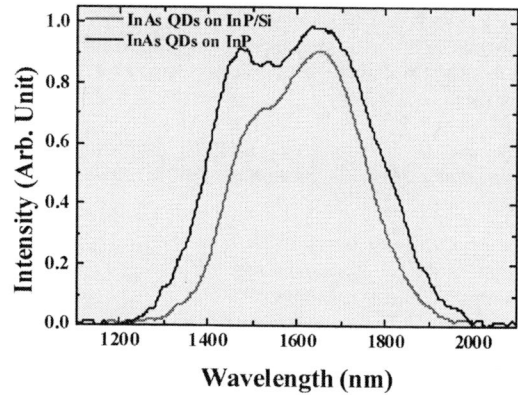

Fig. 4. PL spectra of InAs QDs grown on InP/Si substrate.

TABLE II. Comparison between InAs QDs grown on InP/Si and InP with regard to Wavelength, Intensity and FWHM.

	InAs QDs/InP/Si	InAs QDs/InP
Wavelength (nm)	1650	1646
Intensity (a.u.)	0.91	1
FWHM (nm)	330	420

[1] K. Matsumoto, T. Makino, K. Kimura, and K. Shimomura, "GaInAs/InP MOVPE growth on directly bonded InP/Si substrate", ICMOVPE-XVI, Busan, Korea, MoB2-4, May 2012.

[2] K. Matsumoto, T. Makino, K. Kimura, and K. Shimomura, "Growth of GaInAs/InP MQW using MOVPE on directly-bonded InP/Si substrate", J. Cryst. Growth, in press.

[3] Y. Iwane, F. Kawashima, M. Hirooka, T. Saegusa, and K. Shimomura, "InAs/InP QDs grown by selective MOVPE growth using double-cap procedure for broadband LED improved p-cladding layer", physica status solidi (c), vol.9, pp.210-213, Nov. 2011.

[4] S. Yoshikawa, T. Saegusa, Y. Iwane, M. Yamauchi, and K. Shimomura, "Flat-Topped Emission with Spectral Width above 500nm from InAs/InP Quantum Dot Waveguide Array Light-Emitting Diode", Appl. Phys. Express, vol.5, pp.092103-1-092103-3, Aug. 2012.

MoD4-3 (Oral)
17:15 - 17:30

From surface dimer orientations to bonds at the GaP/Si(100) heterointerface

In situ reflection anisotropy spectroscopy during MOVPE-growth of GaP on Si(100)

Oliver Supplie[1,2], Sebastian Brückner[1,3], Henning Döscher[1,3], Peter Kleinschmidt[1,3,4], and Thomas Hannappel[1,3,4]

[1] Helmholtz-Zentrum Berlin, Institute Solar Fuels, Hahn-Meitner-Platz 1, 14109 Berlin, Germany
[2] Humboldt-Universität zu Berlin, Institut für Physik, Newtonstr. 15, 12489 Berlin, Germany
[3] Technische Universität Ilmenau, Institut für Physik, Postfach 100565, 98684 Ilmenau, Germany
[4] CiS Forschungsinstitut für Mikrosensorik und Photovoltaik, Konrad-Zuse-Str. 14, 99099 Erfurt, Germany
Corresponding author: oliver.supplie@helmholtz-berlin.de

Abstract—Despite intense research in III-V-on-silicon-heteroepitaxy since the 1980ies, ultra-high efficiency opto-electronic devices are yet to be realized. Pseudomorphic growth of GaP films on Si(100) is an adequate model system to study the polar-on-non-polar heterointerface which still is a major source of defects. Generally, in situ control of growth processes is vital in order to understand and, finally, avoid defect formation. While metalorganic vapor phase epitaxy is of big interest regarding device production at large scale, the existence of a process gas limits in situ access to electron-based surface science techniques. Reflection anisotropy spectroscopy (RAS), however, is applicable in vapor phase. Dimerized (100) surfaces of cubic crystals usually exhibit characteristic RA spectra, as known for both monohydride-terminated Si(100) and for P-rich GaP(100). Rotation of an anisotropic structure by 90° flips the sign of its RA spectrum so that dimer orientations are controllable in situ during surface preparation. This is essential, in particular, directly before III-V nucleation since process parameters strongly affect the surface formation. Moreover, the polarity of the GaP film determines the orientation of the P-dimers at the P-rich GaP/Si(100) surface. Knowing both the polarity of the GaP film and the dimer orientation at the Si(100) substrate prior to nucleation, a simplistic model published by Beyer et al. allows to estimate whether Si—Ga or Si—P bonds are preferred at the heterointerface. Our findings are in favor of Si—P bonds which we demonstrate here for both preferentially A-type and B-type Si(100) 2° substrates.

Keywords—*III-V/Si(100); heterointerface; RAS; MOVPE*

I. INTRODUCTION

The polar-on-non-polar heterointerface [1] still is a major challenge in III-V/Si(100) heteroepitaxy in order to produce defect-free epilayers for high-efficiency devices. We study the growth of GaP on Si(100) since the small lattice mismatch allows pseudomorphic growth of thin GaP films making it both an adequate model system for interface analysis [2] and an interesting quasi-substrate for further monolithic III-V integration [3]. We apply in situ reflection anisotropy spectroscopy (RAS) during surface preparation and growth by metalorganic vapor phase epitaxy (MOVPE) [4,5]. RA spectra

This work was supported by the BMBF (project no. 03SF0329C).

of dimerized surfaces are often characteristic, for a basic understanding of its features, however, benchmarking to UHV-based surface science techniques is essential [6].

Single-domain Si(100) surfaces, where dimers on adjacent terraces are aligned in parallel and only even-numbered atomic layer steps occur, are desired in order to avoid anti-phase disorder in the GaP epilayer [1,7]. Detailed knowledge about the RA spectra of Si(100) allowed to study the interaction of silicon with the process gas H_2 in situ [8] and, finally, to prepare almost single-domain Si(100) surfaces [9] of both preferential A-type and B-type [10,11]. Dimers are oriented perpendicular to the step edge on an A-type terrace and parallel to the step edge on a B-type terrace. Their RAS signal is flipped in sign [11]. In situ RAS is of outmost importance since the preparation route strongly varies with offcut [11].

Here, we show that the polarity of the GaP epilayer is dependent on the majority dimer orientation of the Si(100) surface prior to III-V nucleation. This can be understood by a prevalence of either Si—P or Si—Ga bonds occurring identically on both substrate types. Applying a simplified ball-and-stick model [12], we only need the in situ RA spectra of the Si(100) surface prior to III-V nucleation and of the P-rich GaP/Si(100) surface to deduce the preferred bonding at the interface. We find that Si—P bonds are favored on both A- and B-type substrates, which is in contrast to ex situ results in [12].

II. EXPERIMENTAL

All samples were prepared in an AIX-200 MOVPE reactor and monitored with a LayTec EpiRAS 200 RA spectrometer. Benchmarking of the RA spectra via a contamination-free MOVPE-to-UHV transfer [6] to XPS (Specs Focus 500 and Phoibos 100), LEED (Specs ErLEED 100-A) and STM (Specs Aarhus 150) verified the cleanliness and type of surface reconstruction, as well as the sensitivity of the RAS signals to the dimerized surfaces [9,11,13,14]. The Si(100) 2° →[011] substrates were thermally deoxidized and a Si buffer was grown followed by dedicated annealing for the desired surface termination [9,10]. On both substrate types, 40nm GaP were grown identically with TBP and TEGa after pulsed nucleation [10] and, finally, P-rich surfaces were prepared [13].

III. RESULTS AND DISCUSSION

In analogy to P-rich GaP(100) [14-16], the P-rich GaP/Si(100) surface reconstructs (2x1)-like forming buckled P-dimers stabilized by one H atom per dimer [13]. This reconstruction gives rise to a characteristic RA spectrum [13,14,16] and the sign of the signal corresponds to the dimer orientation [7]. Fig.1 shows the RA spectra of P-rich GaP/Si(100) for GaP grown both on A-type and B-type monohydride-terminated Si(100).

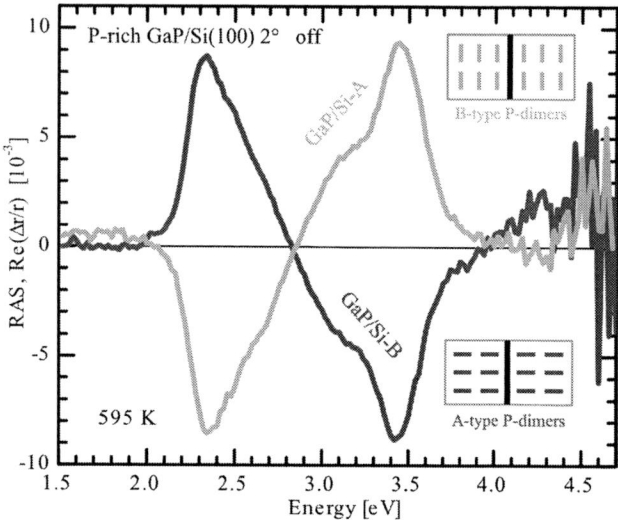

Figure 1. In situ RA spectra of P-rich prepared GaP grown on A-type Si(100) (orange line) and B-type Si(100) (blue line) respectively. The insets indicate the dimer orientations of the majority domain relativ to the step edge.

According to Fig.1, the dimer orientation on the P-rich GaP/Si(100) surface is depending on the dimer orientation of the Si(100) substrate prior to nucleation [10]. We checked the RA spectra of the substrate und thereby the desired surface termination directly before GaP nucleation. A rotation of the P-dimer at the P-rich GaP/Si(100) surface corresponds to a change in GaP polarity due to the zincblende structure. Viewed along [011], consequently, P-polar GaP grew on B-type Si(100) and growth on A-type Si(100) lead to Ga-polar GaP.

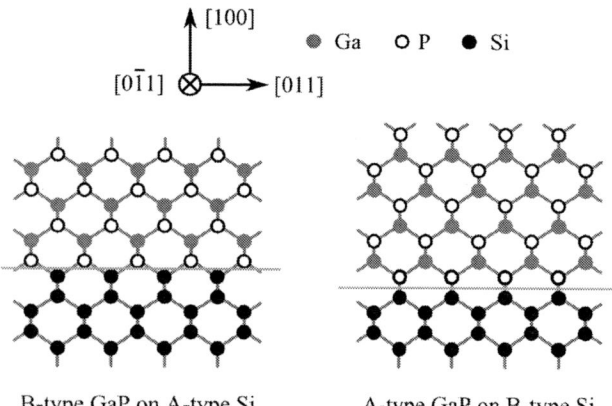

Figure 2. Bonding situation at the GaP/Si(100) heterointerface (green line).

The polarity of the GaP film is a consequence of the binding situation at the GaP/Si(100) heterointerface. Beyer et al. published a simplified ball-and-stick model [12], which, given known the GaP polarity and the dimer orientation of the substrate prior to nucleation, allows to determine if Si—P or Si—Ga bonds are preferred. Applying the same assumptions as in [12] (eg. no rearrangement of the Si atoms during nucleation, charge neutrality at the interface neglected), our RA spectra indicate that Si—P bonds are favored for both type of Si(100) substrates. Once benchmarked to surface science tools, the only input needed here are RA spectra of the Si(100) surface directly prior to nucleation (which is vital on low offcut substrates since annealing in H_2 can lead to monolayer removal of Si atoms [11]) and of the final P-rich GaP/Si(100) surface. Though nucleation is similar, our result is in contrast to [12], where, applying LEED and CBED, Si—Ga bonds were suggested to be favored for Si(001) with 0.1° offcut.

REFERENCES

[1] H. Kroemer, "Polar-on-nonpolar epitaxy", J. Cryst. Growth, vol.81, p.193, 1987.

[2] O. Supplie, T. Hannappel, M. Pristovsek, and H. Döscher, "In situ access to the dielectric anisotropy of buried III-V/Si(100) heterointerfaces", Phys. Rev. B, vol.86, p.035308, 2012.

[3] J. Geisz, J. Olson, D. Friedman, K. Jones, R. Reedy, and M. Romero, "Lattice-matched GaNPAs-on-Silicon Tandem Solar Cells", IEEE PVSC, vol.31, p.695, 2005.

[4] J-T. Zettler, "Characterization of epitaxial semiconductor growth by reflectance anisotropy spectroscopy and ellipsometry", Prog. Cryst. Growth Char. Mat.,vol.35, p.27, 1997.

[5] P. Weightman, DS. Martin, RJ. Cole, and T. Farrell, "Reflection Anisotropy Spectroscopy", Rep. Prog. Phy., vol.68, p.1251, 2005.

[6] T. Hannappel, S. Visbeck, L. Töben, and F. Willig, "Apparatus for investigating metalorganic chemical vapor deposition-grown semiconductors with ultrahigh-vacuum based techniques", Rev. Sci. Instrum, vol.75, p.1297, 2004.

[7] H. Döscher, T.Hannappel, B. Kunert, A. Beyer, K. Volz, and W. Stolz, "In situ verification of single-domain III-V on Si(100) growth via metal-organic vapor phase epitaxy", Appl. Phys. Lett., vol.93, p.122110, 2008.

[8] S. Brückner, H. Döscher, P. Kleinschmidt, and T. Hannappel, „In situ investigation of hydrogen interacting with Si(100)", Appl. Phys. Lett., vol.98, p.211909, 2011.

[9] S. Brückner, H. Döscher, P. Kleinschmidt, O. Supplie, A. Dobrich, and T. Hannappel, "Anomalous double-layer step formation on Si(100) in hydrogen process ambient", Phys. Rev. B, vol.86, p.195310, 2012.

[10] O. Supplie et al., submitted.

[11] S. Brückner et al., submitted

[12] A. Beyer et al., „GaP heteroepitaxy on Si(001): Correlation of Si-surface structure, GaP growth conditions, and Si-III/V interface structure", J. Appl. Phys., vol.111, p.083534, 2012.

[13] H. Döscher, and T. Hannappel, "In situ reflection anisotropy spectroscopy analysis of heteroepitaxial GaP films grown on Si(100)", J. Appl.Phys., vol.107, p.123523, 2012.

[14] L. Töben, T. Hannappel, K. Möller, H. Crawack, C. Pettenkofer, and F. Willig, "RDS, LEED and STM of the P-rich and Ga-rich surfaces of GaP(100)", Surf. Sci., vol.494, p.L755, 2001.

[15] P. Kleinschmidt, H. Döscher, P. Vogt, T. Hannappel, "Direct observation of dimer flipping at the hydrogen-stabilized GaP(100) and InP(100) surfaces", Phys. Rev. B, vol.83, p.155316, 2011.

[16] PH. Hahn, WG. Schmidt, F. Bechstedt, O. Pulci, and R. del Sole, "P-rich GaP(001)(2x1)/(2x2) surface: A hydrogen-adsorbate structure determined from first-principles calculations", Phys. Rev. B, vol.68, p.033311, 2003.

MoD4-4 (Oral)
17:30 - 17:45

in-situ Characterization of MOCVD grown GaAs- and InP-based tunable VCSEL structures

Christian Grasse[2], Yuto Tomita[1*], Peter Wiecha[2], Ralf Meyer[2], Tobias Gründl[2],
Michael Müller[2], and Markus-Christian Amann[2]
[1]LayTec AG, Seesener Str. 10-13, 10709 Berlin, GERMANY
*yuto.tomita@laytec.de
[2]Walter Schottky Institut, Technische Universität München, Am Coloumbwall 3, 85748 Garching, GERMANY

Abstract— **In-situ monitoring of growth parameters such as thickness and quality of InP and GaAs based materials is presented. This is a key technology for the fabrication of optoelectronic devices like VCSELs.**

Keywords— *in-situ monitoring, MOCVD, InP, GaAs, VCSEL*

I. INTRODUCTION

Vertical Cavity Surface Emitting Laser diodes (VCSELs) are semiconductor devices with light emission perpendicular to the chip surface. They are highly attractive for applications in optoelectronics, since they offer several advantages compared to conventional edge-emitting (in-plane) laser diodes, in terms of RF performance, low power consumption, and on-chip testing. The device structure of VCSELs is highly complex (see Figure 1) and the characteristics of the final device strongly depends on the quality of distributed Bragg reflectors (DBRs) which are composed of an alternating stack of thin layers of two different materials with high refractive index contrast. Therefore in-situ monitoring of thickness and composition of layers in complete structures, especially DBR mirrors is a key method to achieve best device characteristics. We demonstrate in-situ monitoring during the growth of various devices such as VCSEL, tunable VCSEL, and quantum cascade laser (QCL).

II. EXPERIMENTAL

All devices were grown on InP substrates using a metal organic chemical vapor deposition (MOCVD) reactor (AIX200/4) with precursors of TMGa, TMIn, TMAl, Arsine, Phosphine, SiH4, and CBr4. For in-situ measurement a LayTec EpiTT sensor was used. Reflectance measurement during the growth was done at 633nm and 950nm which are suitable for InP and GaAs based materials. By measuring Fabry-Pérot oscillations (as seen in Figure 2 and Figure 5), the layer thickness and growth rate can be derived from the reflectance transients. Changes in the reflectance amplitude were used to monitor surface roughening of the films.

III. GROWTH OPTIMIZATION OF VCSEL STRUCTURES

Figure 2 shows an in-situ reflectance measurement taken during the growth of a VCSEL device structure with a DBR

mirror, an InP cladding layer, an active region and a buried tunnel junction (BTJ). By monitoring thickness and growth rate in the DBR mirror, the cavity length was precisely controlled. This is of great significance for laser characteristics to achieve an intense emission at the desired wavelength.

(a) **VCSELs**

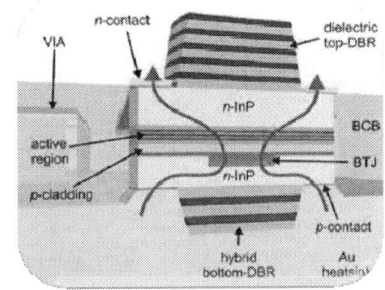

(b) **Tunable VCSELs**

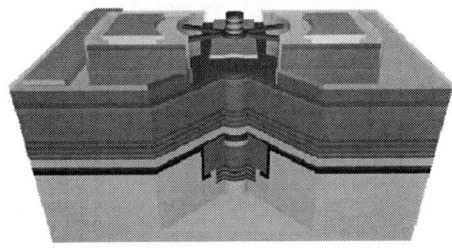

Figure 1. Typical structure of (a) VCSELs and (b) tunable VCSELs

Furthermore the behavior of surface morphology was observed by monitoring the reflectance signals. For an ideal flat surface, the reflectance is stable with a constant oscillation amplitude which was observed during the growth of the InP cladding layer in Figure 3. On the other hand, incident light is scattered on a rough surface, therefore the amplitude of reflectance signal is reduced. The quantum well (QW) layers in the active region were highly strained and its layer surface became rough, therefore a degradation of signal intensity was

978-1-4673-6130-9/13 $31.00 © 2013 IEEE

seen (Figure 3). This result was confirmed by an electron microscope picture in Figure 4. Even these thin layers (ca. 10 nm) in the QWs were able to be visible using the in-situ monitoring method. Since the reflectance signal is very sensitive to thickness, surface roughness, composition ratio and temperature, this technique is useful to check reproducibility from growth to growth and uniformity within a wafer.

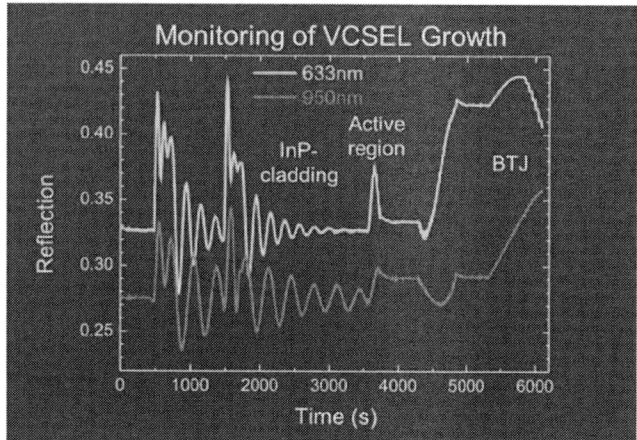

Figure 2. In-situ reflectance measurement of an InP-based VCSEL device structure at 633 nm and 950 nm.

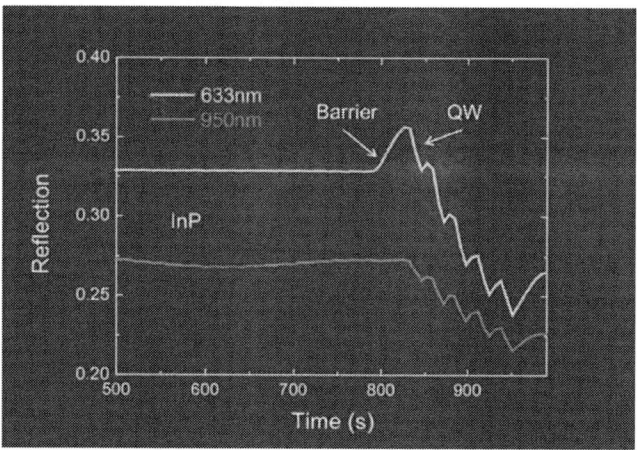

Figure 3. In-situ reflectance measurement during the growth of a strained active region at 633 nm and 950 nm.

Figure 4. A SEM picture of the surface of a QW structure.

Also a complete structure of a tunable VCSEL (ref. Figure 1 (b)) with a movable membrane containing InP, InAsP, and GaInAs layers was monitored with an EpiTT, as shown in Figure 5. Layer thickness and growth rate were precisely controlled and a flat surface could be maintained during the growth of this strained structure after in-situ metrology based growth optimization.

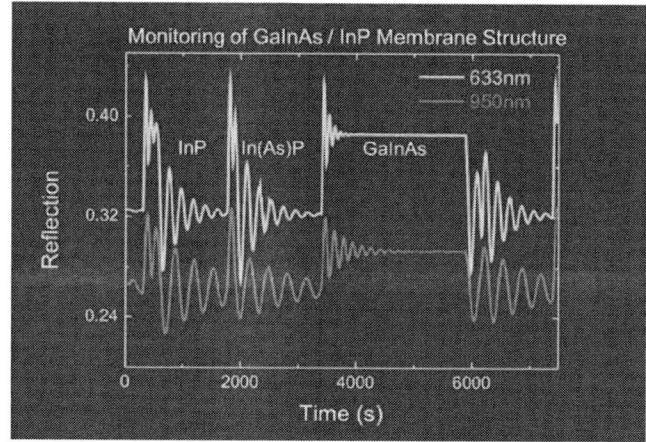

Figure 5. In-situ reflectance measurement of the growth of a tunable VCSEL

IV. CONCLUSION

Real time reflectance was measured in various types of VCSEL device structures during MOCVD. This enabled a thickness and growth rate control with a high accuracy. Precise layer thicknesses even in very thin layers (less than 10nm) have been determined. Utilizing the in-situ 633nm reflectance data the during-growth surface morphology has been optimized for achieving smooth and abrupt MQW interfaces. We demonstrated that quality of tunable VCSEL structures was successfully optimized and controlled by in-situ monitoring.

This technique leads to high reproducibility from growth to growth and is very useful for production use in addition. Further information will be given during the presentation.

MoD4-5 (Oral)
17:45 - 18:00

Preparation of single-domain Si(100) surfaces with in situ control in CVD ambient

Sebastian Brückner[1,2], Oliver Supplie[1,3], Peter Kleinschmidt[1,2,4], Anja Dobrich[1], Henning Döscher[1,2], and Thomas Hannappel[1,2,4]

[1] Helmholtz-Zentrum Berlin, Institute Solar Fuels, Hahn-Meitner-Platz 1, 14109 Berlin, Germany
[2] Technische Universität Ilmenau, Institut für Physik, Postfach 100565, 98684 Ilmenau, Germany
[3] Humboldt-Universität zu Berlin, Institut für Physik, Newtonstr. 15, 12489 Berlin, Germany
[4] CiS Forschungsinstitut für Mikrosensorik und Photovoltaik, Konrad-Zuse-Str. 14, 99099 Erfurt, Germany
Corresponding author: sebastian.brueckner@helmholtz-berlin.de

Abstract—III-V films grown heteroepitaxially on Si(100) substrates by metal-organic chemical vapor deposition (MOCVD) are desired for the combination of optoelectronics with microelectronic devices. Difficulties regarding device quality are related to the formation of the crucial III-V/Si(100) interface, where single-layer steps on the substrate surface induce anti-phase disorder in the epitaxial film. In principle, double-layer steps on the Si(100) substrate prevent the occurrence of anti-phase disorder. While the preparation of silicon surfaces is well-established in UHV, preparation in H$_2$ ambient differs considerably. Considered energetically least favorable on both the clean and the monohydride-terminated Si(100) surface, single domain surfaces with double layer steps in the unusual D$_A$ configuration were recently prepared in MOCVD ambient. The D$_A$ step formation on Si(100) with 2° offcut in CVD ambient is suggested to originate in vacancy generation and diffusion on the terraces accompanied by preferential annihilation at the step edges. Here, we investigate Si removal and vacancy formation on Si(100) substrates with large terraces under CVD preparation conditions. With in situ reflection anisotropy spectroscopy (RAS), we directly observe the domain formation in dependence of the preparation route. Oscillations in transient RA measurements indicate layer by layer Si removal during annealing in hydrogen. Based on scanning tunneling microscopy results, we conclude that vacancy island formation and anisotropic expansion preferentially in parallel to the dimer rows of the terraces explains the layer-by-layer Si removal process.

Keywords—Si(100); single-domain surfaces; RAS; MOVPE

I. INTRODUCTION

High-efficiency III-V-on-silicon devices such as multi-junction solar cells require a suitable Si(100) substrate preparation in order to achieve defect-free high-quality material. The ability to control the material composition combined with scalability of production makes metalorganic chemical vapor deposition (MOCVD) the method of choice for large scale III-V growth. While the present process gas prohibits in situ control by electron-based surface science techniques, reflection anisotropy spectroscopy (RAS) has established as surface sensitive probe of cubic crystals, in particular, regarding dimerized (100) surfaces [1]. The atomic order of the Si(100) substrate is decisive for subsequent III-V

growth since the polar III-V material is to be grown on a non-polar substrate [2]. One fundamental aspect is the avoidance of anti-phase disorder in the III-V epilayers since homopolar bonds would decrease device efficiency. As consequence of the crystal structure, double-layer steps at the heterointerface lead to undisturbed, single-domain III-V growth across the step edges. In recent studies, the surface sensitivity of RAS made it possible to study the interaction of the Si(100) surface with hydrogen as process gas [3]. Hydrogen was found to have a big impact on surface termination and step structure [4]. Characteristic features in the RA spectra of Si(100) clearly correspond to the monohydride-terminated Si(100) surface, which reconstructs forming dimers [5]. The sign of the RAS signal switches its sign if the dimers are rotated by 90° so that domain formation can be studied in situ [4]. While all dimers are aligned in parallel on single-domain Si(100) surfaces forming either A-type (dimer axes perpendicular to the step edges) or B-type (dimer axes parallel to the step edges) terraces, A-type and B-type terraces alternate on single-layer stepped surfaces.

In [4] we showed, that single-domain terraces of type A form on Si(100) with 2° misorientation towards [011] during slow cooling in hydrogen ambient. Though energetically considered unfavored [6,7], D$_A$ steps form as consequence of vacancy generation and diffusion of vacancies to the step edges where vacancies may annihilate [4]. Here, we focus on substrates with larger terraces, namely Si(100) with 0.1° misorientation towards [011]. Annealing at temperatures which were found to be decisive for single-domain formation on 2° offcut samples, in case of larger terraces leads to oscillations in the RAS signal. This indicates layer-by-layer removal of Si atoms implying a rotation of the dimer axis. Based on scanning tunneling microscopy (applied after contamination-free MOVPE-to-UHV-transfer [8]), we assign the removal to formation of vacancy islands and their anisotropic expansion preferentially parallel to the dimer rows.

II. EXPERIMENTAL

We prepared the Si(100) surfaces in an AIX-200 MOVPE reactor equipped with a LayTec EpiRAS 200 RA spectrometer. A contamination-free MOVPE-to-UHV transfer [8] allowed

This work was supported by the BMBF (project no. 03SF0329C).

XPS (Specs Focus 500 and Phoibos 100), LEED (Specs ErLEED 100-A) and STM (Specs Aarhus 150) measurements. The Si(100) 0.1° →[011] substrates were thermally deoxidized, a Si buffer was grown and subsequently annealed at about 1040K in Pd-purified hydrogen (950mbar) [4].

III. RESULTS AND DISCUSSION

RA spectra of the monohydride-terminated Si(100) surface show a characteristic peak at the E_1 transition of silicon which is related to the dimerized surface reconstruction [4,9-11]. For Si(100) samples misoriented 2° we found that the temperature range around 1040K is decisive in order to form stable D_A steps and that annihilation of vacancies at the step edges plays a major role [4]. For low-offcut samples the result differs drastically, as can be seen in the transient RA signals measured during annealing at about 1040K in 950mbar H_2 (Fig.1).

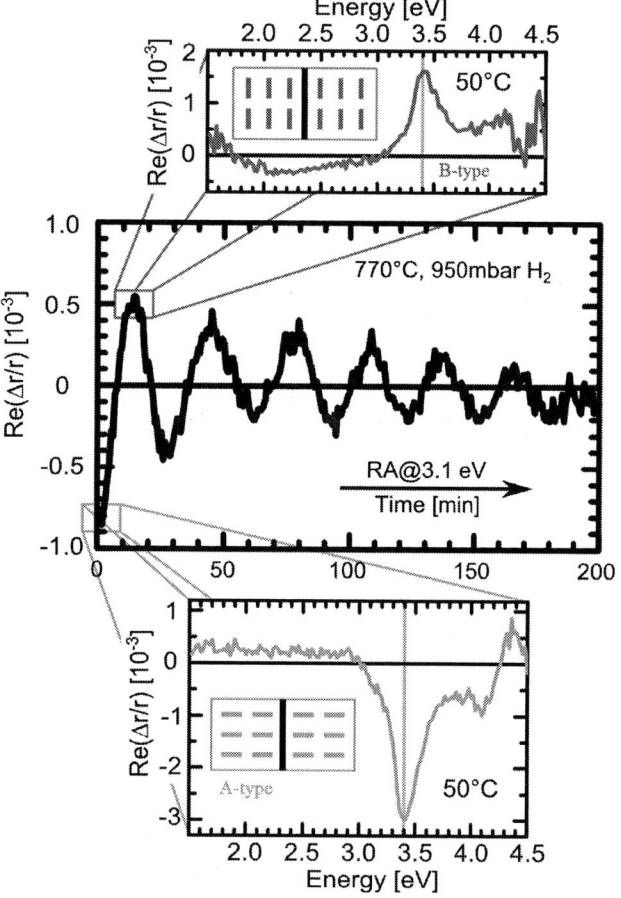

Figure 1. Transient in situ RA measured at 3.1eV during annealing of Si(100) 0.1° → [011] at about 1040K in 950mbar H_2. The insets show RA spectra of corresponding surfaces after optimized preparation and cooling to 320K. The grey lines indicate the thermal shift relative to 3.1eV at 1040K.

Since the sign of the dimer-related peak in the RA spectrum (at 3.1eV at 1040K) corresponds to the dimer orientation, the transient shows a preferential A-type domain decreasing until the contributions of both domains cancel each other out. The

former minority domain then prevails until the opposite development occurs. The prevalence of one domain can be preserved by cooling the sample, the insets in Fig.2 show corresponding RA spectra. The constant period of the oscillation in the transient indicates a uniform transformation. Consequently, Fig.2 shows layer-by-layer removal directly via the dimers which are oriented mutually perpendicular from layer to layer. Based on STM measurements, we conclude that the formation of vacancy islands on the large terraces and their anisotropic expansion, preferentially in parallel to the dimer rows of the terraces, are the reason for the observed layer-by-layer removal. Removal of the subjacent Si layer is limited by the width of the vacancy islands in the first layer and suppressed by refilling with diffusing Si adatoms detached from the step edges of the upper layer [12,13].

The domain formation of Si(100) surfaces in CVD ambient crucially depends on Si removal processes in H_2. Temperature and H_2 pressure highly influence the preparation of nearly exact Si(100) substrates, so that in situ control is essential in order to prepare single-domain substrates for subsequent III-V heteroepitaxy.

REFERENCES

[1] P. Weightman, DS. Martin, RJ. Cole, and T. Farrell, "Reflection Anisotropy Spectroscopy", Rep. Prog. Phy.,vol.68, p.1251, 2005.

[2] H. Kroemer, "Polar-on-nonpolar epitaxy", J. Cryst. Growth, vol.81, p193, 1987.

[3] S. Brückner, H. Döscher, P. Kleinschmidt, and T. Hannappel, „In situ investigation of hydrogen interacting with Si(100)", Appl. Phys. Lett., vol.98, p.211909, 2011.

[4] S. Brückner, H. Döscher, P. Kleinschmidt, O. Supplie, A. Dobrich, and T. Hannappel, "Anomalous double-layer step formation on Si(100) in hydrogen process ambient", Phys. Rev. B, vol.86, p.195310, 2012.

[5] DJ. Chadi, "Stabilities of single-layer and bilayer steps on Si(001) surfaces", Phys. Rev. Lett., vol.59, p.1691, 1987.

[6] TW. Poon, S. Yip, PS. Ho, and FF. Abraham, "Equilibrium structures of Si(100) stepped surface", Phys. Rev. Lett. 65, 2161 (1990).

[7] AR. Laracuente and LJ. Whitman, "Step structure and surface morphology of hydrogen-terminated silicon: (0 0 1) to (1 1 4)", Surf. Sci., vol.545, p.70, 2003.

[8] T. Hannappel, S. Visbeck, L. Töben, and F. Willig, "Apparatus for investigating metalorganic chemical vapor deposition-grown semiconductors with ultrahigh-vacuum based techniques", Rev. Sci. Instrum,. vol 75, p.1297, 2004.

[9] R. Shioda, and J. van der Weide, "Observation of hydrogen adsorption on Si(001) by reflectance difference spectroscopy", Appl. Surf. Sci., vol.130-132, p.266.

[10] M. Palummo, N. Witkowski, O. Pluchery, R. del Sole, and Y. Borensztein, "Reflectance-anisotropy spectroscopy and surface differential reflectance spectra at the Si(100) surface: Combined experimental and theoretical study", Phys. Rev. B,vol.79, p.035327, 2009.

[11] A. Dobrich, P. Kleinschmidt, H. Döscher, and T. Hannappel, "Quantitative investigation of hydrogen bonds on Si(100) surfaces prepared by vapor phase epitaxy", J. Vac. Sci. Technol. B, vol.29, p.04D114, 2011.

[12] B. Poelsema, L. K. Verheij, and G. Comsa, ""Two-Layer" Behavior of the Pt(111) Surface during Low-Energy Ar+-Ion Sputtering at High Temperatures", Phys. Rev. Lett., vol.53, p.2500, 1984.

[13] P. Bedrossian and T. Klitsner, "Anisotropic vacancy kinetics and single-domain stabilization on Si(100)-2×1", Phys. Rev. Lett., vol.68, p.646, 1992.

MoPI-1 (Poster)
18:30 - 20:30

Zn Diffusion in Ruthenium Doped InP with Annealing by Metalorganic Vapor Phase Epitaxy

H. Yamaguchi, T. Nagira, Z. Kawazu, K. Ono and M. Takemi

High Frequency & Optical Device Works, Mitsubishi Electric Corporation
4-1 Mizuhara, Itami, Hyogo 664-8641, Japan
Yamaguchi.Harunaka@dy.MitsubishiElectric.co.jp

Abstract— Ruthenium (Ru) as the semi-insulated doping material for InP has good characteristics in terms of the capacitance and heat dissipation of the current blocking layer for Laser Diodes. However unintentional Zn diffusion from adjacent p-InP into Ru-InP causes the degradation of Laser characteristics such as the output power. In this paper, we fabricated p-InP/Ru-InP/p-InP (p/Ru/p-InP) structure by Metalorganic Vapor Phase Epitaxy (MOVPE) and analyzed the behavior of Zn diffusion from Zn-InP into Ru-InP after annealing by SIMS measurement.

Keywords—Zn; Ruthenium; diffusion; InP; MOVPE;

I. INTRODUCTION

More and more high-capacity and long-haul transmission volume are demanded as the optical communication networks are spreading. From the point of characteristic and cost, semiconductor lasers for the light source of optical fiber communication are required to have high speed and uncooled operation under high temperature. The key process to satisfy high speed operation under high temperature is the growth of semi-insulated current blocking layer at the side of active region of laser device.

Ruthenium (Ru) doped InP has focused on as a new semi-insulated dopant for InP [1]. There are no interdiffusion of Ru with Zn between Ru-InP and Zn-InP, while Fe has strong interdiffusion with Zn if we use Fe instead of Ru. By means of Ru-InP, we can design various current blocking layer structures for Buried-Heterostructure (BH) Lasers. Our group have reported the electrical properties of Ru-InP [2] and applied the Ru-InP as current blocking layer to the AlGaInAs BH Laser [3]. However, we found the unintentional Zn incursion form Zn-InP to Ru-InP at the current blocking layer caused the degradation of Laser characteristics. The fabrication process of BH-lasers needs a few times thermal procedure such as re-growth of the blocking layer. Then we investigated Zn diffusion into Ru-InP layer with annealing process by SIMS analysis. In this paper, we report the behavior of Zn diffusion into Ru-InP and the dependence on the concentration of Ru.

II. EXPERIMENTAL

Ru-InP epitaxial layers were grown by MOVPE apparatus on a n-type InP (100) substrate at 600 °C. The reactor pressure was 100 mbar and the total hydrogen flow rate was 25 slm. Trimethylindium (TMIn) was used as the group-III source, Phosphine (PH_3) was used as the group-V source. Diethylzinc (DEZn) was used for p-type dopant. For Ru doping material to

InP, we selected Bis(2,4-dimethyl 1,3-pentadienyl)Ruthenium. The growth procedure is as follows. First, a 0.75 µm-thick p-InP with the Zn concentration of 1E+18 cm^{-3} was grown on a n-InP substrate. Second, the 1.5 µm-thick Ru-InP or undoped InP layer was grown. The concentration or Ru was 2E+17 and 1E+17 cm^{-3}. Finally the 0.75 µm-thick p-InP with the Zn concentration of 1E+18 cm^{-3} was grown. We added annealing to each sample with the same MOVPE apparatus after we cleaved the grown wafers into four samples. Then we measured SIMS profiles of Zn and Ru in InP.

III. RESULTS AND DISCUSSION

Fig. 1 shows cross-sectional SEM images of Zn-InP / Ru-InP / Zn-InP (p/Ru/p-InP) and Zn-InP / undoped InP / Zn-InP (p/i/p-InP). As seen in Fig. 1, the diffusion length of Zn had no dependence on Ru concentration in InP, and Zn in sample #1, #2 moderately diffused compared with sample #3. Fig. 2 shows Zn and Ru SIMS profiles of #1 and #2. The diffusion length of Zn was similar between #1 and #2. However, compared to the Zn diffusion at the 1.5 µm depth point, the diffusion length of Zn in #2 was slightly longer than that of #1. Fig. 3 shows Zn and Ru SIMS profiles of #1 and #3. It appears the diffusion length of Zn in undoped InP is longer than that in Ru-InP. This result agreed with Fig. 1. In addition, it was accepted in terms of interstitial-substitutional mechanism [4] since Ru-InP could act as hole trap [1]. Compared to the diffusion length of Zn at the interface between Zn-InP and Ru-InP, Zn at the bottom of Zn-InP more diffused than that at the top of Zn-InP. This result indicated Zn diffusion was advanced by the thermal process during the growth of each layer by MOVPE itself.

TABLE I.   DETAILS OF EPITAXIAL LAYER FOR SIMS ANALYSIS

No.	Zn Concentration	Zn-InP Thickness	Ru Concentration	Ru-InP Thickness
#1	1E+18 cm^{-3}	0.75 µm	2E+17 cm^{-3}	1.5 µm
#2	1E+18 cm^{-3}	0.75 µm	1E+17 cm^{-3}	1.5 µm
#3	1E+18 cm^{-3}	0.75 µm	none (undoped)	1.5 µm

MoPI-1 (Poster)
18:30 - 20:30

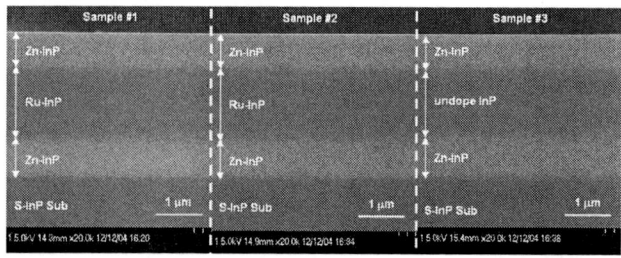

Figure 1. Cross-sectional SEM images of p/Ru/p-InP and p/i/p-InP

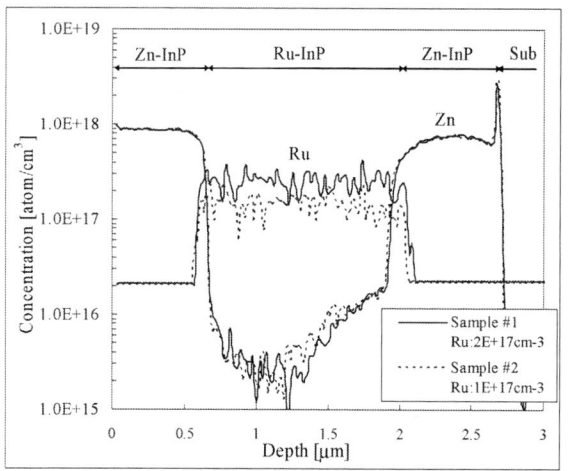

Figure 2. Comparison of Zn SIMS profies for #1 and #2

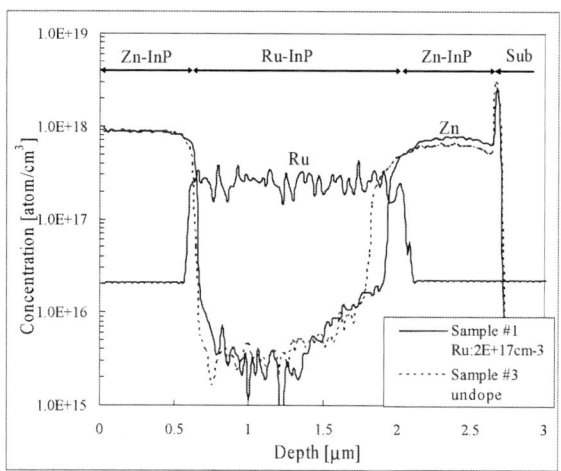

Figure 3. Comparison of Zn SIMS profies for #1 and #3

IV.   CONCLUSION

We investigated Zn diffusion in Ru-InP layer by SIMS analysis in case it was without annealing process so far. We found the Zn diffusion had interesting behavior for the Ru concentration and thermal process. We are going to measure how the Zn diffusion of each sample is advanced by annealing process in MOVPE apparatus.

REFERENCES

[1]   A. Dadgar, O. Stenzel, A. Naser, M. Zafar Iqbal, D. Bimberg, H. Schumann, "Ruthenium: A superior compensator of InP," Appl. Phys. Lett. vol. 73, number. 26, pp. 3878-3880, 1998.

[2]   H. Yamaguchi, T. Nagira, G. Sakaino, K. Ono, M. Takemi, "Electric Characteristics of Ruthenium doped InP and its application for Buried-Heterostructure Lasers" physica status solidi (c) volume. 9, issue. 2, pp. 342–345, 2012

[3]   G. Sakiano, T. Takiguchi, Y. Hokama, T. Nagira, H. Yamaguchi, E. Ishimura, A.Sugitatsu and T. Shimura, "25.8Gbps Direct Modulation AlGaInAs DFB Lasers with Ru-doped InP Buried Heterostructure for 70°C operation" The Optical Fiber Communication Conference and Exposition, OTh3F..3, 2012

[4]   R. L. Longini, "Rapid Zinc Diffusion in Gallium Arsenide" Solid-State Electronics vol. 5, pp. 127-130, 1962

MoPI-2 (Poster)
18:30 - 20:30

The Measurement of Dislocation on InP Wafers

Qingfang Huang[1], Zhiguo Liu[1,2], Ruixia, Yang[2], Xiaolan Li[1], Qiang Wang[1], Xiuwei Tian[1], Jianye Yang[1], Shuai Li[1], Huimin Shao[1], Yanlei Shi[1], Xin Zhang[1], Ning Li[1], Yong Kang[1], Huisheng Liu[1], Tongnian Sun[1], Niefeng Sun[1*]

([1]Science and technology on ASIC Laboratory, Hebei Semiconductor Research Institute, Shijiazhuang, 050051, China)
([2]School of Information Engineering, Hebei University of Technology, Tianjin, 300130, China

Abstract—**We studied the effect of HCl, H$_3$PO$_4$, HBr etchants, temperature, illumination on the display of dislocation pits on <100> InP single crystal wafers, and analyzed the effect of illumination, using the wet chemical etching method. The experimental results show that the etching rate is strongly dependent on the proportion of HBr in the mixed etchant and HBr alone can reveal dislocation pits on <100> InP wafers. Both illumination and higher temperature can increase the etching rate. We also discuss the mechanism of different sizes of dislocation pits.**

Keywords—InP; wet chemical etching method; dislocation

Low dislocation InP have been reported a lot, however, few of them discussed the effects of etching condition[1-4]. In present work, we studied the effects of etchants, temperature, illumination on the display of dislocation pits on <100> InP substrate, and analyzed the influence of illumination, using the wet chemical etching method. The experimental results show that in the mixed etchants, HBr plays a dominant role and HBr alone can reveal dislocation pits on <100> InP wafers. Both illumination and higher temperature can increase the etching rate. We also discuss the mechanism of different sizes of dislocation pits.

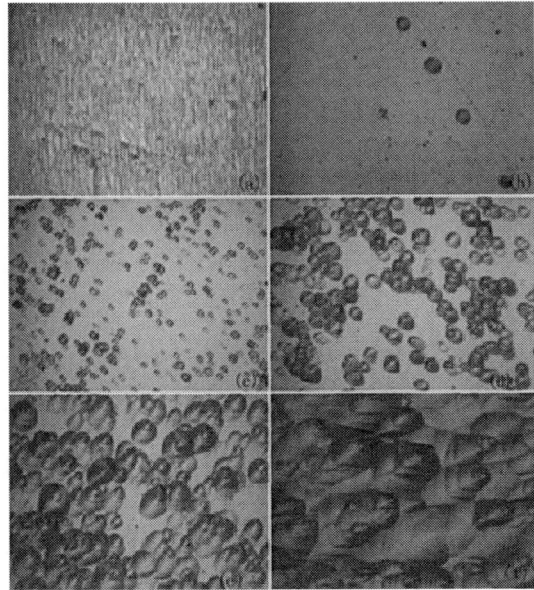

Fig.1 The micrograph of dislocation pits under different etchants.

(a) HCl; (b) H$_3$PO$_4$; (c) HBr: H$_3$PO$_4$=1:1;

(d) HBr: H$_3$PO$_4$=2:1; (e) HBr; (f) HBr: HCl=1:1;

The <100> InP single crystal is grown by home-made puller, using the high pressure liquid encapsulated Czochralski technology (HP-LEC)[5]. In the experiment an epi-ready <100> InP substrate is prepared. After being etched using different etchants under various conditions, the dislocation pits of the wafers are observed with OLYMPUS BX51M metalloscope. Fig.1 shows the micrograph of dislocation pits of <100> InP substrate, being etched 5 mines at 20℃ under different etchants. As shown in Fig.1(a) and (b), sole HCl or H$_3$PO$_4$ can't reveal dislocation pits of <100> InP wafer. As we can see in Fig.1(c), (d), (e), with the increase in the proportion of HBr, the dislocation pits become clearer. HBr plays a dominant role in the mixture etchant and HBr alone can reveal dislocation pits on <100> InP wafers. Over etching may cause neighboring individual etch pits merge together, as is shown in Fig.1(f). HCl is normally used as isotropic etchant for InP. When HBr corrodes a regular square dislocation pits, HCl can deepen and broaden the pits. Therefore, the etching rate is much quicker and the pits are much deeper using HCl and HBr mixed etchant. For well calculating dislocation density, we use Huber etchant[6].

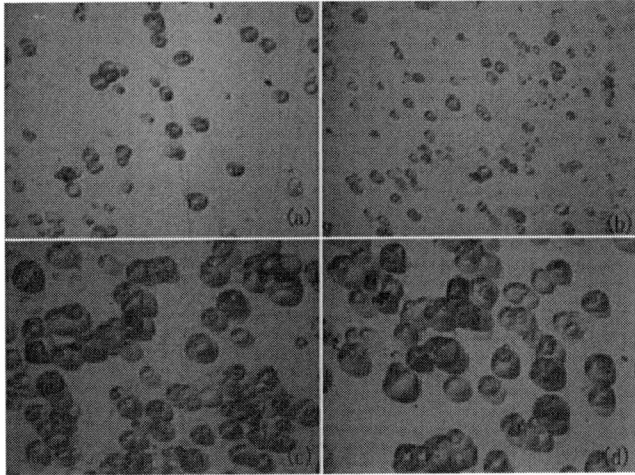

Fig.2 The micrograph of dislocation pits under different temperature.

(a) 2 mines at 20℃; (b) 5 mines at 20℃;

(c) 2 mines at 55℃; (d) 5 mines at 55℃;

Fig.2 shows the micrograph of dislocation pits under different temperature, using mixture etchant of HBr and H$_3$PO$_4$ with volume ratio is 2:1. As we can see from Fig.2, under high temperature, etching rate is much faster than that under low temperature with other etching conditions are the same. According to the law of mass action, such as a chemical reaction process aA + bB + cC → product, the reaction rate was:

$$v = \kappa \, C_A^a \, C_B^b \, C_C^c$$

Where v is the reaction rate, C is the molar concentration of the reactants; a, b and c are coefficients of A, B and C respectively; K is a constant. According to the Arrhenius, K can be written as:

$$\kappa = De^{-E/RT}$$

Where D is an experimental constant, and R is the gas constant, E is the activation energy of the reaction. In the case of increase temperature, the reaction rate constant k is increased rapidly. Therefore, to improve the etching temperature can increase the rate of the reaction markedly.

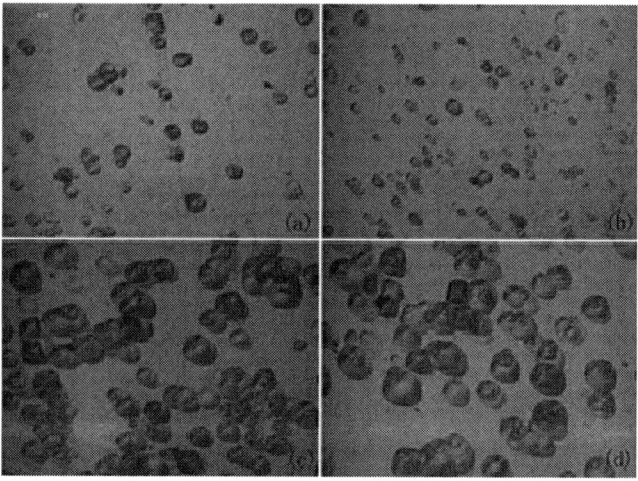

Fig. 3 The micrograph of dislocation pits with illumination or without illumination.

(a) 8 mines with illumination; (b) 8 mines without illumination;

(c) 16 mines with illumination; (d) 16 mines without illumination

The mechanism of the wet chemical etching includes two processes: one is the oxidation-reduction reaction at surface of substrate; the other is the transportation of reactant, product and the transportation of carrier to surface. The illumination can excite electron - hole pairs, accelerating the etching rate. Under a short time the effect is very distinctly, as it is shown in Fig.3(a) and (b). From Fig.3(c) and (d), we can see that the sizes of dislocation pits with illumination and without illumination are almost the same after about 16 mines. In light, HBr deteriorate quickly, the etching rate become slower and slower due to the decrease of HBr.

We also can see from Fig.1 to Fig.3 that in every micrograph of dislocation pits contains different sizes pits. The size of dislocation pits is related to the etching rate which depends on chemical reactions on the wafer surface. The chemical reaction rate can attribute to the different types of threading dislocations, the distortion grade and the inhomogeneous of the substrate surface. The reaction rate is related to the activation energy of dislocation head. There are three kinds of threading dislocations: screw threading dislocations, edge threading dislocations and mixed threading dislocations. The activation energy is different from each kind of threading dislocations, leading to a different etching rate. Different distortion grades can also lead to different activation energy, leading to different size of dislocation pits ultimately. There may be another reason contributes to the size of the dislocation pits, the liquid-solid shape. Due to the temperature gradient and the convection of the melt the liquid-solid shape can be flat, convex, concave or irregular. The substrate is cutting vertically to the growth direction. In this case, the substrate surface isn't at the same altitude until the liquid-solid shape is flat. Due to growing condition changes, the property of the crystal may change, leading to different activation energy around threading dislocation head.

In conclusion, in the mixed etchants, HBr plays a dominant role and HBr alone can reveal dislocation pits on <100> InP wafers. Both illumination and higher temperature can increase the etching rate. The different sizes of the dislocation pits may due to the inhomogeneous of the substrate surface, the different types of threading dislocations and the distortion grade.

ACKNOWLEDGEMENTS

The authors wish to acknowledge valuable financial support from the National Science and Technology Major Project of China and National Natural Science Foundation of China (FSFC) grant No.61076004.

REFERENCES

[1] Akira Noda, Masashi Nakamura, Atsutoshi Arakawa and Ryuichi Hirano, "Growth of dislocation free grade Fe doped InP crystals ", The 16 th IPRM Proceeding, pp.552-553, 2004.

[2] R.Hirano,"Growth of Low Etch Pit Density Homogeneous 2Ⅲ InP Crystals Using a Newly Developed Thermal Baffle", J.Appl.Phys. Vol.38, pp.969-971, August 1999.

[3] R.Hirano and A. Noda, "Growth of low EPD InP crystals", The 13 th IPRM Proceeding, pp.529-532, 2001.

[4] Peter Rudolph, "Present state and future tasks of Ⅲ-Ⅴ bulk crystal growth", The 19 th IPRM Proceeding, pp. 333-338, 2007.

[5] S. Tong-nien, L. Szu-lin and K. Shu-tseng, "The Preparation of Semi-Insulating and Low Dislocation Density InP Single Crystal" The 2nd Conf. on Semi-Insulating Ⅲ-Ⅴ Materials, pp.61-67, 1982.

[6] A. Huber and N. T. Linh, "Révélation métallographique des défauts cristallins dans InP," Journal of Crystal Growth, vol. 29, pp. 80-04, 1975.

MoPI-3 (Poster)
18:30 - 20:30

Optimization of metamorphic buffer layers for extended-InGaAs/InP photodetectors

S. Seifert[1,2], R. Ravash[1], D. Franke[1], F. Wenning[1], D. Zengler[2] and F. Kießling[2]

1. Fraunhofer Institute for Telecommunications, Heinrich Hertz Institute, Einsteinufer 37, 10587 Berlin, Germany
2. TU Berlin, Institut für Optik und Atomare Physik, Straße des 17. Juni 135, 10623 Berlin, Germany

Abstract— **Optimized extended-InGaAs photodetectors were grown on InP substrate using metamorphic buffer layers to achieve a strain relaxation. The samples were investigated by transmission electron microscopy and X-Ray diffraction. The results show that a number of metamorphic buffer layers and their thickness play an important role in material quality and photodetectors performance.**

I. INTRODUCTION

For thermometric measurements of temperatures above 300°C cost-efficient and sufficiently precise devices are well established but this does not apply equally to the lower temperature range. To measure thermal radiation between 40°C and 80°C optical detectors are needed at wavelengths from 2.6 to 2.3 µm. For this purpose, strongly strained extended-InGaAs (E-InGaAs) detectors grown on InP substrate are commonly used. A typical mismatch between the substrate and the active layer in such devices is in the order of $\Delta a/a = +1.5$ % (Fig. 1) which leads to a high density of misfit dislocations. To suppress the formation of these dislocations different metamorphic buffer layers [1-3] are widely used. These buffer layers provide a gradual shift in the lattice constant by varying the layer composition and they often offer opportunities for dislocation lines to eliminate. The achievable overall mismatch for such buffer structures depends on the total thickness of the buffer layers, but can also be improved by using different approaches of grading, relaxation and interface processes in the buffer structure. Reported thicknesses of MOVPE-grown optical detector devices with an acceptable dislocation density at the desired wavelengths are typically in the order of $15 - 20$ µm [4], which implies a time-consuming and therefore expensive growth procedure.

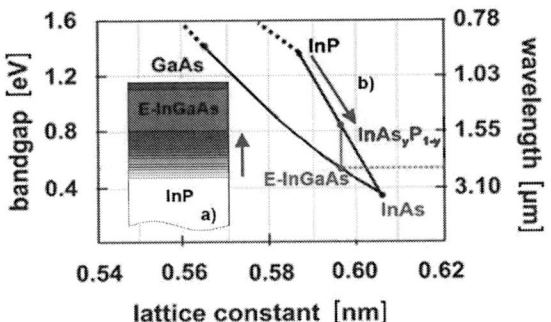

Figure 1. Schematic detector-structure (a). The As-concentration is increased within the buffer layer (b).

In this paper the latest improvement of growth for E-InGaAs/InP photodetectors is presented, Different buffer structures are examined and successively optimized to reduce the overall thickness without decreasing the material quality of the E-InGaAs layers.

II. EXPERIMENTAL

The growth of E-InGaAs detectors on (001) InP substrate was performed using metal-organic vapor phase epitaxy (MOVPE) in a horizontal Aixtron 200 reactor at agrowth temperature of 670°C. trimethylindium (TMIn), trimethylgallium (TMGa), arsine (AsH3) and phosphine (PH3) were used as precursors.

Two types of metamorphic buffer layers were grown: The first one (refered to as Type A) consists of different $InAs_yP_{1-y}$ layers. A variation of the arsenic content y from y = 0.03 to y = 0.47 with each layer transits the lattice constant from a = 0.5868 nm (InP) to a = 0.5956 nm which is matching the lattice of the E-InGaAs used as the active layer. The second type (Type B) overcomes the same range of alteration but the arsenic content y is varied linearly up to the last 0.5µm where it is kept constant at the final composition. On top of both types a 3µm thick E-InGaAs and a 0.5µm lattice matched $InAs_{0.47}P_{1-0.47}$ cap layer is grown. For the comparison of the buffer types two similar samples were grown with this stuctures. Sample A is a stepwise grading with an overall thickness of 11.5 µm, sample B a linear grading with a total thickness of 12.0 µm, whereat the difference in thickness is only due to the composition plateau at the end of the linear grading in sample B.

The samples were then characterized by X-Ray diffraction (XRD) with reciprocal space mapping (RSM) and transmission electron microscopy (TEM). For TEM characterization the samples were prepared in cross-section and investigated with (002) dark-field image under g,3g weak-beam conditions.

A further investigation of the stepwise buffer was carried out varying the number of steps, the alteration of composition per step and the interfaces.

III. RESULTS

Figure 2a shows the RSM of the (115) reflection of sample A and sample B. All measured reflections of sample A are alongside of the relaxation line demonstrating complete relaxation of each composition step. The broadening of the

978-1-4673-6130-9/13 $31.00 © 2013 IEEE

reflections is due to the thickness of the single steps leading to a slightly increased mosaicity (Fig. 2a) [5].

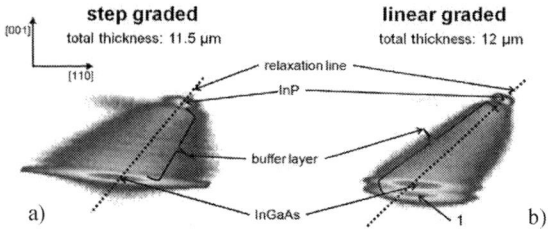

Figure 2. Reciprocal Space Mapping of the (115) reflection of sample A (a) and sample B (b).

The RSM of sample B (Fig. 2b) exhibits a completely relaxed InGaAs. But the reflection of the InAsP composition plateau (marked as 1) indicates that the buffer structure does not provide a fully relaxed platform for further growth. Consequently a step graded buffer is more suitable to achieve lower defect densities in the active device layers without increasing the overall thickness of the structures.

To investigate the material quality of the E-InGaAs layers, TEM examinations were carried out on sample A and B. Figure 3a and b show weak-beam dark-field TEM images [6] of samples A and B. Here, some of the InAsP buffer layers followed by the E-InGaAs active layer are shown.

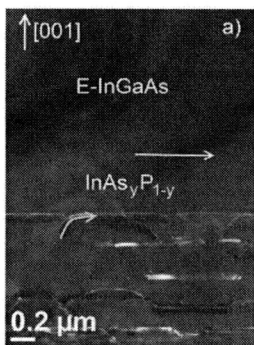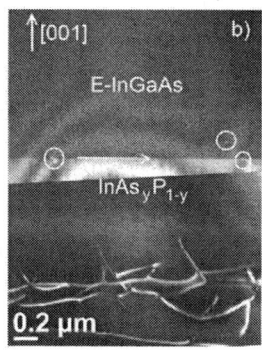

Figure 3. Weak-beam dark-field cross-sectional TEM image of sample A and sample B. The white arrows show the interfaces between the last buffer layer and the E-InGaAs layer (a, b) and the interface between the E-InGaAs layer and the cap layer (b). The yellow arrow (a) shows a bending misfit dislocation. The white circles (b) show defects in the E-InGaAs layer and at the interface.

The misfit strain in the buffer layers results in the nucleation of dislocations on a {111} plane. The bright lines and bright dots in figure 3(a, b) show these misfit dislocations, where some of them bend at the interfaces (Fig. 3a). At the interface region between the last buffer layer and the E-InGaAs active layer no dislocations are observed showing a dislocation free E-InGaAs layer at sample A. In the active layer of sample A the dislocation density is less than $6*10^7$ cm^{-2}. This value shows a good crystal quality of the E-InGaAs layer. In sample B at the interface between the last buffer layer and the E-InGaAs layer defects are observed. In the active layer of sample B the dislocation density is more than $7*10^8$ cm^{-2}. Concluding, the presence of distinct interfaces within the buffer layer is necessary to prevent dislocation lines from advancing throughout the buffer structure towards the device layer.

To derive an ideal relationship between step numbers and layer thickness some step graded buffers were grown reducing the layer thickness stepwise. These samples were analyzed by XRD. Figure 4 shows the required layer thickness, so that the layer is completely relaxed at a certain strain.

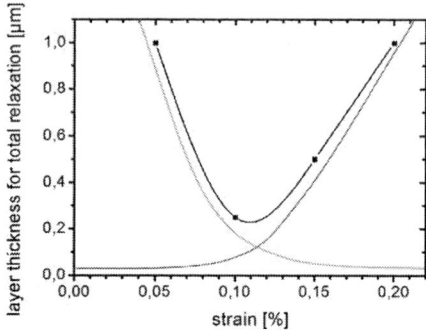

Figure 4. Relationship (shown as black line) between strain and required layer thickness, so that the layer is completely relaxed

The left side of the minimum in figure 4 is influenced by the critical layer thickness [7, 8]. With increasing strain, the critical layer thickness decreases (shown as green line). But with increasing strain the required layer thickness increases by reducing the lattice distortion (shown as blue line), which is shown on the right side of the minimum. It is a minimum layer thickness resulted in a misfit of +0.1%.

IV. CONCLUSION

E-InGaAs detectors were grown on a (001) InP substrate with a nominal misfit of +1.5% applying different step graded and linearly graded buffer layers. TEM and XRD examinations of the different buffer structures show that step graded buffers offer much faster relaxation and are therefore preferable for the thickness reduction of completely relaxed buffer structures. In addition to the advanced relaxation mechanisms it is shown, that the presence of distinct interfaces within the buffer layer prevents dislocation lines from advancing into the device layers. XRD measurements of different step graded buffer structures suggest that there is an optimum balance between the increment of strain and the thickness of each step. This is found to be a balance of the critical layer thickness and the reduction of lattice distortion, and should therefore depend mainly on the elastic parameters of the used material.

For the established InAsP buffer a reduced buffer of less than 4 μm overall thickness was grown with a detector structure on top exhibiting a defect density below $6*10^7$ cm^{-2}.

V. ACKNOWLEGDEMENTS

This work was conducted within the HINT project (grant # 1N10880) funded by the German Federal Ministry of Education and Research under the KMU-Innovation program.

REFERENCES

[1] Ch. Heyn, S. Mendach, S. Loehr, S. Beyer, S. Schnuell, W. Hansen: J. Crystal Growth 251, 832-833 (2003)

[2] D.-S. Kim, S.R. Forrest, M.J. Lange, M.J. Cohen, G.H. Olsen, R.J. Menna and R.J. Pfaff: J. Appl. Phys. 80, 6229 (1996)

[3] M.K. Hudait, Y. Lin, M.N. Palmisiano and S.A. Ringel: IEEE Electron. Device Lett. 24, 538 (2003)

[4] A. Krier, Y. Mao: Infrared Physics & Technology 38, 397-403 (1997)

[5] H. Lin, Y. Huo, Y. Rong, R. Chen, T. I. Kamins, J. S. Harris: J. Crystal Growth 323, 17-20 (2011)

[6] E. P. Butler: Micron, Volume 5, Issue 3, 293-305 (1974-1975)

[7] J.W. Matthews, A.E. Blakeslee: J. Crystal Growth 27, 118 (1974)

[8] R. People, J.C. Bean: Appl. Phys. Lett. 47, 322 (1985); erratum: Appl. Phys. Lett. 49, 229 (1986)

MoPI-4 (Poster)
18:30 - 20:30

Liquid-Phase Electroepitaxy of GaN at atmospheric pressure using ammonia and Ga-Ge solution

D. Kanbayashi, T. Hishida, M. Tomita, H. Takakura, T. Maruyama and S. Naritsuka

Materials Science and Engineering
Meijo University
Nagoya, Japan

Abstract— Liquid-Phase Electroepitaxy of c-plane GaN was tried to perform using NH_3 and a mixed solution of Ga and Ge. Consequently, GaN layer was successfully grown at the atmospheric pressure. The thickness of the grown layer was found to monotonously increase with the current. The thickness of the layer grown with the current of 4 A was more than twice of that of the conventional LPE. The growth thickness shows almost no change by the change of the thickness of the solution. This result strongly suggests the growth was driven mainly by the electromigration, which was caused by the current flow through the solution.

Keywords—Liquid-Phase Electroepitaxy; GaN; Atmospheric Pressure; Electromigration; Growth rate enhancement

I. INTRODUCTION

Group III nitrides have attracted much attention due to their applications in short wavelength light emitting diodes and laser diodes. Usually, these materials are grown by metal-organic chemical vapor deposition (MOCVD) or molecular beam epitaxy (MBE). Liquid phase epitaxy (LPE) is another growth technique, which is simple, easy and low-cost. However, it is not popular to grow GaN by LPE because an extremely high pressure and high temperature are necessary for GaN LPE [1-3], also the addition of Na is required to increase the solubility of nitrogen [5-6]. It is very convenient to grow GaN without the use of Na, which is extremely reactive and hard to handle in the air. Several years ago, Hussy reported on the low-pressure solution growth (LPSG) of GaN, namely a GaN layer was grown at atmospheric pressure without the addition of Na to the solution [7]. The GaN layer grown by LPSG has a fine characteristic but the growth rate was quite low, which was as low as 0.1 μm/h. In this paper, liquid-phase electroepitaxy (LPEE) is tried to apply to the LPSG of GaN to increase the growth rate. The enhancement of the growth rate is expected in LPEE with the help of the electromigration, caused by the current flow. LPEE of GaN was successfully performed using Ga and Ge mixture as a solution under an atmospheric pressure. The thickness dependence of the solution was also studied in order to discuss the growth mechanism.

II. EXPERIMENTAL

Mixture of gallium (75at%) and germanium (25%) is used for the solution of LPEE. The addition of germanium is useful to suppress the parasitic growth and consequently increase the growth rate [7]. Ammonia as a source of nitrogen is transported in the reactor by H_2 carrier gas. The flow rates of the gases are controlled by mass flow controllers, which determine the partial pressure of NH_3. The accurate control of the partial pressure is important to suppress both a parasitic growth of GaN and "creeping" phenomenon of the solution, because they will harm the growth and finally slow down the growth rate.

The sliding boat used in the experiment is illustrated in Fig. 1. Two electrodes are put in the solution to flow the current from bottom to top of the solution. The direction of the current is chosen to cause the electromigration from top to bottom, which helps the transport of the nitrogen, and consequently increases the growth rate. The growth area is determined by the contact area of the solution to the substrate as 8 x 8 mm^2 from the size of the solution sink. A c-plane GaN template of 3.8 μm-thick GaN layer on a sapphire substrate is put in the bottom boat and used as a substrate. The growth temperature, growth time and NH_3 gas (NH_3 is diluted to 1% by H_2) and H_2 flow rates are set as 960 °C, 20 hours, 30 and 70 sccm while the current are changed between 0 and 4 A. In the second experiment, the thickness of the solution is increased from 2.9 to 5.8 mm, to study the driving force of the growth. The morphology and the thickness of the grown layers are studied using scanning electron microscope (SEM). The thickness of the grown layers is averaged over the several points on the sample.

III. RESULTS AND DISCUSSION

The cross-sectional SEM image of a GaN layer grown at 960 °C with the current of 2 A is shown in Fig. 2. It is found that a 3.0 μm-thick GaN LPEE layer was successfully grown on the template substrate. The contrast in the image, which slightly changed between the grown part and the template, is ascribed to the difference in the carrier concentration in these layers. Though some pits with the diameters of about 1 μm are observed on the surface, a flat GaN layer was grown by LPEE at the atmospheric pressure without Na.

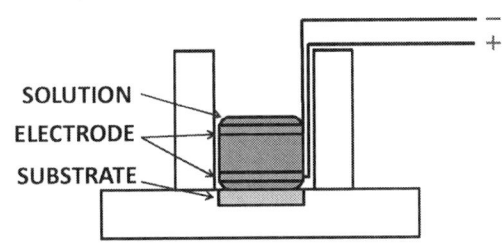

Figure 1. Schematic illustration of LPEE setup.

978-1-4673-6130-9/13 $31.00 © 2013 IEEE

The dependence of the growth thickness of the GaN layer on the current is shown in Fig. 3. In the experiment, the growth temperature, time and the flow rates of the NH_3 and H_2 gasses were fixed at 960 °C, 20 hours, 30 and 70 sccm while the electric current flow was systematically changed. It is found from the figure that the growth thickness of 1.8 µm at the growth without current increases monotonously with the increase of the current. And it finally reached 4.4 µm at the current of 4 A. The growth thickness at 0 A is ascribed to a conventional LPE mode, where the supply of the NH_3 over the solution is the driving force of the growth and the nitrogen atoms are transported in the solution by the diffusion. The thickness also includes the growth during the cooling stage after growth, which is estimated about 1 µm. The increased amount of the growth can be ascribed to LPEE mode. The larger the current became, the thicker the layer grew. The thickness of the growth became more than twice of the conventional LPE mode when the current was 4 A. In LPEE, there are two possible mechanisms of the growth enhancement; one is the electromigration and the other is the temperature difference caused by the Peltier effect. The latter does not correspond to the present case because there is no junction in the current pass. Therefore, there is no place where produces the Peltier effect. In order to clarify the growth mechanism, the following experiment was performed.

The thickness of the solution was changed to study the driving force of the growth. The amount of nitrogen atoms transported by the electromigration is thought to be independent of the thickness of the solution if the current is fixed. Fig.4 shows that the growth thickness dependence on the solution thickness. The growth thickness stayed constant obviously with the thickness change of the solution. This result strongly suggests that the driving force of

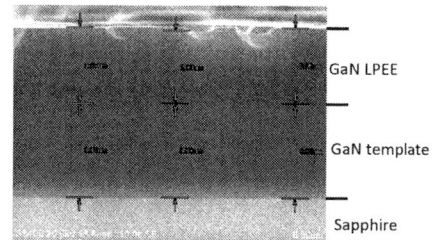

Figure 2. Cross-sectional SEM image of LPEE GaN layer grown at 960 °C for 20 hours with current of 2 A.

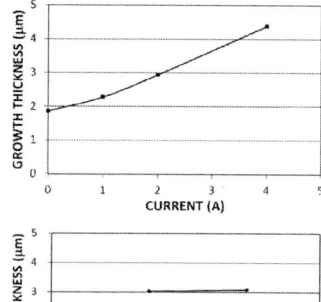

Figure 3. Dependence of growth thickness on the current of LPEE GaN layers grown at 960 °C for 20 hours.

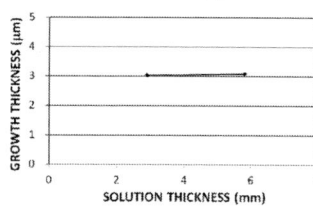

Figure 4. Dependence of growth thickness of LPEE GaN layer on the solution thickness.

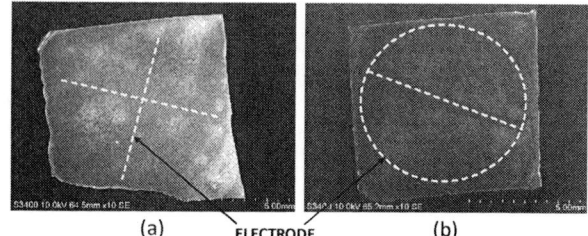

Figure 5. Surface SEM images of LPEE GaN layers grown with different shapes of electrodes in (a) cross-shaped and (b) circle-shaped. Broken white lines show the positions of the electrodes.

the growth is mainly caused by the electromigration. In other words, the growth thickness mainly depends on the current not but the thickness of the solution. The latter possibly change the diffusion condition and, accordingly, the contribution of the growth rate produced by the diffusion. The surface morphology of the grown layers in Fig.5 also supports the discussion. The figure shows a dark contrast near the electrode (shown by the white broken line), which indicates the smooth and thicker growth at the position. That is, the layers grew thicker near the electrodes. This is because the solute was effectively transported by the electromigration especially near the electrodes, and that the growth rate was enhanced.

IV. CONCLUSION

LPEE of c-plane GaN was tried to perform using NH_3 and a mixed solution of Ga and Ge. Consequently, GaN layer was successfully grown at the atmospheric pressure. The thickness of the grown layer was found to monotonously increase with the current. The thickness of the layer grown with the current of 4 A was more than twice of the conventional LPE. The growth thickness shows almost no change by the change of the solution thickness. This result strongly suggests the growth was driven mainly by the electromigration, which was caused by the current flow through the solution.

ACKNOWLEDGMENT

This work was partly supported by a Grant-in-Aid for Scientific Research on Priority Areas B (No. 22360131) from the Ministry of Education, Culture, Sports, Science and Technology of Japan.

REFERENCES

[1] J. A. Van Vechten, Phys. Rev. B7 (1973) 1479.

[2] J. Karpinski, J. Jun, and S. Porowski, J. Cryst. Growth, 66 (1984) 1.

[3] I. Grzegory, M. Bockowski, B. Lucznik, S. Krukowski, Z. Rowanowski, M. Wroblewski, S. Porowski, J. Cryst. Growth, 246 (2002) 177.

[4] H. Yamane, M. Shimada, S. J. Clarke, and F. J. DiSalvo, Chem. Mater. 9 (1997) 413.

[5] M. Yano, M. Okamoto, Y. K. Yap M. Yoshimura, Y. Mori, and T. Sasaki, Jan. J. Appl. Phys. 38 (1999) L1121.

[6] F. Kawamura, M. Morishita, T. Iwahashi, M. Yoshimura, Y. Mori, and T. Sasaki, Jan. J. Appl. Phys. 41 (2002) L1440.

[7] S. Hussy, E. Meissner, P. Berwian, J. Friedrich, and G. Muller, J. Cyst. Growth, 310 (2008) 738.

MoPI-5 (Poster)
18:30 - 20:30

Wide energy level control of InAs QDs using double-capping procedure by MOVPE

Masayuki Yamauchi, Yuto Iwane, Shohei Yoshikawa, Yuta Yamamoto and Kazuhiko Shimomura
Department of Engineering and Applied Sciences, Sophia University
7-1 Kioi-cho, Chiyoda-ku, 102-8554
Tokyo, Japan
e-mail: kshimom@sophia.ac.jp

Abstract—**We have obtained the wide energy level control of InAs QDs structure where the PL peak wavelength were ranged from 1200nm to 1800nm. Stranski - Krastanov InAs QDs were grown by low pressure all metal-organic source MOVPE. We have controlled the InAs QDs energy level by changing the buffer layer composition under the QDs and the height using double-capping procedure and also supply amount of QDs.**

Keywords— Quantum dots, InAs/InP, double-cap procedure, MOVPE

We have investigated the wide energy level control of self-assembled InAs QDs using a double-capping procedure by MOVPE [1]. InAs QDs were grown by the Stranski-Krastanov growth mode using low-pressure MOVPE on an InP substrate. The precursors were trimethyl indium (TMI), triethyl gallium (TEG), tertiary butyl arsine (TBA) and tertiary butyl phosphine (TBP). Fig.1 shows the schematic figures of double-capping process. After the growth of the InAs QDs, a first capping layer (FCL) with a thickness less than the height of the InAs QDs was deposited. Growth interruption was then performed while supplying TBP. During this growth interruption step, As/P exchange occurred and the top of the QDs protruding above the FCL became planarized. After the growth interruption, a second capping layer (SCL) was grown. As a result, we have obtained a wider distribution of peak wavelengths by changing the FCL thickness.

In the higher energy level side control of QDs, we have changed the Ga composition of GaInAs buffer layer. When the Ga composition of GaInAs buffer layer was increased, the PL peak wavelength of InAs QDs was blue shifted because of Ga interdiffusion into InAs QDs from GaInAs buffer layer [2,3]. Fig.2 shows the structure of 2 layer stacked InAs QDs. Firstly, 110 nm thick InP and GaInAs buffer layers were grown on the (100) orientated n-InP substrate at 640 degree-C and a pressure of 100 Torr. Then InAs QDs and 2 nm thick InP FCL were grown at 540 degree-C and 15 Torr. The growth rate of QDs was 0.048 ML/s and the sources supply time was 35 s. During the 8 min growth interruption under TBP flow, the temperature and pressure were raised to 640 degree-C and 100 Torr, and a 22 nm thick InP SCL was grown. In the growth of GaInAs buffer layer, we have grown the Ga composition of 0.7 and 0.47. Fig.3 (a) and (b) show the AFM images of QDs where Ga composition was (a) 0.7 and (b) 0.47, and Fig.4 (a), (b) and (c) show the PL spectrum of QDs at 77K where Ga composition and FCL thickness were (a) 0.7, 2nm, (b) 0.47, 2nm, (c) 0.47, 4nm. As can be seen from Fig.4, the PL peak

wavelength shift of 120 nm was obtained between 0.47 and 0.7 of Ga composition.

In the lower energy level side control of QDs, we have increased the total supply In amount during the InAs QDs and also increased the FCL thickness. By increasing the amount of In of QDs and height of QDs, the volume of QDs is increased and energy level will be shifted to the lower energy. Fig.3 (c) shows the AFM image where the total supply amount was 3.84ML, and Fig.4 (d), (e) show the PL spectrum of InAs QDs where total supply amount of InAs and FCL thickness were (d) 1.68ML, 8nm, (e) 3.84ML, 15nm. Table I summarizes the structure and PL peak wavelength corresponding to Fig.3. By increasing the supply amount of InAs, the PL peak wavelength was clearly red shifted.

By applying these QDs growth layer by layer in the multistacked QDs structure, we can expect the InAs QDs broadband emission devices [4] more than 600nm spectrum width.

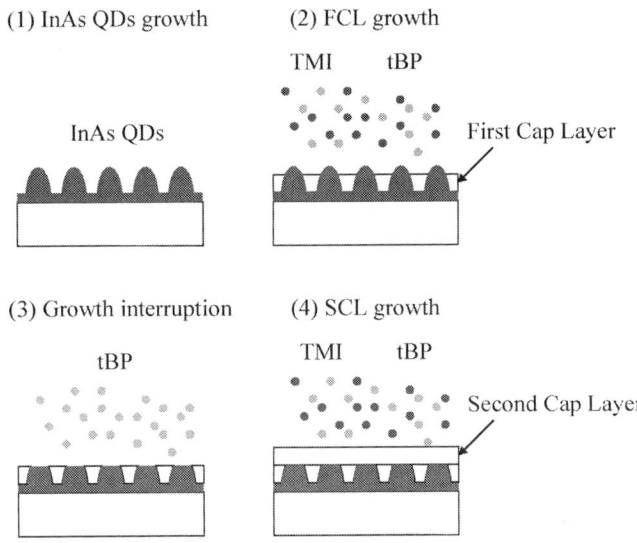

Fig.1 Schematic figures of double-capping process

978-1-4673-6130-9/13 $31.00 © 2013 IEEE

MoPI-5 (Poster)
18:30 - 20:30

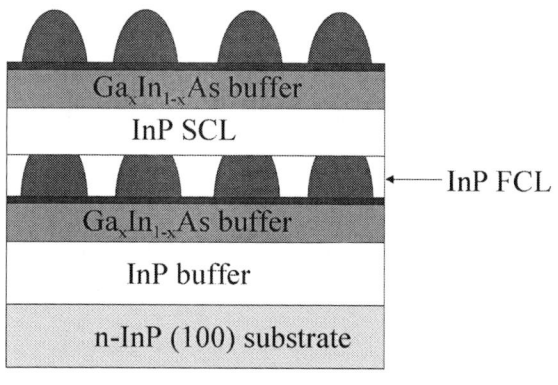

Fig.2 Schematic structure of 2 layer InAs QDs.

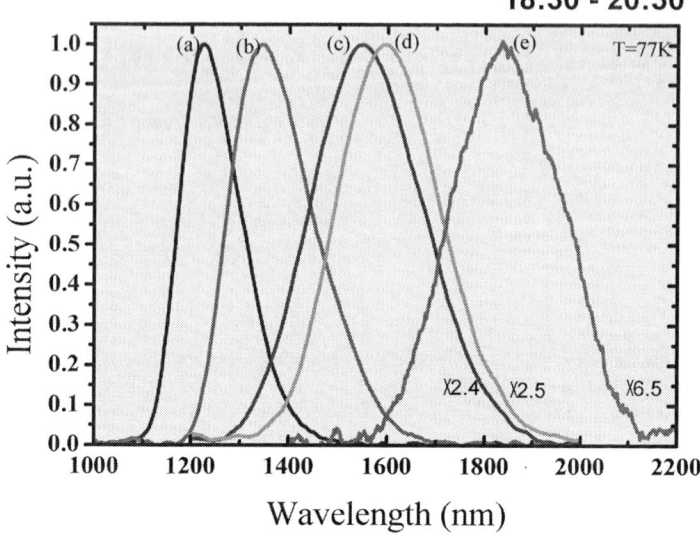

Fig.4 PL spectrum at 77K temperature of 2 stacked InAs QDs.

Table.I Peak wavelength all sumples

	InAs (ML)	FCL (nm)	Ga composition	Peak wavelength (nm)
(a)	1.68	2	0.7	1225
(b)	1.68	2	0.47	1345
(c)	1.68	4	0.47	1551
(d)	1.68	8	0.47	1594
(e)	3.84	15	0.47	1832

REFERENCES

[1] C. Paranthoen, N. Bertru, O. Dehaese, A. Le Corre, S. Loualiche, and B. Lambert, "Height dispersion control of InAsÕInP quantum dots emitting at 1.55 µm" Applied Physics Letters, vol. 78, pp. 1751–1753, March 2001.

[2] S.Barik, H. H. Tan, and C. Jagadish "Selective wavelength tuning of self-assembled InAs quantum dots grown on InP", Applied Physics Letters, vol. 88, p.193112, 2006.

[3] Y.Iwane, T.Saegusa, K.Yoshida, M.Yamauchi, S.Yoshikawa, and K.Shimomura, "V/III ratio of $Ga_{0.7}In_{0.3}As$ buffer layer dependence on InAs/InP QDs structure," ICMOVPE-XVI, Busan, Korea, Wep-49, May 2012.

[4] S. Yoshikawa, T. Saegusa, Y. Iwane, M. Yamauchi, and K. Shimomura, "Flat-Topped Emission with Spectral Width above 500nm from InAs/InP Quantum Dot Waveguide Array Light-Emitting Diode", Applied Physics Express vol.5, pp.092103-1-092103-3, Aug. 2012.

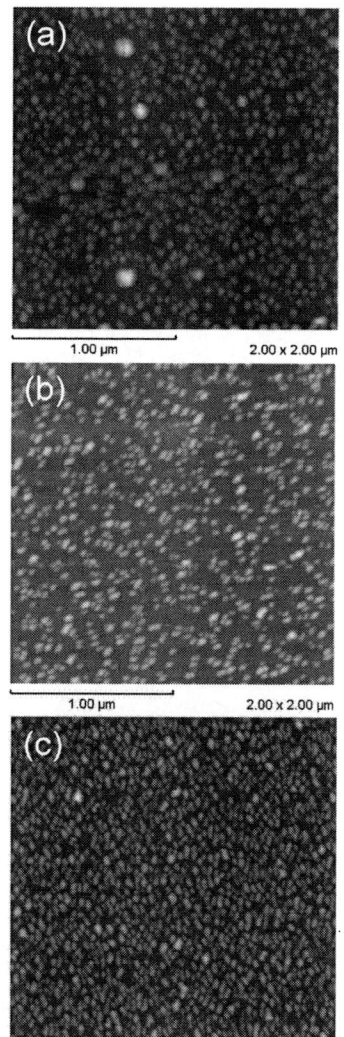

Fig.3 AFM images of InAs QDs (a) 1.68 ML on $Ga_{0.7}In_{0.3}As$ buffer layer (b) 1.68 ML on $Ga_{0.47}In_{0.53}As$ buffer layer (c) 3.84 ML on $Ga_{0.47}In_{0.53}As$ buffer layer

MoPI-6 (Poster)
18:30 - 20:30

Analysis of photoluminescent properties of InAs/InGaAsP/InP quantum dots structure

Rie Sato, Mariya Nakamura, and Hajime Imai

Faculty of Science, Japan Women's University

Faculty of Science, Japan Women's University, JWU
Tokyo, Japan
m1236003sr@gr.jwu.ac.jp

Abstract—We measured photoluminescence (PL) spectra of InAs/InGaAsP/InP quantum dots structures. We examined the PL peak shift changing the polarization of excitation light from p-polarization to s-polarization and the excitation light power. From the results, we estimated that the phonon emission according to the increase in TM component of the p-polarization.

Keywords—InAs/InGaAsP/InP quantum dots structure, Photoluminescence

I. INTRODUCTION

The quantum structures are effective to improve the characteristics of the optical semiconductor devices especially the laser diodes. We studied the luminescent spectrum of the quantum dots structures. We have already reported that the short wavelength shift of PL peak by increasing the excitation light power were caused by the phenomenon like the band filling in the quantum dots. And when we increased the TM component which is perpendicular to the sample surface of the p-polarization, the luminescent efficiency decreased and PL peak shift was not large as compared with the s-polarization. We considered that these phenomena were caused by the phonon emission due to the TM component of the p-polarization. We estimated that the density of the state of the quantum dots when we supposed the sizes of the quantum dots were following Gaussian distribution. And we considered the phenomenon as the band filling of the quantum dots . We analyzed the PL peak shift when we changed the polarization and the excitation light intensity , and examined the result in detail.

II. SAMPLES AND EXPERIMENTAL SET-UP

Experimental set-up

Fig.1 shows the experimental set-up. We used the YAG laser as excitation light whose the wavelength and the power were 1064nm and 1W, respectively. The grating type spectrometer [Jobin Yvon, Triax 320], and the PbS photodiode were used for the PL spectra measurement. The polarization of the excitation light was set at p-polarization or s-polarization. The incident angles of excitation light were set at 30, 50, 70 degrees for each polarization. The temperature of samples was adjusted at 30 ℃ by using the peltier cooler. The excitation light intensity was changed by using the ND filter of 10% or 50% which were set between the excitation light source and the samples.

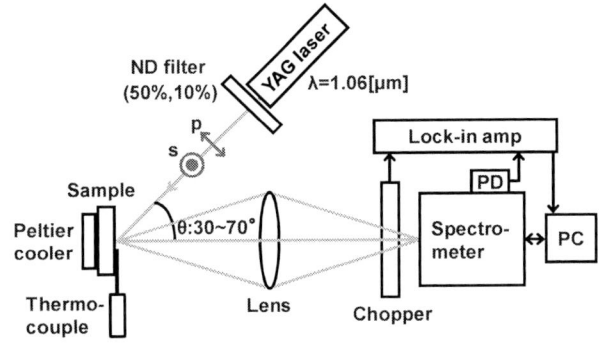

Fig.1. Experimental set-up

Samples

The samples are InAs/InGaAsP/InP quantum dots structures ; the quantum dots was oval about 30nm×40nm×6nm. InAs quantum dots layers and InGaAsP spacer layers consisted of 7 layers.

III. RESULT AND DISCUSSIONS

A. The band filling of the quantum dots

We have already reported that the phenomenon of the short wavelength shift of the PL spectra peak according to the increase in the excitation intensity was seemed to be examined by the band filling effect. We supposed the size of the quantum dots varied within 60%. In this calculation, we considered the size of the quantum dots followed to Gaussian distribution. We estimated the effective density of the state of the quantum dots corresponded the size distribution. Fig.2. shows the relationship between the peak wavelength and the total carrier density of the quantum dots calculated by using this density of the state. This figure shows that the peak wavelength shifted to shorter with the increase in the carrier density and then stable.

MoPI-6 (Poster)
18:30 - 20:30

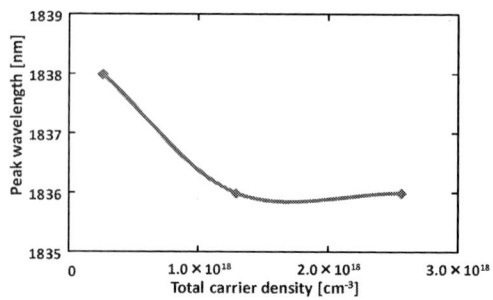

Fig.2 Peak wavelength versus Total carrier density

B. Increasing the TM component and the phonon emission

Fig.3. shows the PL peak shift when the incident angle was set at 50° by changing the excitation light power. This figure shows the short wavelength shift with increase in the absorbed light power. In this range, the PL peak wavelength of the p-polarization was longer than that of the s-polarization for the same value of the absorbed light power. We measured the wavelength difference for each incident angle 30, 40, 50 ° . Using the wavelength difference we estimated the temperature difference. As mentioned above the PL peak became stable according to the increase in the absorbed light power. Therefore we used $\triangle \lambda$ before the saturation. This figure shows that $\triangle \lambda_2$which have more large absorbed light power was larger than$\triangle \lambda_1$.Fig.4. shows the relationship between the $\triangle T$ and the absorbed light power of the TM component including detum for 30° and 70° . The sample temperature increased with the increase in the absorbed light power of the TM component. Therefore we estimated that the phonon emission increased with increase the TM component. The temperature of the sample was controlled by the peltier cooler. However we considered that the temperature of the sample locally increased by the excitation light power.

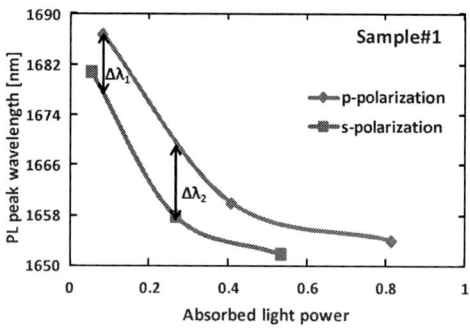

Fig.3.PL peak wavelength versus Absorbed light power

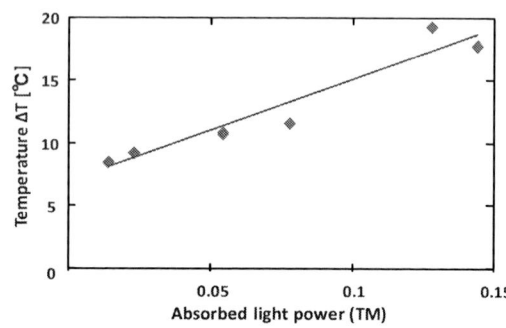

Fig.4. Temperature versus Absorbed light power of TM component

IV. CONCLUSION

We supposed the variation of the size of the quantum dots was Gaussian distribution and we estimated the distribution of the density of the state of the quantum dots. From the results , we considered that the quantum dots caused the phenomenon like the band filling because the peak wavelength shifted to shorter with the increase in the excitation carrier density as shown in the experimental value.

When we changed the excitation light power, we observed the PL peak shift for each polarization. From the result the PL peak wavelength of the p-polarization was longer than that of the s-polarization. When we calculated this wavelength difference and the excitation light power in the TM component of the p-polarization , we confirmed that the temperature difference was increased with the increase in the TM component. Therefore we estimated that there were the phonon emission due to the TM component of the p-polarization.

REFERENCES

[1] H. Imai and A. Motomura, presented at 2007 International Conference on Indium Phosphide and Related Materials, PA4, pp. 103-106 (2007)

[2] H. Imai , M. Esaki, and Y. Saito, presented at 2009 International Conference on Indium Phosphide and Related Materials , WP3 (2009)

[3] M. Esaki, N. Inaba, A. Fukuda, and H, Imai , presented at 2010 International Conference on Indium Phosphide and Related Materials , WeP5(2010)

[4] M. Esaki, A. Fukuda and H, Imai, presented at 2010 in Japanese The Society of Aplplied Physics , 16p-NB-15(2010)

[5] A. Fukuda, M. Esaki, M. Akimoto and H. Imai , presented at 2011 International Conference on Indium Phosphide and Related Materials , (2011)

[6] R. Sato, A. Fukuda, T. Suzuki and H. Imai, presented at 2012 International Conference on Indium Phosphide and Related Materials , (2012)

MoPI-7 (Poster)
18:30 - 20:30

Nature of the optical transition in (In,Ga)As(N)/GaP quantum dots (QDs): effect of QD size, indium composition and nitrogen incorporation

C. Robert[1], C. Cornet[1], K. Pereira da Silva[2], P. Turban[3], S. Mauger[4], T. Nguyen Thanh[1], J. Even[1], J.M. Jancu[1], M. Perrin[1], H. Folliot[1], T. Rohel[1], S. Tricot[3], A. Balocchi[5], P. Barate[5], X. Marie[5], P.M. Koenraad[4], M.I. Alonso[2], A.R. Goñi[2,6], N. Bertru[1], O. Durand[1] and A. Le Corre[1]

[1]Université Européenne de Bretagne, INSA Rennes
CNRS, UMR 6082 FOTON-OHM, 20 avenue des Buttes de Coësmes 35708 Rennes, France
[2]ICMAB-CSIC, Campus UAB, 08193 Bellaterra, Spain
[3]Equipe de Physique des Surfaces et Interfaces, Institut de Physique de Rennes UMR UR1-CNRS 6251
Université de Rennes 1, F-35042 Rennes Cedex, France
[4]Department of Applied Physics, Eindhoven University of Technology,
P.O. Box 513, 5600 MB Eindhoven, The Netherlands
[5]Université de Toulouse, INSA-CNRS-UPS, LPCNO
135 avenue de Rangueil, 31077 Toulouse, France
[6]ICREA, Passeig Lluís Companys 23, 08010 Barcelona, Spain

Abstract—**The structural properties of (In,Ga)As/GaP quantum dots (QDs) are studied by plane view and cross scanning tunneling microscopy. Time-resolved and pressure dependent photoluminescence experiments show a ground optical transition of indirect type. Mixed k.p/tight-binding simulations indicate a possible indirect to direct crossover depending on indium content and QD size. The incorporation of nitrogen in QDs is finally shown.**

Keywords—quantum dots, scanning tunneling microscopy, time-resolved photoluminescence, pressure dependent photoluminescence

I. INTRODUCTION

The pseudomorphic integration of III-V semiconductors through the lattice-matched growth of GaP(N) on silicon has been studied carefully in the last decade, because it may provide defect-free laser devices on silicon [1–4]. In this context, the development of a direct bandgap emitter on the indirect GaP substrate would open the route for efficient laser devices. GaAsPN quantum wells (QWs) on GaP have been proposed but bandstructure is complicated and nitrogen-induced localization effects are not easy to deal with[5]. More recently, (In,Ga)As/GaP QDs have demonstrated promising structural (small size, high density) and optical properties (room temperature photoluminescence (PL)) [6–8]. Finally, InGaAsN/GaP QDs have been proposed to reach wavelengths in the transparency window of silicon (above 1100nm) [9].

II. STRUCTURAL PROPERTIES

Structural properties at the atomic scale of the (In,Ga)As/GaP QDs are first determined. Fig. 1 represents the plane-view STM image of a typical (In,Ga)As/GaP QD. Due to the small QD size, there is a clear contribution of surface energy, and low index facets are visible on these QDs [10].

Surprisingly, QDs have no $\sigma_{(1\text{-}10)}$ plane symmetry as it is usually observed, which result in an overall C_2 symmetry for the QD shape. X-STM images (Fig. 2) allow the determination of the dimensions of the buried QD; i.e. 3.4 nm height and 19.5 nm width. Anisotropic shape of the QD is confirmed, and a thin wetting layer is observed.

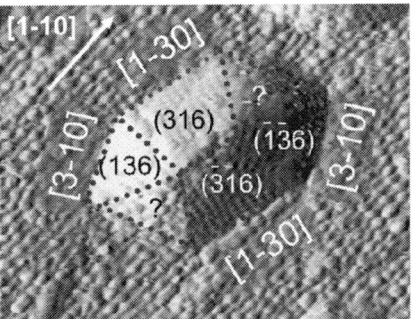

Fig. 1: 30x40nm² STM plane view of (In,Ga)As/GaP QD.

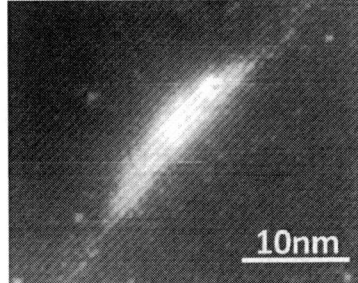

Fig. 2: The cross-sectional STM image of the QDs reveals a thin wetting layer, and anisotropic shape.

III. OPTICAL PROPERTIES

Time-resolved experiments are performed. The QDs exhibit a long time decay of 1 µs when exciting under low power density. Increasing the excitation power density shortens the time decay especially on the high energy side (Fig. 3). A rising time of about 20 ps is measured and demonstrates a favorable carrier capture by the QDs. This can explain the efficient room temperature PL despite the nature of the optical transition which is not the ideal case; i.e. direct and type-I. The room temperature PL is also measured as a function of hydrostatic pressure [11]. It reveals several optical transitions with small negative linear pressure coefficients (Fig. 4). These are related to transitions between X conduction band states and confined hole levels of the QDs [11].

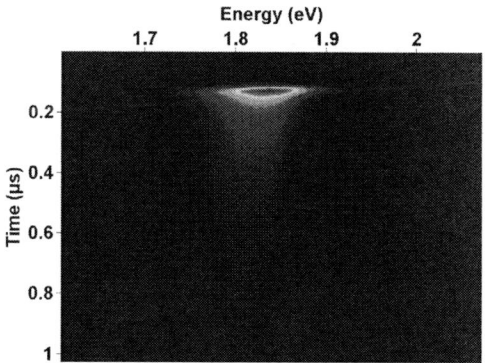

Fig. 3: Streak camera image at 10 K for an excitation density of 5.10^{12} photons.pulse^{-1}.cm^{-2}

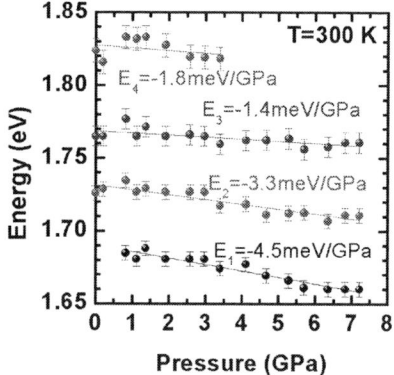

Fig. 4: Pressure dependence of PL peaks

IV. BAND STRUCTURE

Mixed tight-binding / **k·p** calculations are performed indicating the vicinity of both Γ-like and X-like electron levels in the QD and X conduction band of GaP barrier [10]. This competition is dependent of the indium composition inside the QD and the QD size (quantum confinement doesn't act strongly on X-valleys contrary to the Γ-valleys). An indirect to direct crossover is predicted for the largest QDs above an indium content of 30% [10].

V. GROWTH OPTIMIZATION AND NITROGEN INCORPORATION

Increasing the post-growth annealing (and thus the QD size) is found to enhance QD optical properties. The growth temperature is also varied between 500 and 590°C to change the indium sticking coefficient. X-STM images reveal higher QD with lower growth temperature. The optimum of PL emission is found when the growth temperature is around 550°C. Finally, nitrogen incorporation is found to change both structural and optical properties. The QDs areal density is reduced from $1.5.10^{11}$ cm^{-2} to $2.5.10^{10}$ cm^{-2}. The emission wavelength is strongly shifted from 700 nm to 900 nm, in agreement with the huge band gap bowing classically encountered in dilute nitride alloys.

ACKNOWLEDGMENT

This research is supported by "Région Bretagne" through the PONANT project including FEDER funds. The work is also supported through the participation of the SINPHONIC JC JC ANR project N° 2011-JS03-006-01 and the OPTOSI ANR project N°12-BS03-002-02. This work has been performed using HPC resources of GENCI CINES, TGCC/CCRT and IDRIS under the allocation 2013-[x2013096724].

REFERENCES

[1] S. Liebich, M. Zimprich, A. Beyer, C. Lange, D.J. Franzbach, S. Chatterjee, et al., Laser operation of Ga(NAsP) lattice-matched to (001) silicon substrate, Appl. Phys. Lett. 99 (2011) 071109.

[2] K. Yamane, K. Noguchi, S. Tanaka, Y. Furukawa, H. Okada, H. Yonezu, et al., Operation of Monolithically-Integrated Digital Circuits with Light Emitting Diodes Fabricated in Lattice-Matched Si/III–V–N/Si Heterostructure, Appl. Phys. Express. 3 (2010) 074201.

[3] C. Robert, A. Bondi, T. Nguyen Thanh, J. Even, C. Cornet, O. Durand, et al., Room temperature operation of GaAsP(N)/GaP(N) quantum well based light-emitting diodes: Effect of the incorporation of nitrogen, Appl. Phys. Lett. 98 (2011) 251110.

[4] T. Nguyen Thanh, C. Robert, W. Guo, A. Létoublon, C. Cornet, G. Elias, et al., Structural and optical analyses of GaP/Si and (GaAsPN/GaPN)/GaP/Si nanolayers for integrated photonics on silicon, J. Appl. Phys. 112 (2012) 053521.

[5] C. Robert, M. Perrin, C. Cornet, J. Even, J.M. Jancu, Atomistic calculations of Ga(NAsP)/GaP(N) quantum wells on silicon substrate: Band structure and optical gain, Appl. Phys. Lett. 100 (2012) 111901.

[6] T. Nguyen Thanh, C. Robert, C. Cornet, M. Perrin, J.M. Jancu, N. Bertru, et al., Room temperature photoluminescence of high density (In,Ga)As/GaP quantum dots, Appl. Phys. Lett. 99 (2011) 143123.

[7] G. Stracke, A. Glacki, T. Nowozin, L. Bonato, S. Rodt, C. Prohl, et al., Growth of In$_{0.25}$Ga$_{0.75}$As quantum dots on GaP utilizing a GaAs interlayer, Appl. Phys. Lett. 101 (2012) 223110.

[8] Y. Song, M. Larry Lee, Room temperature electroluminescence from light-emitting diodes based on In0.5Ga0.5As/GaP self-assembled quantum dots, Appl. Phys. Lett. 100 (2012) 251904.

[9] F. Fukami, K. Umeno, Y. Furukawa, N. Urakami, S. Mitsuyoshi, H. Okada, et al., Analysis of quantum levels for self-assembled InGaAsN/GaP quantum dots, Physica Status Solidi (c). 8 (2011) 322.

[10] C. Robert, C. Cornet, P. Turban, T. Nguyen Thanh, M.O. Nestoklon, J. Even, et al., Electronic, optical, and structural properties of (In,Ga)As/GaP quantum dots, Phys. Rev. B. 86 (2012) 205316.

[11] A.R. Goñi, C. Kristukat, F. Hatami, S. Dreßler, W.T. Masselink, C. Thomsen, Electronic structure of self-assembled InP/GaP quantum dots from high-pressure photoluminescence, Phys. Rev. B. 67 (2003) 075306.

Optimizing The Double-Cap Procedure for InAs/InGaAsP/InP Quantum Dots by Metal-Organic Chemical Vapor Deposition

Shuai Luo, Hai-Ming Ji, Xiao-Guang Yang and Tao Yang

Key Laboratory of Semiconductor Materials Science
Chinese Academy of Sciences
Beijing, People's Republic of China
tyang@semi.ac.cn

Abstract—**We report the optimization of the double-cap (DC) procedure for InAs/InGaAsP/InP quantum dots (QD) grown by metal-organic chemical vapor deposition. By using a combination of optimized thickness of the first cap layer and elevated growth temperature for the second cap layer, the photoluminescence (PL) linewidth of samples with five QD layers is significantly reduced from 124 meV to 87 meV at room temperature. Furthermore, the uniformity of the PL peak intensity and peak energy on the wafer surface is evidently improved. This distribution improvement is especially beneficial for improving device yield per wafer in device fabrication.**

Keywords—InAs InP quantum dot metal-organic chemical vapor deposition

I. INTRODUCTION

InAs/InP quantum dot (QD) lasers have attracted considerable attention as they can emit in a wide range from 1.4 μm to 2.1 μm, which has promising prospects in optical fiber communication, gas detection and biomedical applications [1-3]. Comparing with InAs/GaAs QD material system, the growth of high quality InAs/InP QDs is more complicated due to the low lattice mismatch and the anion exchange across the interface. The growth of InAs QDs on InP generally leads to a large QD size fluctuation and thereby to a broad photoluminescence (PL) linewidth, typically wider than 110 meV [1]. This would greatly reduce the modal gain for laser application. In order to maximize the gain and to control the QD size distribution, a double cap (DC) procedure where the QD capping growth is divided into two discontinuous steps has been proposed to reduce the large height dispersion [5]. This method successfully controlled the emission wavelength and reduced the full width at half maximum (FWHM) of InAs/InP QDs using gas source molecular beam epitaxy [6] and chemical beam epitaxy [7]. However, for the InAs/InP QDs on (001) substrate grown by metal-organic chemical vapor deposition (MOCVD), the DC procedure has been less successful in controlling QD height inhomogeneity [8]. In this paper, we have investigated the double-cap procedure for InAs/InGaAsP/InP QDs in detail by introducing two growth temperatures to grow the capping layer and found that the PL characteristics of the QDs can be significantly improved, including a reduced PL linewidth from 124 meV to 87 meV, an about 2 times enhanced PL peak intensity, and improved distribution uniformity in PL peak intensity and peak wavelength on wafer surface by using a combination of optimized thickness of the first cap layer (FCL) and elevated growth temperature for the second cap layer (SCL).

II. EXPERIMENTAL DETAIL

Samples used in the study were grown on InP (001) substrates using an Aixtron 3×2 FT MOCVD system. Trimethylindium (TMIn), triethylgallium(TEGa), arsine(AsH$_3$) and phosphine (PH$_3$) were used as the precursors. A 400 nm InP buffer and a 100 nm lattice matched InGaAsP quaternary alloy matrix emitting at 1.1 μm (referred as 1.1Q hereafter) were deposited on a InP substrate at 645 °C, and then the substrate was cooled down to 470 °C for the growth of InAs QDs. A 3.3-monolayer of InAs was deposited followed by a 5 second growth interruption for full migration of indium. After that, the QDs were capped with 1.1Q. The capping process for the QDs was divided into two steps. The FCL with the thickness of h_1 was deposited at the same growth temperature as the QDs. Subsequently a 5 minute growth interruption in a flow of phosphine was used to set and stabilize the substrate temperature prior to the growth of the SCL. For the growth of multilayer QDs, the total thickness of the capping layers between the QD layers was 40 nm.

III. RESULTS AND DISCUSSION

The influence of the growth temperature of the SCL on the PL characteristics of QDs was first investigated. To do this, three samples consisting of five QD layers were grown. Figure 1 (a) presents the PL spectra measured at RT from the three samples. It can be clearly seen that the PL intensity is markedly enhanced for the samples grown at the higher temperatures of 510 and 540 °C. The inset in Figure 1 (a) shows the FWHM dependence of the PL spectra on the growth temperature of the SCL. The FWHM decreases significantly from 124 meV to 87 meV as the SCL growth temperature increases from 470 to 540 °C. This reduction in FWHM can be attributed to the effective control of QD height dispersion using a DC procedure that introduces a higher temperature during the growth interruption. At elevated temperature, the enhancement of both the As/P exchange process and indium migration benefits the planarization process of the DC procedure so that a more uniform QD height can be achieved, and thereby a narrower PL linewidth is obtained. In Figure 1

MoPI-8 (Poster)
18:30 - 20:30

(b) the statistical of the PL peak intensity and of the peak energy are shown. It is found that for the sample with an SCL grown at 470 °C, the PL peak energy has a large fluctuation range of about 31.5 meV, whereas for the samples with the SCL grown at the higher temperatures, 510 and 540 °C, this range is reduced markedly to 15.7 and 2.5 meV, respectively. The fluctuation of the peak intensity also diminishes as the growth temperature increases. These improvements in PL peak intensity and peak energy are ascribed to the improved 1.1Q InGaAsP material quality that results from the more efficient decomposition of the precursors and from the enhanced migration of adatoms at higher temperatures.

Figure 2 plots the PL FWHM as a function of the FCL thickness (h_1). As the h_1 decreases from 10 nm to 2.7 nm, the FWHM decreases from 110 meV to 87 meV. This decrease in FWHM is mainly attributed to the unprotected part of the InAs QDs being removed, resulting in an improved QD height uniformity. As the h_1 further decreases from 2.7 nm to 1.8 nm, the FWHM is found to increase. This increase in FWHM is not due to the broadening of QD size distribution during PH_3 annealing, but rather due to a higher quantum confinement of the smaller QDs that enhance the effect of the size fluctuation on PL linewidth.

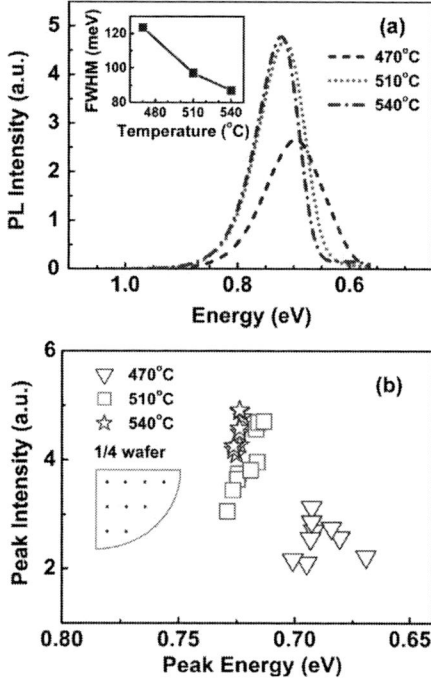

Fig.1 Influences of SCL growth temperature on PL characteristic of samples with five layers of QDs. (a) RT PL spectra from the samples with different SCL growth temperatures. The inset in Fig.1 (a) shows the evolution of FWHM with SCL growth temperature. (b) Statistical distribution of PL peak energy and peak intensity. The inset of quarter circle in Fig.1 (b) schematically shows a quarter of a 2 inch wafer and the spots above indicate the PL measurement sites.

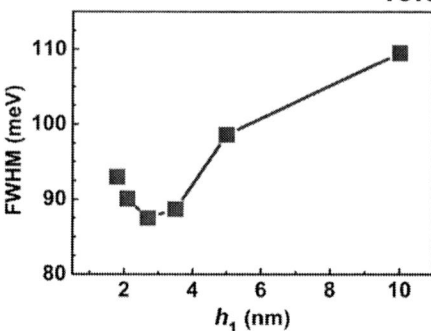

Fig.2 RT PL FWHM as a function of the FCL thickness, h_1. The growth temperatures of the FCL and SCL for the samples are 470 and 540 °C, respectively.

ACKNOWLEDGMENT

This work was supported by the National Natural Science Foundation of China (Nos. 61176047, 61076050, 61204057 and 61204076) and the National Basic Research Program of China (No.2012CB932701).

REFERENCES

[1] S. Anantathanasarn, R. Nötzel, P. J. van Veldhoven, F. W. M. van Otten, Y. Barbarin, G. Servanton, T. de Vries, E. Smalbrugge, E. J. Geluk, T. J. Eijkemans, E. A. J. M. Bente, Y. S. Oei, M. K. Smit, and J. H. Wolter, "Lasing of wavelength-tunable 1.55 μm region InAs/InGaAsP/InP (100) quantum dots grown by metal organic vapor-phase epitaxy," Appl. Phys. Lett. vol. 89, pp. 073115, 2006.

[2] J. Kotani, P. J. van Veldhoven, and R. Nötzel, "What Is the Longest Lasing Wavelength of InAs/InP (100) Quantum Dots Grown by Metal Organic Vapor Phase Epitaxy", Appl. Phys. Express vol. 3, pp. 072101, 2010.

[3] W. Lei and C. Jagadish, "Lasers and photodetectors for mid-infrared 2–3 μm applications," J. Appl. Phys. vol. 104, pp. 091101, 2008.

[4] K. Kawaguchi, M. Ekawa, A. Kuramata, T. Akiyama, H. Ebe, M. Sugawara, and Y. Arakawa, "Fabrication of InAs quantum dots on InP(100) by metalorganic vapor-phase epitaxy for 1.55 μm optical device applications," Appl. Phys. Lett. vol. 85, pp. 4331, 2004.

[5] C. Paranthoen, N. Bertru, O. Dehaese, A. Le Corre, S. Loualiche, B. Lambert, and G. Patriarche, "Height dispersion control of InAs/InP quantum dots emitting at 1.55 μm," Appl. Phys. Lett. vol. 78, pp. 1751, 2001.

[6] G. Elias, A. Létoublon, R. Piron, I. Alghoraibi, A. Nakkar, N. Chevalier, K. Tavernier, A. Le Corre, N. Bertru and S. Loualiche, "Achievement of High Density InAs/GaInAsP Quantum Dots on Misoriented InP (001) Substrates Emitting at 1.55μm," Jpn. J. Appl. Phys. vol. 48, pp. 070204, 2009.

[7] P. J. Poole, K. Kaminska, P. Barrios, Z. Lu, and J. Liu, "Growth of InAs/InP-based quantum dots for 1.55 μm laser applications," J. Cryst. Growth. vol. 311,pp. 1482, 2009.

[8] Y. Sakuma, M. Takeguchi, K. Takemoto, S. Hirose, T. Usuki, and N. Yokoyama, "Role of thin InP cap layer and anion exchange reaction on structural and optical properties of InAs quantum dots on InP (001)," J. Vac. Sci. Technol. B. vol. 23, pp. 1741, 2005.

MoPI-9 (Poster)
18:30 - 20:30

Junction Field-Effect Transistor Based on GaAs Core-Shell Nanowires

O. Benner, A. Lysov, C. Gutsche, G. Keller, C. Schmidt, W. Prost, F. J. Tegude
Solid State Electronics Department
University Duisburg-Essen
Duisburg, Germany
oliver.benner@uni-due.de

Abstract— **Nanowire FETs with all-around Junction-Gate are demonstrated using GaAs core-shell nanowires. The electrical properties of the n-channel Junction FET were determined by DC measurements. The radial pn-junctions show diode-type I-V characteristics. The output and transfer I-V characteristics exhibit good pinch-off, and hysteresis-free transient behavior. First devices with 190 nm nanowire channel diameter show a drain current of I_D = 260 nA and a transconductance of g_m = 300 nS.**

Keywords—nanowire, JFET, GaAs

I. INTRODUCTION

Nanowire transistors are routinely formed as core-shell structures consisting of the nanowire channel core and the surrounding gate metal isolated by a dielectric shell. The vertical nanowire with all-around-gate geometry allows for an optimum field-effect control of the channel carriers and provides the most important advantage of the nanowire FET design [1]. Today, InAs is the preferential III/V nanowire channel material because the Fermi level at the InAs surface is pinned within the conduction band. This enhances the nanowire conductivity down to very low diameters and simplifies ohmic contact formation substantially such that an impressive performance may be possible [2]. However, a highly sophisticated three-dimensional contact technology is required, dielectric/InAs interface states may cause instabilities, and the limited on/off ratio of the small band gap InAs severely affects digital applications. GaAs offers a wider band gap and may improve the on/off ratio and the mircroware power efficiency of nanowire transistors. However, the bare GaAs surface exhibits a huge trap density and its interface to the high-k dielectric will cause interface states and severe I-V instabilities. Therefore, despite the huge number of publications on GaAs nanowires there are almost no reports on transistors [3-5].

In order to avoid surface and dielectric interface effects the replacement of the gate dielectric by a semiconductor material is a promising approach. Recently, GaAs core-shell p-i-n nanowires were grown for optoelectronic purposes [6-7] that may also enable the fabrication of a radial pn-junction gate field-effect transistor. Here, we report on the demonstration of n-GaAs nanowire transistor with an all-around epitaxially grown p-GaAs junction gate. For the ease of fabrication based on selective wet-chemical etching a GaInP interlayer is inserted

as an intershell allowing for precise removal of the shell prior to nanowire channel contact formation.

II. FABRICATION

A. Processing of NW-JFET

The n-GaAs/i-GaInP/p-GaAs core-shell nanowires schematically shown in Fig. 1a were grown by VLS mode MOVPE. Details of growth are described elsewhere [7]. The nanowires were mechanically removed from the growth substrate and poured into isopropyl alcohol. The nanowire solution was redeposited on a semi-insulating GaAs host substrate that was covered with a 150 nm SiNx layer for improved isolation and a reduced effective dielectric constant. Electron beam lithography and lift-off technique were used to apply contacts to the nanowire. For ohmic contact of the gate electrode to the p-doped shell of the nanowire Pt/Ti/Pt/Au

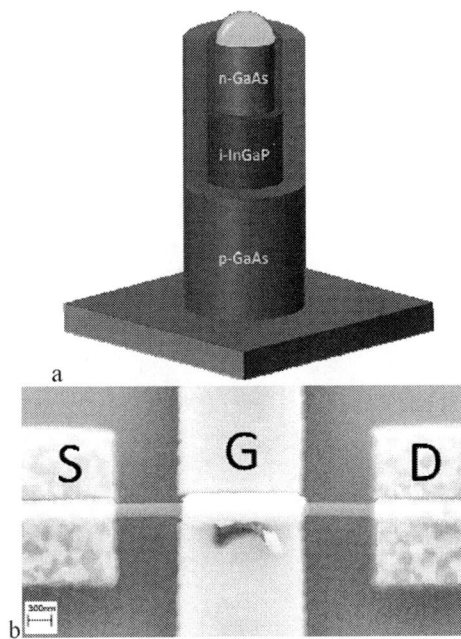

Figure 1. GaAs core-multi-shell nanowire (**a**) Schematic of the as grown p-i-n core-multishell nanowire and (**b**) preliminary fabrication of a JFET using redeposited nanowries on a host substrate.

metallization was used, followed by thermal annealing for 30 s at 360 °C. After the exposure of drain and source contacts, the outer multi-shells were removed by selective etching, performed with phosphoric and hydrochloric acids. Pd/Ge/Au was evaporated to form drain and source contacts to the n-doped GaAs nanowire core, followed by thermal annealing for 30 s at 280 °C. Figure 1b shows a SEM image of the NW-JFET. The diameter of the nanowire is 190 nm.

III. CHARATERIZATION

The electrical properties of the NW-JFET were determined by DC measurement technology. The I-V characteristic of the radial pn-junctions is shown in Fig.2a in a semi-log plot. A clear rectifying behavior is observed though a relatively large reverse current is present attributed to the high doping levels used. A substantial increase of the diode current towards the micro ampere range shall be possible [7]. Fig. 2a also shows the measured transfer characteristic of the device at fixed drain-source voltage of $V_{DS} = 1,5$ V. The transfer characteristic shows a typical FET type parabolic increase of the drain current. The open-channel resistance of the device is here 6.6 MΩ, which leads to a current flow in the 200 nA range. A transconductance of $g_m = 300$ nS has been determined. Measurements with different integration times are free of hysteresis or transient behavior.

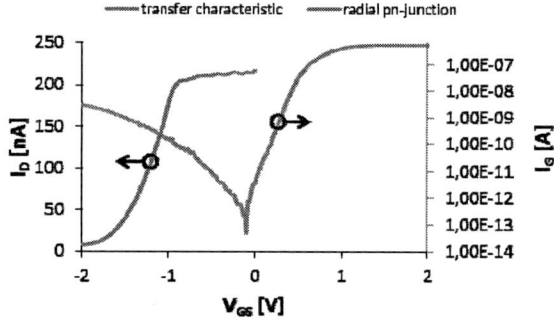

a

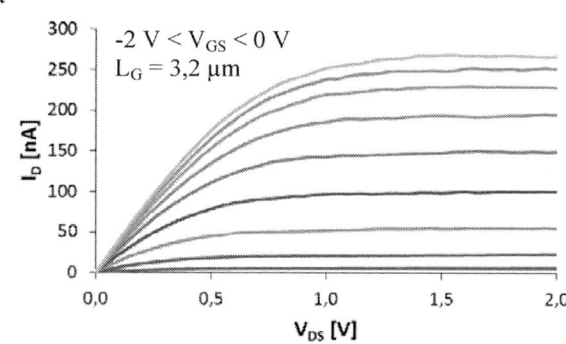

b

Figure 2. I-V characteristic of the NW-JFET: (**a**) transfer characteristic/gate-source and (**b**) Ouput characterstics.

IV. SUMMARY

GaAs core-shell nanowires have been grown via MOVPE and NW-JFETs have been fabricated. The I-V characteristics of the radial pn-junction show diode behavior. The output and transfer characteristic prove the all-around junction gate transistor behavior of the core-shell structure. A transconductance of $g_m = 300$ nS can be determined. Measurements with different integration times indicate that there is no hysteresis. A maximum drain current of $I_D = 260$ nA can be measured. This report shows that pn core-shell structures are suitable for hysteresis-free nanowire transistors.

ACKNOWLEDGMENT

The authors gratefully acknowledge financial support of the Deutsche Forschungsgemeinschaft with (i) Japanese/German cluster on Nanoelectronics, project "Nanowire/CMOS Heterogeneous Integration for Next-Generation Communication Systems", and (ii)the Priority Programme 1616.

REFERENCES

[1] Bryllert, T; Wernersson, LE; Froberg, LE; Samuelson, L; "Vertical high-mobility wrap-gated InAs nanowire transistor", IEEE Electron Dev. Lett., vol. 27, no. 5, pp. 323-325, 2006.

[2] Jansson, Kristofer; Lind, Erik; Wernersson, Lars-Erik, "Performance Evaluation of III-V Nanowire Transistors", IEEE Trans. Electron. Dev, vol. 59, LECTRON DEVICES Volume: 59(9) pp. 2375-2382, 2012.

[3] Fortuna SA, Li XL; "GaAs MESFET With a High-Mobility Self-Assembled Planar Nanowire Channel", IEEE Electron Dev. Lett., 30 (6) p.593, 2009.

[4] Muramatsu, Toru; Miura, Kensuke; Shiratori, Yuta; et al.; "Characterization of Low-Frequency Noise in Etched GaAs Nanowire Field-Effect Transistors Having SiNx Gate Insulator" Jap. J. Appl. Phys. 51(6), 06FE187, 2012.

[5] X. Miao, X. Li; "Scalable Monolithically Grown AlGaAs–GaAs Planar Nanowire High-Electron-Mobility Transistor", IEEE Electron Dev. Lett., 32 (9) p.1227, 2011.

[6] C. Colombo1, M. Heiβ1, M. Grätzel2, and A. Fontcuberta i Morral1, "Gallium arsenide p-i-n radial structures for photovoltaic applications", Appl. Phys. Lett. 94, 173108 (2009).

[7] C Gutsche, A Lysov, D Braam, I Regolin, G Keller, Z-A Li, M Geller, M Spasova, W Prost, F-J Tegude; "n-GaAs/InGaP/p-GaAs Core-Multishell Nanowire Diodes for Efficient Light-to-Current Conversion", Adv. Funct. Mater. 2012, 22, 929–936.

MoPI-10 (Poster)
18:30 - 20:30

MOVPE-preparation of Si(111) surfaces for III-V nanowire growth

Matthias Steidl[1,2], Agnieszka Paszuk[1,2], Weihong Zhao[1,2], Sebastian Brückner[1,2], Anja Dobrich[2], Oliver Supplie[2,3], Johannes Luczak[2], Peter Kleinschmidt[1,2,4], Henning Döscher[1,2] and Thomas Hannappel[1,2,4]

[1] Technische Universität Ilmenau, Institut für Physik, Postfach 100565, 98684 Ilmenau, Germany
[2] Helmholtz-Zentrum Berlin, Institute Solar Fuels, Hahn-Meitner-Platz 1, 14109 Berlin, Germany
[3] Humboldt-Universität zu Berlin, Institut für Physik, Newtonstr. 15, 12489 Berlin, Germany
[4] CiS Forschungsinstitut für Mikrosensorik und Photovoltaik, Konrad-Zuse-Str. 14, 99099 Erfurt, Germany
Corresponding author: matthias.steidl@tu-ilmenau.de

Abstract—We studied the preparation of the clean Si(111) surface in H_2 ambient with in situ reflection anisotropy spectroscopy and UHV-based surface science tools after contamination-free transfer. X-ray photoelectron spectroscopy confirmed complete oxide removal after high-temperature annealing. In situ RAS enabled observation of the oxide removal in dependence of process temperature. Monohydride termination was verified by Fourier transform infrared spectroscopy which agrees with a (1x1) surface reconstruction we observed by scanning tunneling microscopy and low energy electron diffraction. By atomic force microscopy analysis of the morphology, we found that wet-chemical pretreatment has an impact on the different silicon surfaces we have prepared, including homoepitaxy and termination of silicon with arsenic.

Keywords— MOVPE, silicon, arsenic, morphology, XPS, FTIR, LEED, STM

I. INTRODUCTION

III-V nanowires are one example for new solar cell concepts aiming at high efficiency [1]. Compared to III-V wafers, silicon substrates benefit from low cost and mature high quality manufacturing. Typically, {111} surfaces are considered for nanowire solar cell concepts where the wires are grown perpendicular to the surface. In order to prevent difficulties associated with different lattice constants and the different nature of the non-polar Si and the polar III-V material, III-V buffer layers are commonly grown on Si(111) prior to nanowire growth. However, there are two possibilities for the subsequent growth of III-Vs, either starting with Si-III or Si-V bonds at the heterointerface. As a requirement for successful growth of nanowires, type B buffer layers (Si-V bonds) are necessary [2]. While direct III-V growth on Si(111) results in type A buffer layers, the polarity of the epitaxial layer can be changed by terminating the Si surface with one monolayer of arsenic. In contrast to more established surface preparation in UHV, we prepare our samples in metal organic vapor phase epitaxy (MOVPE) ambient. Since the presence of hydrogen as a carrier gas limits access to most surface science techniques, we apply a contamination-free MOVPE-to-UHV transfer system to correlate in situ reflection anisotropy spectroscopy (RAS) to results from UHV-based surface science techniques.

II. EXPERIMENTAL

All samples were n-type Si(111) with 0.1° or 6° misorientation and were prepared in a modified Aixtron AIX-200 MOVPE reactor. We studied three different types of surface preparation: On all Si(111) samples, oxides were removed by thermal annealing at 1000 °C under flow of purified hydrogen at nearly atmospheric pressure (950 mbar). On some samples, a homoepitaxial buffer was grown using silane and some surfaces were terminated with arsenic by annealing in TBAs for 10 minutes at 420°C and 670°C, respectively. A dedicated sample transfer from MOVPE to UHV combined with a mobile UHV shuttle [3] enabled contamination-free access to several remote surface analysis systems, such as x-ray photoelectron spectroscopy (XPS), low energy electron diffraction (LEED), Fourier transform infrared spectroscopy (FTIR) in an attenuated total reflection (ATR) configuration and scanning tunneling microscope (STM). The morphology of the surfaces was analyzed ex situ by atomic force microscopy (AFM).

III. RESULTS AND DISCUSSION

After annealing at 1000°C, we transferred the samples to XPS to confirm that the surfaces are free from oxides and other contaminations. The results are shown in Fig.1. Before the annealing, oxygen is present as can clearly be seen in both the chemically shifted side peak of Si2p photoemission line, which is assigned to silicon oxide, and the O1s photoemission line. Both oxygen related signals vanished after annealing at 1000°C for 30 minutes in hydrogen ambient. Thus, temperature and time are sufficient to completely deoxidize the samples (within the detection limits of XPS). We found no traces of other contaminations.

MoPI-10 (Poster)
18:30 - 20:30

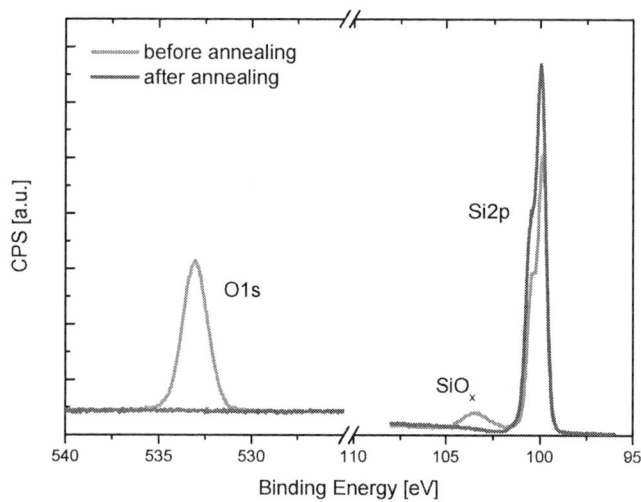

Figure 1. In XPS measurements before and after annealing it can be seen that the O1s as well as the SiO$_x$ peak vanish completely, confirming the removal of oxide.

During annealing, we observe a change in the RA spectrum, as depicted in Fig. 2 for a sample with 6° misorientation, where the changes are more pronounced than for 0°. We attribute the final shape of the spectrum to the oxide-free surface which enables in situ control over the deoxidation process.

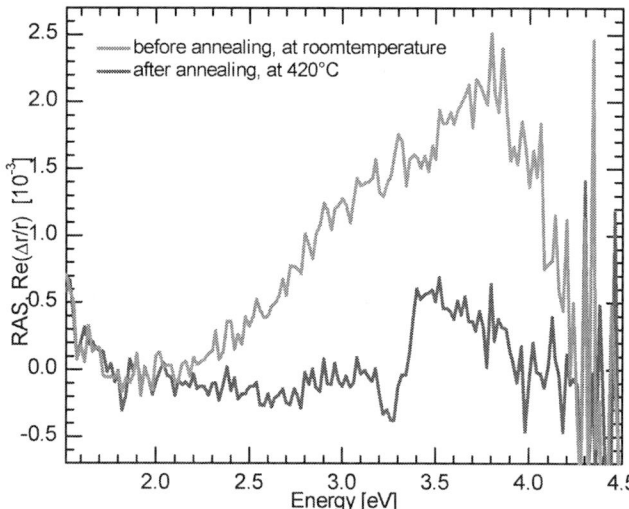

Figure 2. In situ RA spectra for Si(111) with 6° offcut before (green line) and after (red line) oxide removal.

In order to check if the Si(111) preparation in hydrogen ambient leads to a hydrogen terminated surface (in analogy to the Si(100) surface [4]), we performed FTIR measurements in ATR configuration. As can be seen in Fig. 3, we observe a strong absorption band at 2083 cm^{-1}, which corresponds to the stretching vibration mode of Si-H monohydride [5]. From measurements with light polarized perpendicular and parallel to the plane of incidence, we conclude that the Si-H bonds are

perpendicular to the surface and the hydrogen is bonded to the uppermost layer of Si atoms.

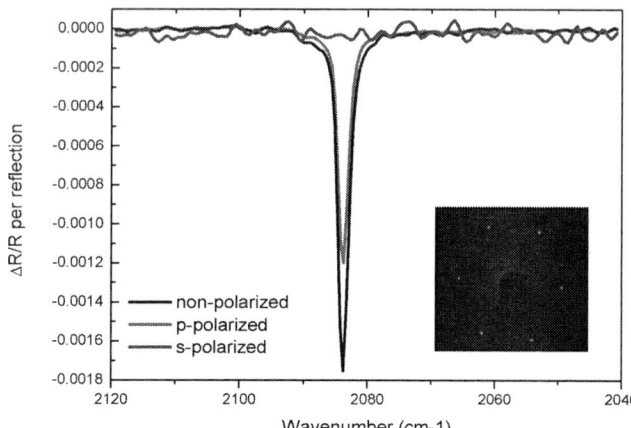

Figure 3. IR spectra of Si(111) after oxide removal including non-polarized, perpendicular polarized and parallel polarized radiation. Strong absorption at 2083 cm^{-1} is assigned to Si-H bonds which are perpendicular to the surface. The inset shows a LEED pattern of 1x1 Si(111) surface reconstruction.

Both LEED and STM (not shown here) measurements confirmed that the surface is 1x1 reconstructed, which agrees with a monohydride termination.

By AFM we analyzed the influence of oxide removal, homoepitaxial growth and arsenic termination on the surface morphology of either etched or non-etched substrates. We found that a chemical pretreatment in HF and NH$_4$F [6] reduces the roughness and atomically flat surfaces can be achieved after homoepitaxy. Similar behavior was observed for the treatment with TBAs. Moreover, arsenic strongly influences the surface morphology, changing the shape of the steps from smooth waves to sawtooth-type step edges.

REFERENCES

[1] M. T. Borgstroːm, J. Wallentin, M. Heurlin, S. Faːlt, P. Wickert, J. Leene, M. H. Magnusson, K. Deppert, and L. Samuelson, "Nanowires With Promise for Photovoltaics," IEEE Journal of Selected Topics in Quantum Electronics, vol. 17, no. 4, pp. 1050–1061, Jul. 2011.

[2] J. H. Kang, Q. Gao, H. J. Joyce, H. H. Tan, C. Jagadish, Y. Kim, D. Y. Choi, Y. Guo, H. Xu, J. Zou, M. a Fickenscher, L. M. Smith, H. E. Jackson, and J. M. Yarrison-Rice, "Novel growth and properties of GaAs nanowires on Si substrates," Nanotechnology, vol. 21, no. 3, p. 035604, Jan. 2010.

[3] T. Hannappel, S. Visbeck, L. Toːben, and F. Willig, "Apparatus for investigating metalorganic chemical vapor deposition-grown semiconductors with ultrahigh-vacuum based techniques," Review of Scientific Instruments, vol. 75, no. 5, p. 1297, 2004.

[4] H. Döscher, A. Dobrich, S. Brückner, P. Kleinschmidt,T. Hannappel, "Si(100) surfaces in a hydrogen-based process ambient," Appl. Phys. Lett. 97, p.151905, 2010.

[5] G. S. Higashi, Y. J. Chabal, G. W. Trucks, and K. Raghavachari, "Ideal hydrogen termination of the Si (111) surface," Applied Physics Letters, vol. 56, no. 7, p. 656, 1990.

[6] M. Lublow, T. Stempel, K. Skorupska, a. G. Muñoz, M. Kanis, and H. J. Lewerenz, "Morphological and chemical optimization of ex situ NH$_4$F (40%) conditioned Si(111)-(1×1):H," Applied Physics Letters, vol. 93, no. 6, p. 062112, 2008.

978-1-4673-6130-9/13 $31.00 © 2013 IEEE

MoPI-12 (Poster)
18:30 - 20:30

Bandgap Wavelength Shift in Quantum Well Intermixing using Different SiO$_2$ masks for Photonic Integration

Jieun Lee[1], Yoshiaki Yamahara[1], Mitsuaki Futami[1], Takahiko Shindo[2],
Tomohiro Amemiya[2], Nobuhiko Nishiyama[1], and Shigehisa Arai[1,2]

[1]Department of Electrical and Electronic Engineering, [2]Quantum Nanoelectronics Research Center
Tokyo Institute of Technology, Meguro-ku, Tokyo 152-8552, Japan
Email address: lee.j.aj@m.titech.ac.jp

Abstract—As a photonic integration method of semiconductor membrane structure, quantum-well-intermixing (QWI) process using O$_2$-sputtered SiO$_2$ mask was investigated by comparing the photoluminescence peak wavelength shift between two sections with/without the SiO$_2$ mask. As the result, a large bandgap wavelength difference of 80 nm (47 meV) was obtained while quite large transient region (90 μm) was observed. Since this fact was considered to be attributed to the temperature gradient along the masked and window regions during the rapid thermal annealing (RTA) process, we deposited CVD SiO$_2$ (which has smaller vacancies than O$_2$-sputtered SiO$_2$ hence the bandgap wavelength shift is smaller) on the entire surface after forming O$_2$-sputtered SiO$_2$ mask pattern and successfully reduced the transient region length to less than 5 μm.

Keywords—quantum-well intermixing (QWI); photonic integration; membrane photonic integrated circuits

I. INTRODUCTION

To realize the optical interconnections on a Si-LSI, we proposed the concept of the membrane photonic integrated circuit (PIC) [1] as shown Fig. 1, and reported its components such as a membrane laser [2], a photo-detector [3] and an InP-based waveguide [4]. As for the photonic integration method of active and passive devices, quantum-well-intermixing (QWI) is expected to be versatile technique by increasing in the bandgap wavelength and changing in the refractive-index partially, with only single epitaxial growth [5]. A great deal of intermixing techniques have been reported; impurity induced disordering (IID) [6], photoabsorption-induced disordering (PAID) [7], and impurity-free vacancy-enhanced disordering (IFVD) [8] in multiple applications. In this paper, we employ the IFVD mixing method, which relies on creating vacancies in the III-V semiconductor using SiO$_2$ mask during annealing process. In regard to the bandgap wavelength shift for formation of passive region, the SiO$_2$ mask margin and surface cover mask are considered as the key for wide bandgap wavelength shift value and sharp shift gradient on the boundary of active and passive regions.

II. QWI PROCESS

For the QWI process, the base structure shown in Fig. 2 (a) was grown on an n-InP wafer using organo-metallic vapor-phase epitaxy (OMVPE). The base structure consists of five 6-nm-thick quantum-wells and six 10-nm-thick barriers, sandwiched by 55-nm-thick 1.55Q (GaInAsP; λ$_g$ = 1.55 μm) layers. Above the core layer, a 10-nm-thick InP was followed by 10-nm thick GaInAs cap layer designed to cover semiconductor surface during the heating process.

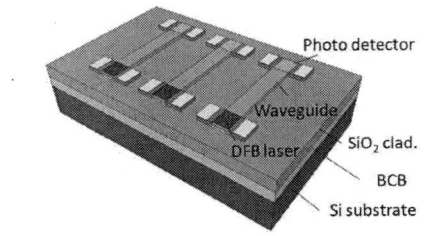

Fig. 1 Schematic representation of membrane photonic integrated circuits.

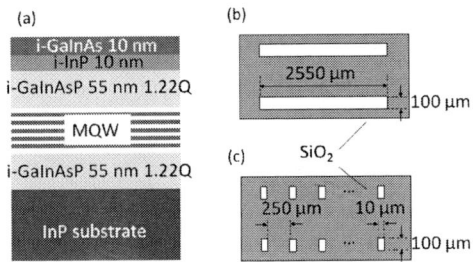

Fig. 2 (a) Epitaxial base structure with 5 quantum-well active region of 90 nm, (b) QWI-SiO$_2$ masks margin stripe type, and (c) island type.

The fabrication process is started with formation of photo-resist mask pattern of active region on the base structure by photolithography. Then, 200-nm-thick SiO$_2$ mask was deposited on the wafer by O$_2$-sputtering (5 sccm) with a pressure of 1.0×10^{-4} Torr at room temperature. Then the SiO$_2$ mask on the active region was removed by a lift-off process for remaining the mask margin and for QWI process. Finally, the rapid thermal anneal (RTA) was held at 750°C for 180 seconds in nitrogen atmosphere for creating vacancies in the III-V semiconductor. When the dielectric mask formed on III-V wafer was heated in the RTA process, Gallium atoms migrated into the SiO$_2$ mask layer created group III vacancies and inter-diffused through the quantum-well, then result in QWI.

III. QWI CHARACTERIZATION

In this study, 2 types of SiO$_2$ mask patterns were prepared; stripe pattern with 100×2550 μm^2 and island pattern with 100×10 μm^2 as shown in Figs. 2(b) and 2(c) to confirm the effect of mask margin to QWI. After the RTA was carried out, photoluminescence (PL) mapping with 3×3 μm^2 resolution was done. Figures 3(a) and 3(b) show the PL mapping results of stripe and island type patterns, respectively. In the stripe type, the bandgap wavelength shifts were 135 nm (76 meV) in the passive section (covered with SiO$_2$ mask) and 55 nm (29 meV) in the active section (without SiO$_2$ mask), hence the difference of 80 nm (47 meV) was obtained. The full width at half

978-1-4673-6130-9/13 $31.00 © 2013 IEEE

MoPI-12 (Poster)
18:30 - 20:30

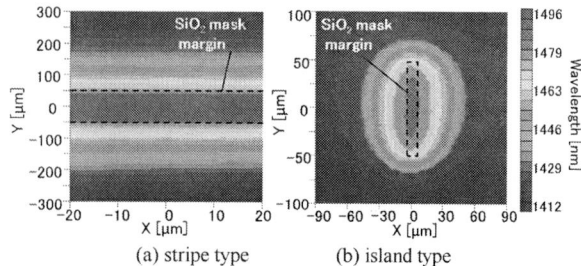

(a) stripe type   (b) island type

Fig. 3 PL mapping results of peak wavelength.

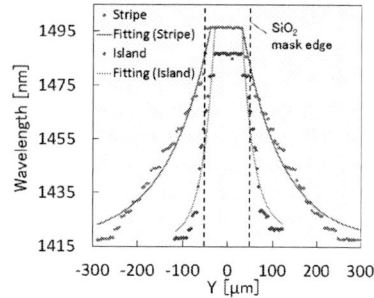

Fig. 4 Peak wavelength distribution depending on the distance from the mask center.

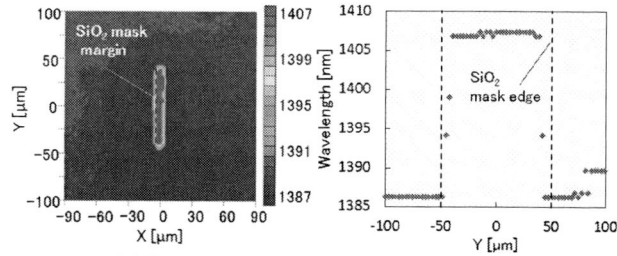

(a) SiO$_2$ covered island type   (b) PL distribution

Fig. 5 PL measurement results of peak wavelength in the SiO$_2$ covered island type.

maximum (FWHM) was 41 meV in the passive section and was slightly wider than 35 meV in the active section. In the island type mask case, the shifts in the passive and the active sections were 130 nm (73 meV) and 60 nm (32 meV), respectively, and the difference was 70 nm (41 meV). The FWHM was 40 meV in the passive section and was 38 meV in the active section.

Although the wavelength shift difference between the two sections in the stripe pattern was wider than that in the island pattern, the transient region length was also wider. Figure 4 indicates the peak wavelength distribution depending on the distance from the center (Y = 0) of the mask. Since the PL peak wavelength shifted exponentially from the mask edge, we defined the transient region length as the distance where the PL wavelength shift becomes 1/e of the maximum wavelength shift such as diffusion constant. The transient region length was 90 µm for the stripe pattern and was 25 µm for the island pattern. From these results, we considered that this gradient was caused by the temperature gradient during the RTA process since the open window region is exposed to the ambient N$_2$ while the SiO$_2$ masked region is insulated.

In order to confirm the above mentioned assumption and to reduce the transient region length, a supplementary experiment was performed so as to decrease thermal conductivity difference between these two regions by covering the entire surface with additional SiO$_2$ film deposited by plasma enhanced chemical vapor deposition (PECVD) which has less defects than O$_2$-sputtered SiO$_2$. Figure 5(a) indicates the PL mapping result of peak wavelength, and Fig. 5(b) shows the peak wavelength distribution along the vertical direction. The transient region length was reduced to less than 5 µm. The bandgap wavelength shifts were 164 nm (95 meV) in the passive section and 143 nm (81 meV) in the active section, and the FWHMs were 47 meV (passive) and 49 meV (active). Even though the wavelength shift difference was reduced to only 21 nm (14 meV) in this experiment, it is reasonable to assume that the wavelength shift total distance can decrease to 30 µm from 84 µm when the wavelength shift was supposed to 70 nm proportionally with same transient region as island pattern. Therefore, taking care of surface temperature distribution is very important to achieve sharp division for QWI.

IV. CONCLUSION

In conclusion, we investigated QWI process using SiO$_2$ mask for photonic integration and obtained the transient region length of less than 5 µm by combining 2 different SiO$_2$ masks to avoid the temperature gradient during the RTA process.

ACKNOWLEDGMENT

This work was supported by JSPS KAKENHI (Grant numbers 24246061, 24656046, 22360138, 21226010, 23760305, and 10J08973), the Council for Science and Technology Policy (CSTP) under the FIRST program, and the Ministry of Internal Affairs and Communications under the SCOPE program.

REFERENCES

[1] S. Arai, N. Nishiyama, T. Maruyama, and T. Okumura, "GaInAsP/InP membrane lasers for optical interconnects," *IEEE J. Sel. Top. Quantum Electron*, vol. 17, no. 5, pp. 1381-1389, Oct. 2011.

[2] M. Futami et al., "Low-threshold operation of LCI-Membrane-DFB lasers with Be-doped GaInAs contact layer," *Proc. Int. Conf. Indium Phosphide and Related Materials*, Th-2C.5, Aug. 2012.

[3] Y.Yamahara et al., "Characterization of GaInAsP lateral junction waveguide type membrane photodiode," the Institute of Electronics, Information and Communication Engineer, C-4-15, Sep. 2012.

[4] J. Lee et al., "Low-loss GaInAsP wire waveguide on Si substrate with benzocyclobutene adhesive wafer bonding for membrane photonic circuits," *Jpn. J. Appl. Phys.* vol. 51, no. 4, pp. 042201-1-042201-5, Apr. 2012.

[5] J. W. Raring et al., "Demonstration of widely tunable single-chip 10-Gb/s laser-modulators using multiple-bandgap InGaAsP quantum-well intermixing," *IEEE Photon. Technol. Lett.*, vol. 16, no. 7, pp. 1613-1615, July 2004.

[6] D. Deppe and N. Holonyak, Jr., "Atom diffusion and impurity-induced layer disordering in quantum well III-V semiconductor heterostructures," *J. Appl. Phys.* vol. 64, no. 12, pp. 93-113, Dec. 1988.

[7] B. Qui, A. Bryce, R. De La Rue, and J. Marsh, "Monolithic integraion in InGaAs-InGaAsP multiquantum-well structure using laser processing," *IEEE Photon. Technol. Lett.*, vol. 10, no. 6, pp. 769-771, June 1988.

[8] S. K. Si, D.H. Yeo, K. H. Yoon, and S. J. Kim, "Area selectivity of InGaAsP-InP multiquantum-well intermixing by impurity-free vacancy duffusion," *IEEE J. Sel. Topics Quantum Electron.*, vol. 4, no. 4, pp. 619-623, June/Aug. 1988.

MoPI-13 (Poster)
18:30 - 20:30

Carrier-transport, Optical and Structural Properties of Large Area ELOG InP on Si Using Conventional Optical Lithography

H. Kataria, W. T. Metaferia, M. Nagarajan, C. Junesand, Y. Sun, S. Lourdudoss*

Laboratory of Semiconductor Materials, KTH- Royal Institute of Technology, Electrum 229, 16440, Kista, Sweden
*e-mail: slo@kth.se

Abstract— We present the carrier-transport, optical and structural properties of InP deposited on Si by Epitaxial Lateral Overgrowth (ELOG) in a Low Pressure-Hydride Vapor phase epitaxy (LP-HVPE). Hall measurements, micro photoluminescence (μ-PL) and X-ray diffraction (XRD) were used to study the above-mentioned respective properties at room temperature. It is the first time that electrical properties of ELOG InP on Si are studied by Hall measurements. Prior to ELOG, etching of patterned silicon dioxide (SiO_2) mask leading to a high aspect ratio, i. e. mask thickness to opening width >2 was optimized to eliminate defect propagation even above the opening. Dense high aspect ratio structures were fabricated in SiO_2 to obtain ELOG InP on Si, coalesced over large area, making it feasible to perform Hall measurements. We examine this method and study Hall mobility, strain and optical quality of large area ELOG InP on Si.

Keywords— HVPE, ELOG, InP, Hall

INTRODUCTION

In the recent years there has been an increased demand of miniaturized and compact integrated circuits (ICs). It started with the conventional electronic ICs, but in the last decades the demand for photonic integrated circuits (PICs) has increased exponentially. An efficient combination of these two circuitries, in recent years have resulted in faster networks, higher data speeds etc. A viable option to satisfy the ever increasing demand on large bandwidth is the integration of both electronic and photonic components on a single chip. Many successful demonstrations have been made using bonding techniques [1] [2], which however are not yet fully exploited in large scale fabrication. As an alternative, ELOG has been proposed as one of the candidates to enable monolithic integration [3]. Here, the active material is grown selectively through openings made in dielectric mask. Despite several studies on ELOG, its carrier-transport properties have not been addressed so far. In this study, we carry out Hall measurements along with PL and XRD of unintentionally doped (u.i.d) ELOG InP (hereafter called only as ELOG InP) on semi-insulating InP:Fe (SI-InP:Fe) layer on Si substrate and compare the results with the reference samples.

EXPERIMENT

In this experiment, 3 samples are used. Details of those samples are given in table I. To measure the carrier concentration of the ELOG InP on Si, SI-InP:Fe (nominal resistivity $\geq 1x10^8$ $\Omega\cdot$cm) is grown to isolate the conducting n-InP seed layer from the ELOG InP.

Table I.

Sample ID	Description
Sample A	ELOG InP/ELOG InP:Fe/n-InP seed/Si substrate
Sample B	ELOG InP/ELOG InP:Fe/InP:Fe substrate
Sample C	u.i.d. InP/InP:Fe substrate

The ELOG InP on SI-InP:Fe on Si (sample A) is prepared as follows: (SI) InP:Fe (~ 2 μm thick) is grown by hydride vapor phase epitaxy (HVPE) at 600°C for 15 minutes with a V/III ratio of 10, on patterned (001) Si substrate 4° off-oriented toward <111> precoated with n-InP seed, enough to fill the openings in the mask. This resulted in partially coalesced ELOG InP:Fe layer. To obtain sample B, InP:Fe substrate is patterned and grown in the similar manner as sample A. Subsequently a 5 μm thick ELOG InP is grown at 610 °C for 20 minutes with a V/III ratio of 10, on these samples and studied. For comparative studies a 5 μm thick u.i.d InP on a planar InP:Fe substrate, sample C is also considered. For conducting ELOG, a SiO_2 mask of thickness 2 μm is deposited using plasma enhanced chemical vapor deposition (PECVD) on the respective substrates of samples A and B, which are patterned using standard optical lithography. The photomask used in these experiments consists of 1.25x1.25 cm^2 pattern field with 1.25 cm long parallel and 1 μm wide openings with a separation of 3 μm. Line openings 30° off [110] were made on the samples A and B. Dense high aspect ratio patterns were then fabricated using reactive ion etching (RIE). The growth of ELOG InP resulted in well-connected coalesced layer although some voids of a few μm^2 were observed on both samples A and B, probably due to uneven growth rate from different openings due to uneven coalescence of the preceding ELOG InP:Fe layer. But the u.i.d InP layer on sample C was void free. Hall, μ-PL and XRD studies were undertaken on A, B and C at room temperature.

RESULTS AND DISCUSSION

Electrical measurements: Table II shows the measured Hall data of all the three samples which exhibited n-type conductivity. With respect to the planar sample C, samples A and B have lower mobility values which can be attributed to the corresponding increase of the carrier concentrations in A and B with respect to C. Since u.i.d. InP was grown on the partially coalesced layers of ELOG SI-InP:Fe on Si substrate (sample A) and on InP:Fe substrate (sample B), the non-uniformity of the ELOG layers might have caused the formation of micro-facets at certain regions leading to different levels of enhanced auto-doping [4]. Since the coalesced layer of sample A is the least uniform, it is also but natural that its carrier concentration is the highest and its mobility the lowest.

TABLE II.

Sample ID	Hall mobility (cm²/Vs)	Carrier (n) Concentration (cm⁻³)
Sample A	815	1.8×10^{17}
Sample B	1950	1.8×10^{15}
Sample C	2850	1.9×10^{14}

Hall mobility and carrier concentration of Samples A, B and C

Optical Measurements: Fig. 1 shows the µ-PL-spectra of samples A, B, C. Comparable PL intensity and FWHM are observed on all the 3 samples, which indicates the good optical quality of the investigated layers. A red shift of approximately 4 nm is observed for sample A which could be due to certain residual strain.

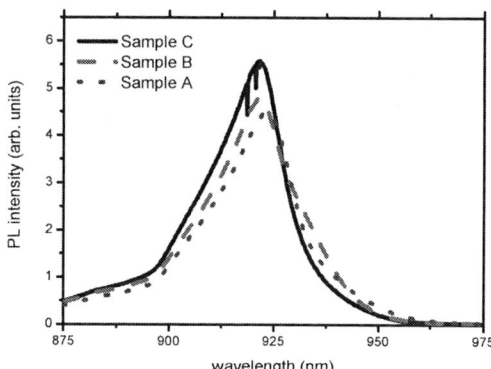

Fig 1. PL-spectra of Sample A, Sample B and Sample C

Structural Measurements: Fig. 2 shows the XRD curve of sample A. The peaks due to the Si substrate, GaAs buffer layer (from the seed layer) and InP are all distinctly visible. The broader GaAs peak actually indicates that it is somewhat relaxed. Some thermally induced strain remains in the GaAs and InP layers on Si [5] [6]. This residual strain is also evident in the PL-spectra of sample A, where a shift of 4 nm is observed. Table III compares the measured XRD data of all the samples. We notice a shift in the InP peak of sample A, which corresponds to the same residual tensile strain reflected in the PL measurements, Fig. 1. The trend of FWHM is also in accordance with that of the Hall measurements.

Table III.

Sample ID	FWHM (ω)(deg)	ω (deg)
Sample A	0.054	31.48
Sample C	0.026	30.2
InP Bulk	0.012	30.06

XRD data measured from Sample A, Sample C and Bulk InP

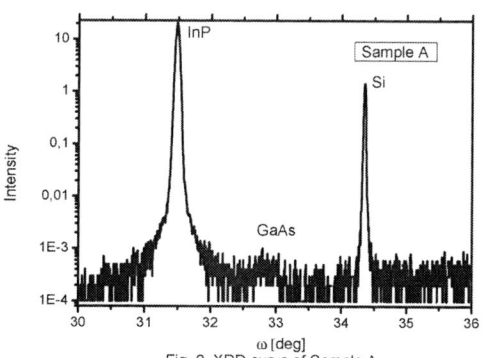

Fig. 2. XRD curve of Sample A

CONCLUSIONS

Carrier-transport, optical and structural properties of ELOG InP on Si are studied. It is the first time that electrical properties of ELOG InP on Si are reported. High optical and structural quality InP on Si is obtained using conventional lithography. Large area ELOG on InP is optimized, which makes it possible to fabricate broad area active devices monolithically integrated on Si. More uniform morphology with void-free growth of the ELOG layer is expected to further improve the electrical properties.

ACKNOWLEDGMENT

The work was supported by Swedish Research Council (VR), Swedish Foundation for Strategic Research (SSF) and INTEL Corporation through URO program.

REFERENCES

[1] Alexander W. Fang et al.,"Electrically pumped hybrid AlGaInAs-siliconevanescent laser," OPTICS EXPRESS 9203, Vol. 14, No. 20, 2 October 2006.

[2] Joris Van Campenhout et al."Design and Optimization of Electrically Injected InP-Based Microdisk Lasers Integrated on and Coupled to a SOI Waveguide Circuit," JOURNAL OF LIGHTWAVE TECHNOLOGY, VOL. 26, NO. 1, JANUARY 1, 2008.

[3] Sebastian Lourdudoss, "Heteroepitaxy and selective area heteroepitaxy for silicon photonics," Current Opinion in Solid State and Materials Science, 16 (2012) 91–99.

[4] Anand S., Sun Y. T., Lourdudoss S., Xu W. W., Vandervorst W.,"High Resolution Electrical Characterization of Laterally Overgrown Epitaxial InP," 15th International conference on indium phosphide and related materials, 563-566, IEEE, 2003.

[5] Sugo M., Uchida N., Yamamoto A., Nishioka T., Yamaguchi M.," Residual strains in heteroepitaxial III-V semiconductor films on Si-(100) substrates," J. Appl. Phys. 1989, 65, 591-595.

[6] Fang S. F., Adomi K., Iyer S., Morkoc H., Zabel H.," Gallium arsenide and other compound semiconductors on silicon," J. Appl. Phys. 1990, 68, 31-58.

MoPI-14 (Poster)
18:30 - 20:30

Low Power Consumption Operation of Light Sources for Inter-chip Optical Interconnects

Nobuaki Hatori[1,2], Takanori Shimizu[1,2], Makoto Okano[2,3], Masashige Ishizaka[1,2], Tsuyoshi Yamamoto[1,2], Yutaka Urino[1,2], Masahiko Mori[2,3], Takahiro Nakamura[1,2], and Yasuhiko Arakawa[2,4]

1: Photonics Electronics Technology Research Association
Tsukuba, Japan, n-hatori@petra-jp.org
2: Institute for Photonics-Electronics Convergence System Technology,
3: National Institute of Advanced Industrial Science and Technology, 4: Institute of Industrial Science, The University of Tokyo

Abstract — **In this work, we present a multi-channel configuration for the low power consumption operation of light sources on a silicon substrate in a photonics-electronics convergence system. We simulated the power consumed by the entire system with a focus on the characteristics of a laser diode used as a light source and the total loss of the system. Simulation results showed that lower power consumption per channel can be achieved with a single-LD multi-channel branching configuration. We found that using a hybrid integrated light source on a Si substrate with this branching configuration is suitable for photonics-electronics convergence systems.**

Keywords—silicon photonics, inter-chip optical interconnects, light sources on Si substrate, hybrid integrated light sources

I. INTRODUCTION

Optical interconnection with silicon (Si) photonics has been developed to overcome bandwidth bottlenecks in LSI chips because of the intrinsic properties of optical signals, including wide bandwidth, low latency, low power consumption, and low mutual interference [1]. We have previously proposed a photonics-electronics convergence system in which photodetectors (PDs), optical modulators, and light sources are linked by optical waveguides on a Si substrate [2]. It is necessary to use a huge number of light sources to ensure multi-channel operation when the scale of integration density increases. In terms of the light sources, an increase in power consumption with an increase in the scale of the optical circuit integration is a concern. In this study, we investigated the introduction and optimization of an optical branching structure for the low-power operation of a multi-channel light source on a Si substrate.

II. HYBRID INTEGRATED LIGHT SOURCES ON A Si SUBSTRATE

Much research attention has been devoted to the formation of light sources on Si substrates [3]–[5]. We previously developed a hybrid integrated light source by passive alignment on a Si substrate [5]. This structure has several advantages: it is possible to efficiently use a laser diode (LD), which is a light-emitting device; high optical output power operation is possible by introducing a spot-size converter (SSC) with a low coupling loss; and the light source has superior temperature characteristics due to the relatively high thermal conductivity of the Si substrate to which the mounted LDs connect. Precise alignment between the Si platform and the LD is required, which can be tricky but is possible by using modern passive alignment mounting technology. Its low operating voltage and high light output characteristics make the hybrid integrated light source the dominant contender as a light source for large-scale integrated devices, especially from the viewpoint of low-power operation. Multi-channel operation and reduced footprint per channel have already been achieved using a 1 × 4 branch configuration [5].

III. POWER CONSUMPTION IN MULTI-CHANNEL OPERATION

Here, we discuss the power consumption of a light source in two multi-channel configurations using a hybrid integrated light source. Figure 1 (a) shows a configuration using an optical branch from a single-mounted LD and (b) shows the configuration including multi LDs equal to the number of channels with no optical branch.

Figure 2 shows the LD characteristics used in the hybrid integrated light source at 25°C. The LD was a Fabry-Perot laser diode with a 1.55-μm lasing wavelength. The cavity length was 400 μm. A spot-size convertor was formed in the cavity and the spot-size of the LD was around 3 μm. We used the current-voltage and current-output characteristics to clarify the relationship between the light output power and the power consumption of the LD. The light output power required at the LD facet corresponds to the sum of the light intensity required for receiving at a PD and the total loss from coupling to the Si waveguide up to the PD. This loss has already shown in Ref. 2. In this simulation, we assumed that excess loss by branching was constant regardless of the number of branches. We calculated the total power dissipation with and without the

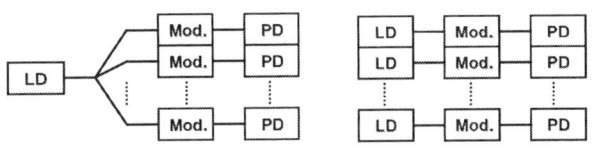

(a) Single LD mounting configuration with branching

(b) Multi LD mounting configuration with no branching

Figure 1. Multi-channel configuration.

branched structure and estimated the power consumption of the LD against the received power at the PD by referencing the relationship of the LD output power at the facet and the power consumption. The results are shown in Fig. 3.

Simulation results showed that the power consumption per channel with the single-mounted LD branching configuration could be decreased compared to that with the multi-LD configuration when the received power was less than a certain degree. The LD output can be obtained after injecting a current and power higher than the threshold current. In Fig. 2(b), the light output was radiated after input power of about 10 mW was injected. In the case of the multi-LD configuration with no branching, power corresponding to the number of multiple LDs is required, while in the case of the single-LD configuration with branching, the power consumption per channel is decreased because the power required for lasing is divided by the number of channels. In this simulation, there are advantages even in the case of 16-branching configuration below about -10 dBm of received power, and power consumption per channel can be reduced.

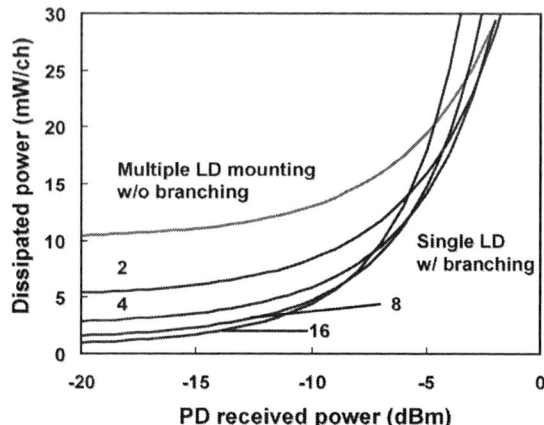

Figure 3. Dissipated power of multi-channel operation.

IV. CONCLUSION

We investigated the low power consumption operation of light sources for large-scale inter-chip optical interconnects. Simulation results showed that lower power consumption per channel could be achieved with a single LD multi-channel configuration. We conclude that the light source using a hybrid integrated light source on a Si substrate with a branching configuration is suitable for photonics-electronics convergence systems.

ACKNOWLEDGMENTS

This research is granted by the Japan Society for the Promotion of Science (JSPS) through the "Funding Program for World-Leading Innovative R&D on Science and Technology (FIRST Program)," initiated by the Council for Science and Technology Policy (CSTP).

REFERENCES

[1] I. A. Young, E. M. Mohammed, J. T. S. Liao, A. M. Kem, S. Palermo, B. A. Block, M. R. Reshotko, and P. L. D. Chang, "Optical technology for energy efficient I/O in high performance computing," IEEE Commun. Mag., vol. 48, no. 10. p.184, (2010.

[2] Y. Urino et. al., "Demonstration of 12.5-Gbps optical interconnects integrated with lasers, optical splitters, optical modulators and photodetectors on a single silicon substrate," Opt. Exp., vol. 20, p. B256, 2012.

[3] R. E. Camacho-Aguilera, Y. Cai, N. Patel, J. T. Bessette, M. Romagnoli, L. C. Kimerling, and J. Michel., "An electrically pumped germanium laser," Opt. Exp., vol. 10, p. 11316, 2012

[4] M. N. Sysak, H. Park, A. W. Fang, J. E. Bowers, R. Jones, O. Cohen, O. Raday, and M. Paniccia, "Experimental and theoretical analysis of a hybrid silicon evanescent laser," Opt. Exp., vol. 15, no. 23, p.15041, 2007.

[5] T. Shimizu, N. Hatori, M. Okano, M. Ishizaka, T. Yamamoto, Y. Urino, M. Mori, T. Nakamura, and Y. Arakawa, "High Density Hybrid Integrated Light Source with a Laser Diode Array on a Silicon Optical Waveguide Platform for Inter-Chip Optical Interconnection," Digests 8th Group IV Photonics, ThB5, p.181, 2011.

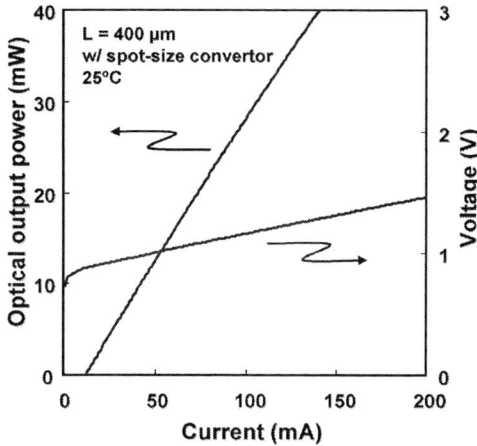

(a) Current-output power and current-voltage characteristics

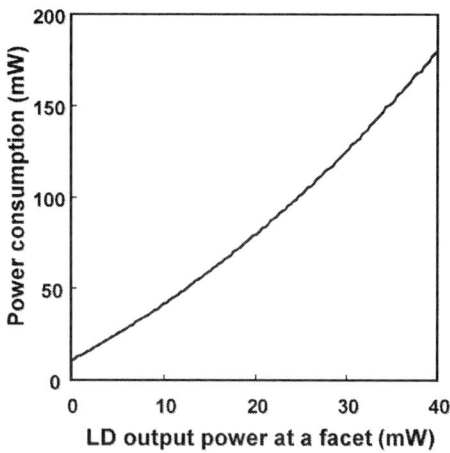

(b) Power consumption dependence on light output
Figure 2. LD characteristics.

MoPI-15 (Poster)
18:30 - 20:30

1540 to 1645 nm continuous VCSEL emission based on quantum dashes

C. Paranthoen[1], C. Levallois[1], J. P. Gauthier[2], F. Taleb[1], N. Chevalier[1], M. Perrin[1], Y. Leger[1], O De Sagazan[2] and A. Le Corre[1]

1 Université Européenne de Bretagne, France INSA, CNRS UMR 6082 FOTON
F-35708 Rennes, France
2 GM-IETR UMR-CNRS 6164, Université de Rennes1,
F-35042 Rennes Cedex, France
cyril.paranthoen@insa-rennes.fr

Abstract—We report on an optically excited InAs quantum dash vertical cavity surface emitting lasers (VCSEL) on InP substrate. By introducing a wedge microcavity design, we obtain a spatial dependence of the resonant wavelength along the wafer, enabling us to monitor the gain material bandwidth. In this paper we show a continuously variable VCSEL emission from 1645 down to 1540 nm all across the wafer, a consequence of the important and wide gain afforded by the use of optimized quantum dashes.

Keywords—VCSEL, quantum dashes, wide gain.

I. INTRODUCTION

Wide band gain material (WBGM) has always been essential for the development of a large variety of applications : telecommunication with large gain amplification in semiconductor optical amplifier (SOA), sensing with large wavelength tuning in external cavity lasers, and ultra high bit rate modulation with increasing the number of longitudinal modes in mode-locked lasers. Because of their important inhomogeneous broadening, nanostructures such as quantum dot (QD) or quantum dashes (QDHs) gain in interest as being a WBGM. Recent demonstrations have shown a 300 nm wide wavelength amplification with QDH [1], and a 200 nm wide wavelength tuning with QDs [2]. Inserted in vertical cavity surface emitting lasers (VCSEL), such nanostructures are very attractive for the development of compact and potentially widely wavelength tunable lasers [3], provided that the gain satisfies lasing conditions in such demanding devices. In this paper, we demonstrate a VCSEL based on optimized quantum dash nanostructures operating at telecommunication wavelength with a 105 nm continuous wavelength emission.

II. GROWTH AND PROCESSING

Optically excited VCSEL device are grown by molecular beam epitaxy on a 2 inch diameter InP(001) substrate. Active layers consist of three groups of QDHs, located at the position of the microcavity antinode field, each groups correspond to six InAs QDH layers separated with 15 nm $Ga_{0.2}In_{0.8}As_{0.435}P_{0.565}$ barriers [4]. In order to maximize the overall gain, a specific attention has been paid to compensate the natural wavelength shift related to internal strain field, and

to obtain exactly the same wavelength emission from each layer and each group of QDH.

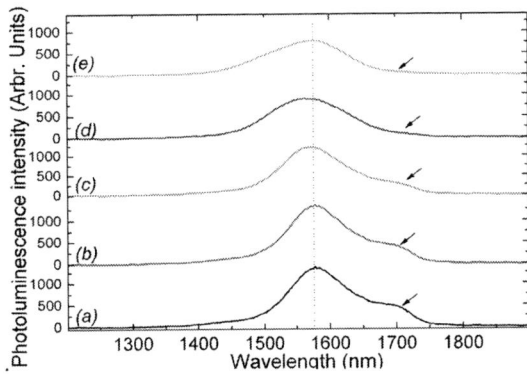

Figure 1. Photoluminescence spectra (293K, 10W/cm²) measured at different positions from the center of the substrate : 5, 7.5, 10, 12.5 and 20 mm corresponding to spectrum (a), (b), (c), (d) and (e) respectively). The vertical dotted line is only an eye guided centred at 1580 nm.

Prior to the active layers gowth, while growing the 1500 nm thick InP phase layer, the sample rotation has been interrupted, in order to create a voluntarily InP graded thickness from the center up to the edges of the wafer. This wedge has been estimated from past calibration to be around 10 %. Fig.1 presents photoluminescence (PL) spectra mapping measured on the 2 inch substrate, prior to processing. Note that the arrow indicates a PL shoulder which is not ascribed to QDH but related to the underneath InGaAs stop etch layer. The QDHs wavelength emission is constant and centered at 1580 nm. The linewidth slightly increases from 116 to 130 nm, and the overall integrated intensity decreases only by a factor of 1.6. Considering those slight variation, we may consider that all across the wafer, the QDH properties are nearly identical.

The processing of the optically excited VCSEL consists in several steps. First a large stop-band (>600 nm) dielectric mirror constituted of six pairs of amorphous Si and SiN is deposited on the wafer, followed by metallic layers of Ti and Au. Then a 50 μm thick copper film is electro-plated on the wafer, to form a metallic substrate. The whole InP substrate is then removed by mechanical polishing and selective chemical

978-1-4673-6130-9/13 $31.00 © 2013 IEEE

etching, Finally, the InGaAs stop-etch layer is also removed by chemical etching, and a second DBR is deposited.

III. VCSEL CHARACTERIZATION BELOW THRESHOLD

VCSEL are characterized at room temperature (293 K), and pumped with a 980 nm CW laser diode, on a 15 μm diameter focused spot. Fig.2 presents the spontaneous emission from the VCSEL, for an excitation power below threshold (5 mW), as a function of the position on the wafer; from 5 up to 20 mm from the center of the wafer.

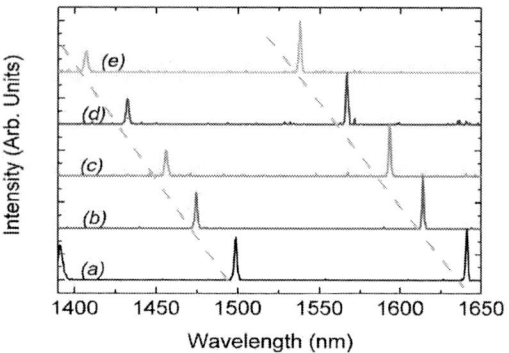

Figure 2. Spontaneous emission from VCSEL measured below threshold (CW, 5 mW), and at different positions from the center of the substrate : 5, 7.5, 10, 12.5 and 20 mm corresponding to spectrum (a), (b), (c), (d) and (e) respectively. Dotted lines are guides for the eye.

The resonant wavelength continuously decreases from 1640 down to 1535 nm, and the free spectral range decreases from 143 down to 130 nm, from the center up to the edge of the wafer respectively. Comparing those experimental data with simulations, we deduced a 153 nm decrease of the InP phase layer thickness, close to the 10% expected variation.

IV. VCSEL CHARACTERIZATION ABOVE THRESHOLD

At several positions on the wafer, laser experiments have been carried out, on CW mode. Optical output power as a function of incident pump power (LI curves), and output spectra have been monitored with an optical spectrum analyzer. Fig. 3 shows the VCSEL output spectra, measured at different position on the wafer, at a 1O kW/cm² constant power density above threshold. Close to the center of wafer (5 mm), laser emission operates at 1645 nm. As the distance from the center increases, the laser wavelength continuously decreases, and reaches 1540 nm 20 mm further. Note that as previously demonstrated for such a QDH based VCSEL (not shown), the output VCSEL intensity exhibits polarization ratio greater than 25 dB along the [1-10] crystallographic at all wavelength [4]. From LI curves (not shown), we measured variation of maximal output power from 170 down to 30 μW, and threshold power from 15.5 up to 20.5 mW. Minimum threshold and maximum output power have been obtained at a position of 10 mm away from the center, corresponding to a lasing wavelength of 1620 nm. This position corresponds to the optimal matching between the QDH maximal afforded

gain and the cavity resonance. We can thus conclude that the QDH VCSEL emission covering the [1540-1645] nm range, is mainly limited by the microcavity resonance itself, as it covers the low wavelength side of the QDH gain spectrum. This means that we can expect an overall QDH based VCSEL emission covering more than the measured 105 nm.

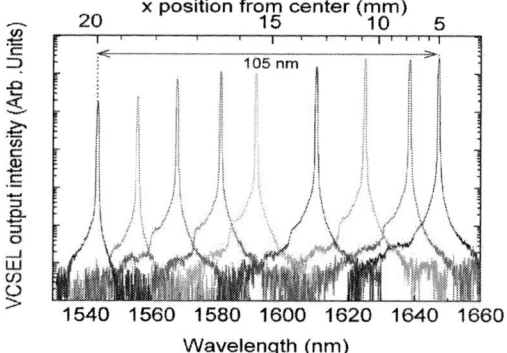

Figure 3. VCSEL output spectra (293 K, CW) measured above threshold at 10 kW/cm², for different position across the wafer (from 5 up to 20 mm, from right to left respectively). The upper scale relies on the dependance of the position from the center of the wafer and the cavity resonant wavelength.

CONCLUSION

We demonstrated that a QDH active medium is well suited to achieve laser emission in VCSEL, covering a wavelength range as high as 105 nm. Combining this active medium with a wavelength tunable VCSEL process [5-6], we can thus expect to achieve very large tuning, for a large variety of applications. The authors acknowledge the INSCOOP ANR project n° 11-NANO-012 for financial supports.

REFERENCES

[1] A. Somers, W. Kaiser, J. P. Reithmaier, A. Forchel, M. Gioaninni ans I Montrosset, "Optical gain properties of InAs/InAlGaAs/InP quantum dash structures with a spectral gain bandwidth of more than 300 nm", Appl. Phys. Lett. Vol 89, pp 061107, 2006

[2] K. Fedorova, M. Cataluna, I. Krestnikov, D. Livshits, and E. Rafailov,"Broadly tunable high-power InAs/GaAs quantum-dot external cavity diode lasers," Opt. Exp., vol. 18, pp. 19438–19443, 2010.

[3] Alexander R. Albrecht, Andreas Stintz, Felix T. Jaeckel, Thomas J. Rotter, Pankaj Ahirwar, Victor J. Patel, Christopher P. Hains, Luke F. Lester, Kevin J. Malloy and Ganesh Balakrishnan, "1220–1280-nm Optically Pumped InAs Quantum Dot-Based Vertical External-Cavity Surface-Emitting Laser", IEEE Jounr. Of . Select. Top. In Quant. Electron., vol 17, pp 1787-1793, november 2011

[4] J.-P. Gauthier, C. Paranthoën, C. Levallois, A. Shuaib, J.-M. Lamy, H. Folliot, M. Perrin, O. Dehaese, N. Chevalier, O. Durand, and A. Le Corre, "Enhancement of the polarization stability of a 1.55 μm emitting vertical-cavity surface-emitting laser under modulation using quantum dashes", Opt. Exp. Vol 20, pp 16832-16837, July 2012

[5] C. Gierl, T. Gruendl, P. Debernardi, K. Zogal, C. Grasse, H. A. Davani, G. Bohm, S. Jatta, F. Kuppers,P. Meißner, and M.-C. Amann , "Surface micromachined tunable 1.55 μm-VCSEL with 102 nm continuous single-mode tuning", Opt. Exp. vol 19, pp 17336-17343, August 2011

[6] O. Castany, L. Dupont, A. Shuaib, J. P. Gauthier, C. Levallois, and C. Paranthoen, "Tunable semiconductor vertical-cavity surface-emitting laser with an intracavity liquid crystal layer", Appl. Phys. Lett vol 98, pp 161105 (2011)

MoPI-16 (Poster)
18:30 - 20:30

InAs/InP quantum dot mode-locked lasers grown on (113)B InP substrate

K. Klaime[1], C. Calò[2], R. Piron[1], C. Paranthoen[1], D. Thiam[1], T. Batte[1], O. Dehaese[1], J. Le Pouliquen[1], S. Loualiche[1], A. Le Corre[1], K Merghem[2], A. Martinez[2] and A. Ramdane[2]

[1] *UEB INSA-RENNES, CNRS UMR6082 FOTON, FRANCE*

[2] *CNRS LPN MARCOUSSIS, FRANCE*

Abstract — We report for the first time the passive mode-locking of single section Fabry-Perot (FP) lasers based on InAs quantum dots grown on (113)B InP substrate. Devices under study are a 1 and 2 mm long laser diodes emitting around 1.58 μm. Self-starting pulses with repetition rates around 39 and 23 GHz and pulse widths down to 1.5 ps are observed after propagation through a suitable length of single-mode fiber for intracavity dispersion compensation.

Index Terms — Mode-locking, quantum dots (QDs), semiconductors laser.

I. INTRODUCTION

Mode-locked laser (MLL) diodes have been regarded for the last two decades as the centre of interest for a large range of photonics and optical telecommunication applications. Owing to their capability of generating short pulses at high repetition rates, these sources are well suited for ultra-high bit rate optical telecommunications. In addition, MLLs may serve as building blocks for the development of all optical systems for clock distribution, clock recovery, radio over fiber signal generation and optical sampling so as to overcome the electronic bandwidth limitation at reasonable cost [1]. For applications in the 1.3 μm telecommunication window, QD-lasers have already outperformed the quantum well (QW) lasers [2]. On the other hand for operation in the C and L telecommunication band, much attention is devoted to InAs nanostructures grown on InP substrates. Stranski-Krastanov growth of InAs nanostructures on InP by Molecular Beam Epitaxy (MBE) leads to the formation of Quantum Dashes (QDashes) on nominal InP substrate ((001)InP) or QDs on the (113)B InP substrate. QDashes obtained by MBE growth on (001)InP have already demonstrated good results for mode-locking in single or two section devices [3][4]. In contrast, despite the high density (10^{11}/cm^2) of small QDs achieved on (113)B InP substrate using MBE and the good laser results with low threshold current density[5], no mode-locking results have been reported on this substrate. In fact, InAs/InP QD based MLL have been demonstrated on InP nominal substrate using chemical beam epitaxy (CBE) [6] or on misoriented InP substrate using MBE [7]. In this work, we report, for the first time to our knowledge, the fabrication and characterization of a mono-section InAs/InP QD MLL on (113)B substrate emitting around 1.58 μm.

II. FABRICATION AND STATIC CHARACTERISATIONS

The active region (AR) consists of 9 layers of QDs embedded in a GaInAsP (1.18 μm) barrier (Fig. 1. (a)) grown by Gas Source MBE on an n doped (113)B InP substrate using the double cap technique[8]. As shown in the inset of Fig. 1. (b), QD exhibits a photoluminescence spectrum centred at 1.52 μm. with a full width at half maximum of 177 nm determined by the size dispersion of InAs QDs within the same layer and between different layers. Measurements on broad area lasers processed from such epitaxial structure show an internal efficiency of about 46%, and a modal gain and internal losses of about 25 cm^{-1} and 6 cm^{-1}, respectively. Single section FP ridge lasers have been fabricated by photolithography and inductively coupled plasma (ICP) reactive ion etching. Benzocyclobutene (BCB) is used for laser passivation and planarization. A ridge width of 2 μm has been chosen to ensure single transverse mode operation of the lasers. FP Lasers with cavity lengths of 1 and 2 mm, as cleaved facets and threshold currents of about 60 and 80 mA respectively, have then been mounted on copper bases for thermal management purposes.

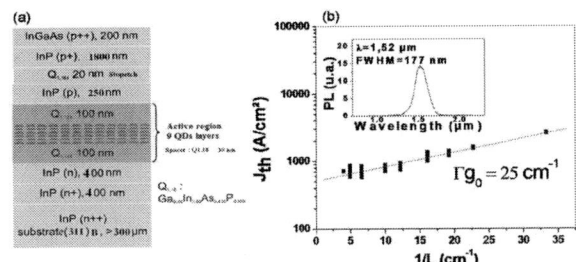

Fig. 1. (a) Schematic of the epitaxial structure of the InAs/InP QDs MLL under study. (b) Threshold current density evolution versus reciprocal cavity length for the 9 QDs layer laser, inset: photoluminescence measured at 300K for the investigated 9 QDs layers sample.

III. MEASUREMENTS AND DISCUSSIONS

Mode-locking performances of the devices described above are evaluated in this session. The lasers are operated in CW regime and temperature controlled at 20°C using a Peltier cooler. The laser signal is collected by an antireflection coated lensed fiber followed by an optical isolator to prevent feedback from reflections into the laser cavity and analysed using an optical spectrum analyser, an RF electrical spectrum analyser with a 50 GHz bandwidth InGaAs photodiode and an autocorrelator.

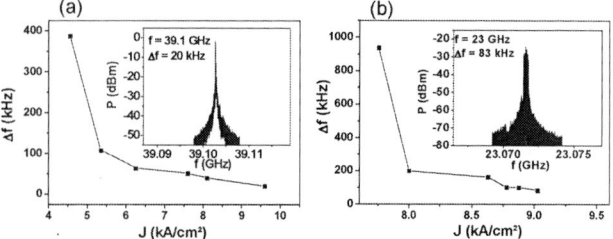

Fig. 2. RF peak width Δf versus the injection current density J of the 1 mm (a) and 2 mm (b) cavity length devices, (inset): RF spectrum for a gain current of 192 and 361 mA respectively.

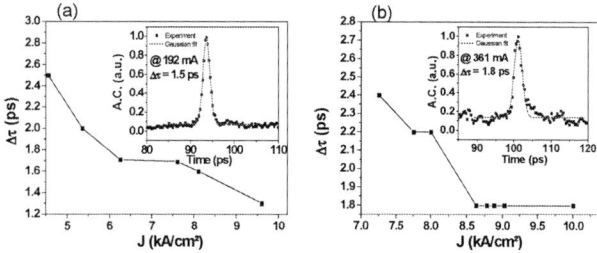

Fig. 3. Optical pulse width $\Delta\tau$ versus the injection current density J of the 1mm (a) and 2 mm (b) cavity length devices, (inset): Measured autocorrelation trace, for a gain current of 192 and 361 mA and after propagation in 220 and 230 m of SMF respectively.

Measurements were taken at 192 and 361 mA respectively for the 1 and 2 mm devices. These operating points were selected by choosing the optimal RF spectra in term of power and width Fig. 2. shows the variation of the RF peak width Δf versus the injection current density J for the 1 mm (a) and 2 mm (b) cavity length devices. Insets of the figure show the RF spectra for the both devices. The RF peak width decrease with increasing the injection current. The repetition rates are about 39 and 23 GHz and the optimal RF peak width at -3 dB is around 20 and 83 kHz respectively for the two devices. The optical spectrum width increases with injection current and reaches a maximum value of 4.0 nm and 4.8 nm respectively. . Self-starting pulses are observed by autocorrelation after propagation through 220 and 230 m of SMF respectively to fully compensate the group delay dispersion of the laser. Fig. 3. exhibits the optical pulse width $\Delta\tau$ versus the injection current density of the 1 mm (a) and 2 mm (b) cavity length devices. The autocorrelation traces are presented in the inset of the figure. Similarly, the pulse width decreases with the current and reach its minimum value for the optimal current presenting the minimum value of the RF peak width. Assuming a Gaussian fit, we deduce a pulse width of about 1.5 ps for the 1 mm long device and 1.8 ps for the 2 mm long one. Recently, a shorter pulse has been reported on a single section QD MLL using CBE [6]. Here we report for the first time a mode-locking regime on an InAs QD laser elaborated

on InP (113)B substrate using gas source MBE. The advantage of this type of substrate is the capability to obtain a high density of QD by layer. This increase of the QD density takes us to an increase of the modal gain and then to decrease threshold current for the ML. We note that the pulse width decrease with increasing the gain current. This process is observed for the monosection device MLL [3][7] and has been attributed to a compensation of the dispersion by nonlinear effects enhanced by higher carrier density and larger intracavity laser fields [3].

IV. CONCLUSION

Mode-locking without the presence of saturable absorber is demonstrated for the first time for single-section QD lasers based on (113)B InP substrates. Lasers with repetition frequencies of 39 and 23 GHz exhibit self-starting pulses with pulse widths down to 1.5 ps after intracavity dispersion compensation by light propagation through suitable lengths of SMF.

Acknowledgement: This work was supported by the French National Research Agency through the project TELDOT.

REFERENCES

[1] T. Ohno, K. Sato, R. Iga, Y. Kondo, I. Ito, T. Furuta, K Yoshino, and H. Ito, "Recovery of 160 GHz optical clock from 160 Gbit/s data stream using mode locked laser diode" Electron. Lett., vol. 40, pp. 265-267, 2004.

[2] E. U. Rafailov, et al., "High-power picosecond and femtosecond pulse generation from a two-section mode-locked quantum-dot laser"Appl. Phys. Lett. 87, 081107-1 (2005)

[3] R. Rosales, S.G. Murdoch, R.T. Watts, K. Merghem, A. Martinez, F. Lelarge, A. Accard, L.P. Barry, A. Ramdane, "High performance mode locking characteristics of single section quantum dash lasers", Opt. Express, vol. 20, 8649-8657 (2012)

[4] M. Dontabactouny, R. Piron, K. Klaime, N. Chevalier, K. Tavernier, S. Loualiche, A. Le Corre, D. Larsson , C. Rosenberg , E. Semenova , K. Yvind , " 41 GHz and 10.6 GHz low threshold and low noise InAs/InP quantum dash two-section mode-locked lasers in L band" Journal of Applied Physics, 111, 023102 (2012)

[5] P. Caroff, et al., "High-gain and low-threshold InAs quantum-dot lasers on InP" Appl. Phys. Lett. 87, 243107 (2005).

[6] Z.G. Lu, J.R. Liu, P.J. Poole, Z.J. Jiao, P.J. Barrios, D. Poitras, J. Caballero, X.P. Zhang, "Ultra-high repetition rate InAs/InP quantum dot mode-locked lasers, Optics Communications 284, 2323-2326 (2011)

[7] K. Klaime, R. Piron, C. Paranthoen, T. Batte, F. Grillot, O. Dehaese, S. Loualiche, A. Le Corre, R. Rosales, K. Merghem, A. Martinez and A. Ramdane " 20 GHz to 83 GHz single section InAs/InP quantum dot mode-locked lasers grown on (001) misoriented substrate", (Oral) IPRM Santa Barbara (U.S.A.) 2012

[8] C. Paranthoen, N. Bertru, O. Dehaese, A. Le Corre, S. Loualiche, B. Lambert, and G. Patriarche, "Height dispersion control of InAs/InP quantum dots emitting at 1.55 µm" Appl. Phys. Lett. 78, 1751 (2001)

MoPI-17 (Poster)
18:30 - 20:30

Analysis of Uni-Traveling-Carrier Photodetectors (UTC-PDs) with Dipole-Doped Interface

Q. Q. Meng[1*], C. Y. Liu[1], H. Wang[1,2], K. S Ang[1], K. Manoj[1], T. X. Guo[1], and B. Gao[1,3]

1.Temasek Laboratories (TL@NTU) , Nanyang Technological University,
50 Nanyang Drive, Singapore 637553
2.School of Electrical and Electronic Engineering, Nanyang Technological University,
Nanyang avenue, Singapore, 639798
3. School of Electronic and Information Engineering, Xi'an Jiaotong University, No.28, Xianning West Road, Xi'an, Shaanxi, China,710049
*Corresponding author, Phone: (65)97930516, E-mail: qqmeng@ntu.edu.sg

Abstract—A uni-traveling-carrier photodetector (UTC-PD) with dipole-doped structure at the InGaAs/InP interface has been designed, fabricated and characterized. The device transit time delay and RC time delay was extracted using an equivalent circuit model. A transit time delay time less than 4.5 ps was obtained with a junction reverse bias lager than 4 V. The results suggest that the current-blocking at InGaAs/InP interface can be effectively suppressed by the dipole-doped interface.

Keywords—uni-traveling-carrier photodetector, dipole doping, equivalent circuit model, time delay.

I. INTRODUCTION

In recent years, the uni-traveling-carrier photodetector (UTC-PD) has attracted much research attention. Compared to conventional photodiodes, such as PIN diodes [1,2], the UTC-PD has a unique mode of operation where only electrons are the active carriers [3-9]. The overshoot velocity of electrons is one order of magnitude larger than that of holes, and this helps shorten the carrier transit time, as well as reduce the space charge effect, which is a common issue in any surface illuminated photodiode. These factors result in high-speed and high saturation-output-powers simultaneously obtained from UTC-PDs [3]. On the other hand, the current blocking at the InGaAs/InP absorber-collector interface may result in a large carrier transit delay and compromise the device bandwidth. Different InGaAs/InP interface designs were used to minimize the current blocking effect. In this work, a uni-traveling-carrier photodetector (UTC-PD) with dipole-doped structure at the InGaAs/InP interface to suppress the current blocking is characterized and analyzed.

II. EXPERIMENTAL DETAILS

A dipole-doped interface structure is introduced between the InGaAs absorber and InP collector for our devices to minimize the current-blocking at InGaAs/InP interface [7]. Simulations results have suggest the effectiveness of minimized conduction band barrier at the InGaAs/InP heterojunction.

Top-illuminated UTC-PDs with cylindrical mesas of different diameters were fabricated using standard processing techniques. The mesa formation was based on wet etching processes to minimize the surface damage. The mesa was connected to the coplanar waveguide for RF probing with BCB planarization.

The DC and high frequency performance of UTC-PDs with different diameters were characterized. Fig.1 shows the dark current characteristics of the UTC-PDs with different diameters. The devices shows low junction leakage (10^{-9} to 10^{-8} A) and series resistance (48 ohm). Fig.2 shows the typical frequency response characteristics measured from a UTC-PD with a diameter of 12 µm under the constant optical power of 150 mW. The reverse bias is varied from 1 V to 6 V. The 3dB bandwidth of the devices is around 29GHz .

The 3dB bandwidth of UTC-PD is limited by both the carrier transit time and the junction RC delay. An equivalent circuit model shown in Fig.3 [10] is used to model the UTC-PDs and extract the model parameters. The carrier transit time delay can thus be estimated by the product of R_t and C_t.

III. RESULTS AND DISCUSSION

Fig.4 compares the modeling results of the 12µm device at the reverse bias voltage of 1 V, 2 V and 6 V with the experimental results. It can be seen that the modeling S_{21} parameters match well with the measured ones in the frequency range from 10 MHz to 20 GHz. The extracted parameters of 12 µm diameter device at different reverse bias voltage from 1 to 6 V at a injection power of 150 mW are summarized in Table 1. It can be seen that, at the low reverse biases, the device bandwidth is limited by the carrier transit time, with could be attributed to the low carrier velocity and possible carrier pile-up at the InGaAs/InP absorber-collector interface With the increase in biasing voltage, a drastic decrease in the carrier transit is observed at $V_{RB} \geq$ 4 V). A transit delay smaller than 4.5 ps, which is less than the junction RC time is obtained. The results suggest that the current blocking at the InGaAs/InP interface

MoPI-17 (Poster)
18:30 - 20:30

could be effectively suppressed by the dipole-doped interface.

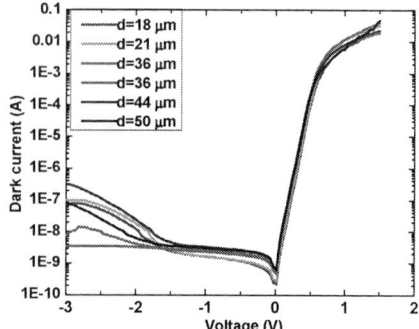

Fig.1 Dark I-V characteristics of the UTC-PD.

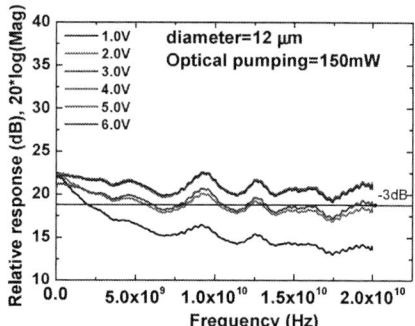

Fig.2 Frequency response of UTC-PD with a diameter of 12 μm at optical power of 150mW.

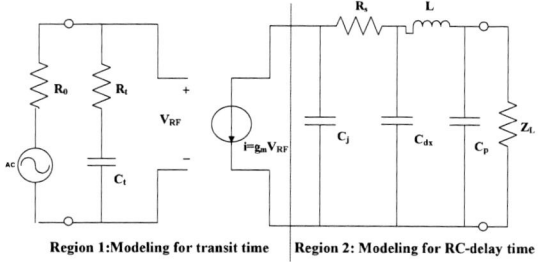

Fig.3 Equivalent-circuit model of the UTC-PD.

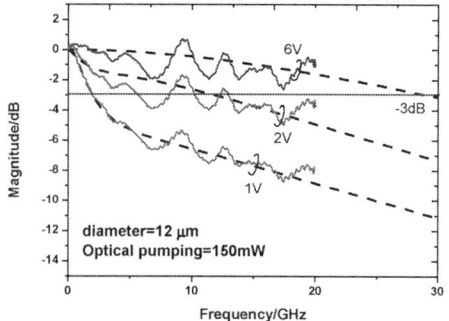

Fig.4 Measured and fitted S_{21} magnitude of the device with a diameter of 12 μm and reverse bias voltages of 1V, 2 V and 6 V. The solid lines are the experimental results, and the dashed lines are the modelling curves.

TABLE.1 EXTRACTED MODEL PARAMETERS

V_{RB} (V)	C_j (pF)	R_s (Ω)	C_{dx} (fF)	L (nH)	C_p (fF)	R_t (Ω)	C_t (fF)	$\tau_t = R_t C_t$ (ps)	g_m (S)
1	0.45	48	70	0.16	12	40	1500	60	0.76
2	0.3	48	70	0.16	10	180	750	135	0.76
3	0.18	48	70	0.16	5	170	300	51	0.76
4	0.12	48	70	0.16	4	90	50	4.5	0.76
5	0.12	48	70	0.16	4	90	40	3.6	0.76
6	0.12	48	70	0.16	4	90	30	2.7	0.76

IV. CONCLUSION

In summary, a 29 GHz 3 dB bandwidth UTC-PD (12 μm diameter) with dipole-doped interface is demonstrated. The carrier transit time delay is extracted using an equivalent circuit model. Small carrier transit time demonstrated at $V_{RB} \geq 4$ V suggests the effectiveness of the suppression of current blocking at the InGaAs/InP interface using the dipole-doped structure.

REFERENCES

[1] J. E. Bowers, C. A. Burrus, and R. J. McCoy, "InGaAs PIN photodetectors with modulation response to millimetre wavelengths," Electronics Letters, vol. 21, pp. 812-814, 1985.

[2] L. Y. Lin, M. C. Wu, T. Itoh, T. A. Vang, R. E. Muller, D. L. Sivco, and A. Y. Cho, "High-power high-speed photodetectors - Design, analysis, and experimental demonstration," IEEE Transactions on Microwave Theory and Techniques, vol. 45, pp. 1320-1331, Aug 1997.

[3] T. Ishibashi, T. Furuta, H. Fushimi, S. Kodama, H. Ito, T. Nagatsuma, N. Shimizu, and Y. Miyamoto, "InP/InGaAs uni-traveling-carrier photodiodes," Ieice Transactions on Electronics, vol. E83C, pp. 938-949, Jun 2000.

[4] N. Li, X. W. Li, S. Demiguel, X. G. Zheng, J. C. Campbell, D. A. Tulchinsky, K. J. Williams, T. D. Isshiki, G. S. Kinsey, and R. Sudharsansan, "High-saturation-current charge-compensated InGaAs-InP uni-traveling-carrier photodiode," IEEE Photonics Technology Letters, vol. 16, pp. 864-866, Mar 2004.

[5] D. A. Tulchinsky, X. W. Li, N. Li, S. Demiguel, J. C. Campbell, and K. J. Williams, "High-saturation current wide-bandwidth photodetectors," IEEE Journal of Selected Topics in Quantum Electronics, vol. 10, pp. 702-708, Jul-Aug 2004.

[6] H. Ito, S. Kodama, Y. Muramoto, T. Furuta, T. Nagatsuma, and T. Ishibashi, "High-speed and high-output InP-InGaAs unitraveling-carrier photodiodes," IEEE Journal of Selected Topics in Quantum Electronics, vol. 10, pp. 709-727, Jul-Aug 2004.

[7] H. Wang and S. Mao, "High speed InP/InGaAs uni-traveling-carrier photodiodes with dipole-doped InGaAs/InP absorber-collector interface," in Compound Semiconductor Week (CSW/IPRM), 2011 and 23rd International Conference on Indium Phosphide and Related Materials, 2011, pp. 1-3.

[8] X. Wang, N. Duan, H. Chen, and J. C. Campbell, "InGaAs-InP photodiodes with high responsivity and high saturation power," IEEE Photonics Technology Letters, vol. 19, pp. 1272-1274, Jul-Aug 2007.

[9] M. Chtioui, A. Enard, D. Carpentier, S. Bernard, B. Rousseau, E. Lelarge, F. Porrimereau, and M. Achouche, "High-power high-linearity uni-traveling-carrier photodiodes for analog photonic links," IEEE Photonics Technology Letters, vol. 20, pp. 202-204, Jan-Feb 2008.

[10] Gang Wang, Tsuneo Tokumitsu, Ikuo Hanawa, Yoshihiro Yoneda, Keiji Sato, and Masahiro Kobayashi, "A Time-Delay Equivalent-Circuit Model of Ultrafast p-i-n Photodiodes," IEEE Transactions on Microwave Theory and Techniques, vol. 51, No. 4, April 2003.

978-1-4673-6130-9/13 $31.00 © 2013 IEEE

Multi-regrowth steps for the realization of buried single ridge and μ-stripes quantum cascade lasers

O. Parillaud, G-M de Naurois, B. Simozrag, V. Trinité, G. Maisons, M.Garcia, B. Gérard, and M. Carras

III-V Lab
1, Avenue Augustin Fresnel, 91767 Palaiseau, France
olivier.parillaud@3-5lab.fr

W. Metaferia, C. Junesand, H. Kataria, Y. Sun, and S. Lourdudoss

KTH – Royal Institute of Technology
Electrum 229, Isafjordsgatan 22-24, 16440 Kista, Sweden

Abstract—We report on the realization of buried single ridge and μ-stripes quantum cascade lasers using HVPE and MOVPE regrowth steps of semi-insulating InP:Fe and Si doped layers. We present here the preliminary results obtained on these devices. The reduction of the thermal resistance achieved using semi-insulating InP:Fe for regrowth planarization and μ-stripe arrays approaches are shown and performance perspectives are addressed.

Keywords— Quantum Cascade Lasers; Thermal resistance; μ-stripes arrays; MOVPE and HVPE.

I. INTRODUCTION

The interest for efficient and high power, single and multi-mode Quantum Cascade Lasers (QCLs) for applications such as gas sensing or mid-IR counter measures has increased continuously in the past years, following the device performance. InP based structures emitting at 4.6 μm have now reached output power of 5W with above 20% Wall Plug Efficiency (WPE) [1]. Among all the improvements achieved, for example in the active region design or in the device mounting optimization, the use of buried ridge structures into InP:Fe has been critical in order to reduce the thermal resistance R_{th} [2], [3]. This allows the increase of the output optical power, using higher electrical injection. We have recently proposed the use of μ-stripes QCLs arrays to achieve high beam quality and high power devices [4] [5]. We combine here the use of this device geometry with the InP:Fe regrowth by Hydride Vapor Phase Epitaxy (HVPE). This epitaxial technique has been successfully used for many applications requiring highly selective and high anisotropic growth behavior, such as III-V growth on Si [6][7], planarization of near IR laser diodes [8][9], and more recently epitaxial lateral overgrowth of GaN [10].

II. DEVICES FABRICATION AND CHARACTERISTICS

We used MBE, MOVPE and HVPE epitaxial techniques in this work for the growth of AlInAs/GaInAs active region, InP:Si cladding and InP:Fe semi-insulating layers, respectively. Dielectic mask deposition, optical lithography, and ICP etching technological steps have been inserted in between the growth runs in order to define the ridge and stripes geometry.

A. Single ridge FP lasers

As a first step, we have realized and characterized single FP lasers. The QCL active region is based on a shallow well design from [1]. Fig. 1 presents the V(I) and P(I) characteristics of a 14 μm wide, 5 mm long ridge waveguide laser, buried in semi-insulating InP:Fe. The mirrors facets are HR coated / cleaved and the device exhibits a total output optical power of 2.4 W at 2.5 A.

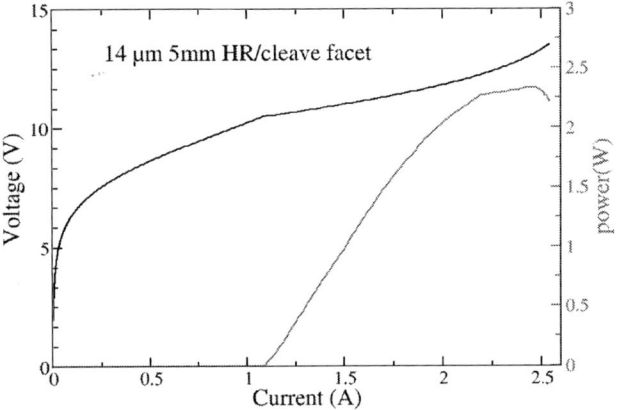

Figure 1 : V(I) and P(I) characteristics of a 14 μm wide, 5 mm long ridge waveguide laser emitting at 4.9μm.

B. μ-stripes arrays

As mentioned above, we have recently realized μ-stripe QCL devices with various geometries, up to 2 μm wide x 32 emitters [4] [5]. This configuration enhances the lateral thermal dissipation as shown in the simulation presented in fig. 2. The temperature elevation ΔT is shown for three different geometries. The good beam quality control is conserved thanks to the phase-locking provided by evanescent coupling between adjacent ridges.

MoPI-18 (Poster)
18:30 - 20:30

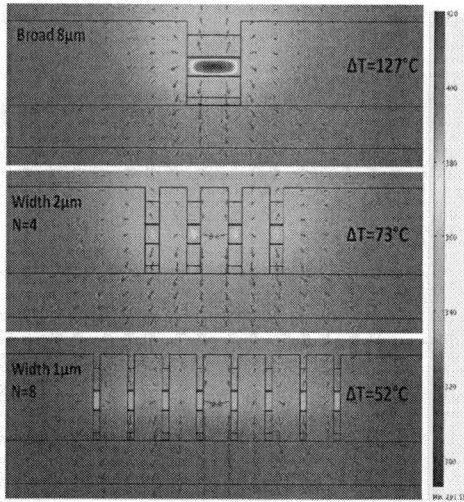

Figure 2 : Cross-section simulation using a thermo-electrical model (COMSOL). ΔT is the thermal elevation in the active region. The total active region width of 8μm is kept constant for the 3 configurations

While the semi-insulated InP:Fe regrowth is quite well established for high power emitting devices, the fabrication of the buried μ-stripes QCLs is still challenging and requires a perfect control of the selective and anisotropic InP:Fe layers regrowth. We present in Fig. 3 and Fig.4 the cross section of the device at different stages of the fabrication. The first SEM picture has been realized after ICP etching of the μ-stripes arrays. On the second one, we can observe the active region and the perfect InP:Fe planarization achieved using HVPE. It has to be noticed that the growth duration was less than 10 minutes, taking advantage of the highly anisotropic growth mode of the technique. The last image shows the cross section after the mask removal and overall regrowth of the InP:Si cladding layer by MOVPE.

Figure 3 : SEM tilted view after ICP etching of the μ-stripes array

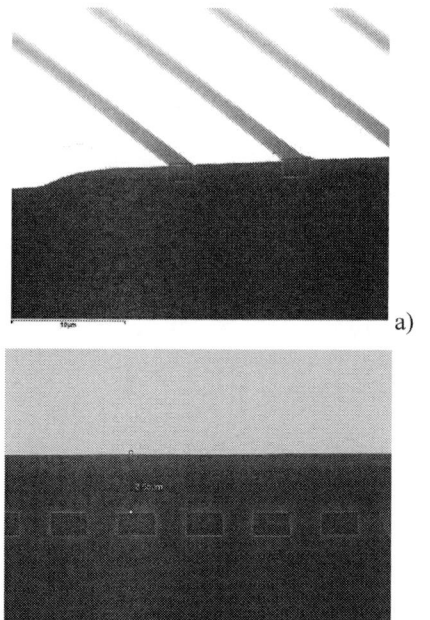

Figure 4: SEM tilted views at the various stages of the process (AlInAs/GaInAs active region has been revealed).
a) after HVPE InP:Fe planarization in between the stripes and before removing the dielectric masks
b) after InP:Si cladding regrowth by MOVPE

III. CONCLUSION

We have realized both single ridge waveguide and μ-stripe QCLs using MBE, HVPE and MOVPE epitaxial steps. Very promising results have been obtained for single ridge lasers exhibiting output power up to 2.4 W. The realization of μ-stripe devices is now under processing. The combination of all these approaches should lead to higher power emitting QCLs with good beam quality.

REFERENCES

[1] Y. Bai, N. Bandyopadhyay, S. Tsao, S. Slivken, and M. Razeghi, Appl. Phys. Lett. **98**, 181102 (2011)

[2] M. Beck, D. Hofstetter, T. Aellen, J. Faist, U. Oesterle, M. Ilegems, E. Gini, and H. Melchior, Science **295**, 301 (2002)

[3] M. Razeghi, IEEE J. Sel. Top. Quantum Electron. **14**, 3 (2009)

[4] G-M. de Naurois, M. Carras, G. Maisons and X. Marcadet, Opt. Lett. 37, 3 (2012)

[5] G-M. de Naurois, B. Simozrag, G. Maisons, V. Trinité, F. Alexandre and M. Carras, Appl. Phys. Lett. 101, 041113 (2012)

[6] O. Parillaud, E. Gil-Lafon, B. Gérard, P. Etienne, D. Pribat, Appl. Phys. Lett. **68** , 2654 (1996)

[7] S.Lourdudoss and O.Kjebon, IEEE J. Sel. Topics in Quant. Electron., vol. 3(3), pp. 749-767 (1997).

[8] F. Alexandre, O. Parillaud, D. C. Nguyen, R. Azoulay, M. Quillec, S. Bouchoule, G. Le Mestreallan, M. Juhel, G. Le Roux and E. V. K. Rao, J. of Crystal Growth, 187, pp 347-354 (1998)

[9] O.Kjebon, S.Lourdudoss, B.Hammarlund, S.Lindgren, M.Rask, P.Ojala, G.Landgren and B.Broberg, Appl. Phys. Lett., 59(3), pp. 253-255 (1991).

[10] A Usui, H Sunakawa, A Sakai, AA Yamaguchi, Jpn. J. Appl. Phys. **36** , L899 (1997)

MoPI-19 (Poster)
18:30 - 20:30

IPRM2013, May 19 - 23, 2013, Kobe, Japan
The 25th International Conference on Indium Phosphide and Related Materials

Mid-infrared photodetectors with InAs/GaSb type-II quantum wells grown on InP substrate

H.Inada[1], K.Miura[1,2], Y.Iguchi[1], Y.Kawamura[2]
J. Murooka[3], H. Katayama[3], S. Kanno[4], T. Takekawa[4], M. Kimata[3,4]

[1]Transmission Devices R&D Laboratories, Sumitomo Electric Industries, Ltd.
Yokohama, Japan
E-mail: inada-hiroshi@sei.co.jp
[2]Frontier Science Innovation Center, Osaka Prefecture University
Sakai, Japan
[3]Japan Aerospace Exploration Agency (JAXA)
Tsukuba, Japan
[4]Department of Mechanical Engineering, Ritsumeikan University
Kusatsu, Japan

Abstract—Infrared photodetectors with InAs/GaSb type-II quantum wells on InP substrate was fabricated and evaluated. Dark current density was $0.1mA/cm^2$ at 112K. Quantum efficiency at 5μm was 10%. This results show that InAs/GaSb quantum wells on InP substrate has potential for infrared image sensor.

Keywords—Infrared, Photodetector, type-II, InAs/GaSb, InP

I. INTRODUCTION

Mid-infrared photodetectors with cutoff wavelength over 3μm are expected for precise sensing of environmental gases, toxic gases and so on. In particular, focal plane arrays (FPAs) are useful for image sensing of distribution of gases. Type-II InAs/GaSb quantum wells (QWs) are attractive materials for absorption layer of photodetectors because they are expected to realize lower dark current and better controllability of cutoff wavelength compared to state-of-the-art HgCdTe [1][2][3]. GaSb substrates are generally used for the epitaxial growth of this type-II QWs. However, GaSb substrate has some problems. For example, its large absorption coefficient in mid-infrared region diminishes the external quantum efficiency of photodetectors with back-illuminated type such as FPAs. Also, the difference of thermal expansion coefficient between GaSb substrate and silicon CMOS read-out integrated circuit (ROIC), which are bonded to each other with indium bumps, makes the reliability of bonding poor, because the FPAs are used by cooling down to the temperature lower than 100K. These problems can be resolved by removing GaSb substrate and leaving only epitaxial layers. But this method is technically difficult. Recently, the type-II InAs/GaSb SLs grown on GaAs substrates with higher transparency is proposed and demonstrated [4]. But the epitaxial growth is difficult owing to the large lattice mismatch between GaAs and GaSb (7.8%). And the problem of the large difference of thermal expansion coefficient between GaAs substrate and ROIC still remains. InP substrates seem to be more favorable because of not only high transparency in mid-infrared region

but also smaller lattice mismatch. Moreover, InP has a closer thermal expansion coefficient to that of Si than GaAs and GaSb[5][6].
In this work, we demonstrated the InAs/GaSb type-II photodetector on InP substrate.

II. EXPERIMAENTARL PROCEDURE

A. MBE growth

InAs/GaSb quantum well structures were grown on a Fe doped (100) InP substrates by solid source molecular beam epitaxy (MBE) method. Prior to growth, the InP substrates were thermally cleaned. $0.15μm$-thick $In_{0.53}Ga_{0.47}As$ layers were grown in order to smoothen the surfaces of the substrates. 4.5μm-thick GaSb buffer layers were followed. The growth temperature was 480°C. Type-II InAs/GaSb QWs, which consist of totally 100 pairs of 3.6nm-thick InAs and 2.1nm-thick GaSb, were grown. To form pin structure, Be doped GaSb for p-type region, and Si doped InAs for n-type region used, respectively. The growth temperature of the QWs was 450°C. Finally, 20nm-thick InAs cap layer was grown. To characterize device characteristics, samples grown on GaSb substrate were also prepared. Tetramer As_4 and monomer Sb_1 were used for group V beam sources. X-ray diffraction and room-temperature photoluminescence (PL) were utilized to characterize the as-grown samples.

B. Device process and evaluation

Mesa type pin-photodiode(PD)s as shown in figure 1 were fabricated. Wet etching with mixture of citric and phosphoric acid, was conducted to form mesa isolation. 300nm-thick SiO_2 for passivation was deposited by PECVD. The temperature of deposition was 150°C. P-electrodes were placed at the top of mesa and n-electrodes were at the bottom around the mesa, respectively. After contact window opening, non-alloy electrode of Au/Pt/Ti was evaporated with lift-off method. Current-voltage characteristics and quantum efficiency at the low temperature were evaluated.

MoPI-19 (Poster)
18:30 - 20:30

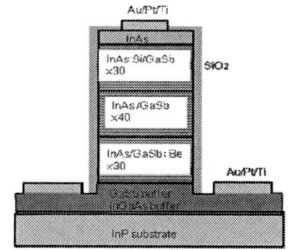

Figure 1. Schematic device structure

III. RESULT AND DISCSSION

Crosshatch was not observed on the surfaces. The thick GaSb showed a diffraction peak with narrow full width half maximum, indicating good crystalline quality. PL peak was the wavelength from 6.6 μm to 6.9 μm. Current-voltage measurement was conducted at the temperature of 112-205K. From the diode characteristics, pin structure was confirmed. As shown in figure 2, dark current density of specimen on InP substrate was nearly same to one on the conventional GaSb substrae in the low bias region at temperatue of 205K. However, decreasing the temperature or incresing the reverse bias, dark current density incresing compare to the one of GaSb substrate. The idial factor n calculated from temperature dependence of dark current was 1.78 for specimen on InP substrate and 1.34 for GaSb substrate, respectively. These results indicate that generation-recombination, which related to some defects, is dominant component rather than the specimen on GaSb substrate. To reduce dark current, it is necessary to optimize the growth condition. Moreover,

external quantum efficiency was estimated as shown in figure 3. At wavelength of 5μm, quantum efficiency is beyond 10 % at the temperature of 20K. This value showed InAs/GaSb QWs on InP substrate has the potential for the absorption layer of mid-infrared FPAs.

IV. SUMMURY

We have successfully realized photodetectors with InAs/GaSb type-II QWs grown on InP substrate. Dark current density is comparable to conventional InAs/GaSb QWs pin-PD on GaSb substrate. And Responsivity at the wavelength from 3μm to 6μm is confirmed.

REFERENCES

[1]   M. Razeghi, Y. Wei, A.Hood, D .Hoffman, B.M.Nguyen, P.Y.Delaunay, E.Michel, R.McClintock "Type-II Superlattice Photodetectors for MWIR to VLWIR Focal Plane Arrays", Proc. of SPIE Vol.6206 62060N-1

[2]   P-Y Delaunay, B.M. Nguyen, D.Hofman, M.Razeghi,"Substrate removal for high quantum efficiency back side illuminationed type-II InAs/gaSb photodetectors", Appl. Phys. Lett. Vol.91, 231106 (2007)

[3]   B.-M. Nguyen, D. Hoffman, E. K. Huang, S. Bogdanof, P. -Y. Delaunay, M.Razeghi, and M. Z. Tidrow,"Demonstration of midinfrared type-II InAs/GaSb superlattice photodiodes grown on GaAs substrate", Appl. Phys. Lett., Vol. 94, 223506 (2009)

[4]   P. M. Thibado, B. R. Bennett, M. E. Twigg, B. V. Shanabrook, and L. J. Whitman , "Evolution of GaSb epitaxy on GaAs(001)c(4×4)", J. Vac. Sci. Technol. A Vol. 14, 885 (1996)

[5]   I. Vurgaftman, J. R. Meyer and L. R. Ram-Mohan, "Band parameters for III–V compound semiconductors and their alloys" J. of Appl. Phys., Vol.   89,   5815   (2001)

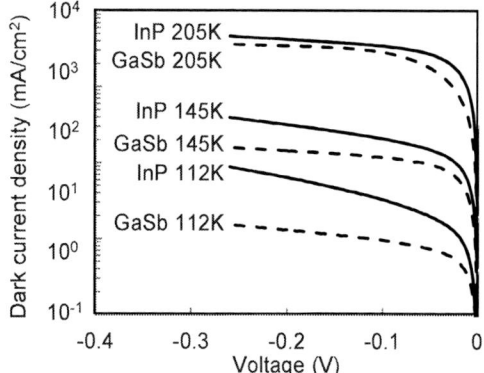

Figure 2. Current-Voltage characteristics

Figure 3. Wavelength dependence of external quantum efficiency

MoPI-20 (Poster)
18:30 - 20:30

Cryogenic DC Characterization

of InAs/Al$_{80}$Ga$_{20}$Sb Self-Switching Diodes

Andreas Westlund[1#], Giuseppe Moschetti[1], Per-Åke Nilsson[1], Jan Grahn[1], Ludovic Desplanque[2], Xavier Wallart[2]

[1]Department of Microtechnology and Nanoscience (MC2)
Chalmers University of Technology, Göteborg, Sweden
#andreas.westlund@chalmers.se

[2]Institute of Electronics, Microelectronics and Nanotechnology
UMR CNRS 8520, University of Lille, CS 60069,
59652 Villeneuve d'Ascq, France

Abstract—DC characterization of an InAs/Al$_{80}$Ga$_{20}$Sb self-switching diode for THz detection is presented at 300 K and 6 K. Compared to 300 K, an enhancement of the diode I-V non-linearity and associated responsivity was observed under zero-bias conditions at 6 K. The intrinsic responsivity was estimated to 490 V/W at 300 K and 4400 V/W at 6 K.

Keywords—InAs; THz; detector; diode; self-switching; SSD

I. INTRODUCTION

InAs-based self-switching diodes (SSDs) have been predicted as a promising terahertz detector by Monte Carlo simulations [1]. GaAs SSDs have been demonstrated as a detector at 1.5 THz [2]. Monte Carlo simulations predict that InAs SSDs may operate at even higher frequencies, well above 2 THz [1]. Furthermore, the responsivity is shown to stay constant from DC to more than 1 THz [1].

In this work, the DC-characteristics of an InAs SSD are analyzed at 300 K and 6 K. The consequences for RF detection performance are discussed.

II. FABRICATION

A non-intentionally doped InAs/Al$_{80}$Ga$_{20}$Sb quantum well was used to form a two-dimensional electron gas (2DEG), see Figure 1. A metamorphic AlSb buffer was grown in order to accommodate the lattice mismatch with the GaAs substrate. In the upper part of the buffer as well as in the barrier, AlSb was replaced with Al$_{80}$Ga$_{20}$Sb, known to be less prone to oxidation than AlSb [3].

Hall measurements were performed to investigate the transport properties of the 2DEG. The contact layer of the Hall sample was removed by wet etching followed by deposition of 180 nm silicon nitride (SiN$_x$) passivation. In this way, only the transport properties in the 2DEG were investigated. Table I shows how the sheet resistance R_{sh} is virtually unaffected upon cooling from 300 K to 77 K whereas the sheet carrier concentration n_s is halved and mobility μ doubled.

TABLE I. HALL MEASUREMENTS OF THE HETEROSTRUCTURE WITH THE CAP LAYER REMOVED THROUGH WET ETCHING.

	R_{sh} [Ω/sq]	n_s [cm^{-2}]	μ [cm^2/Vs]
300 K	181	1.3×10^{12}	26 000
77 K	175	6.3×10^{11}	57 000

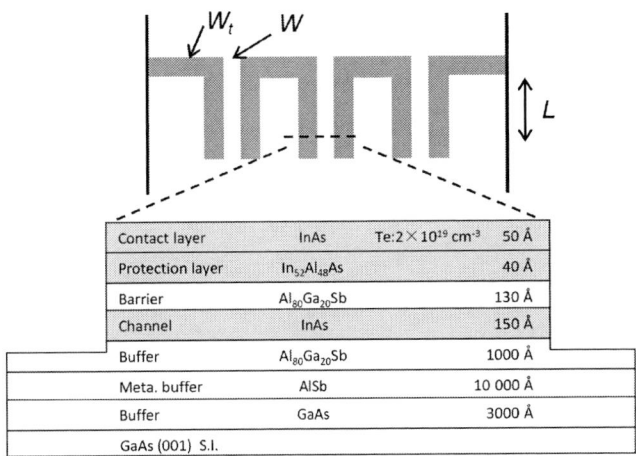

Figure 1. Layout (above) of three parallel SSDs with nominal trench width W_t=500 nm, channel width W=300 nm and channel length L=3.8 μm. Below, the InAs/Al$_{80}$Ga$_{20}$Sb heterostructure used in fabrication of InAs SSDs.

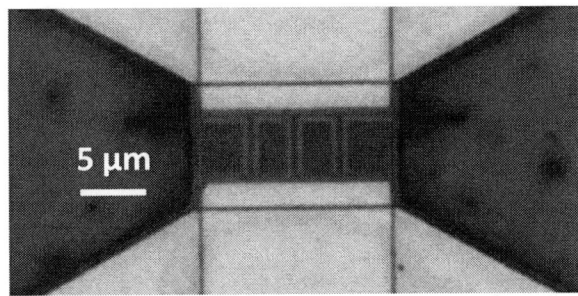

Figure 2. Micrograph of the fabricated 3-channel InAs/AlSb SSD.

SSD fabrication was started by the formation of Pd/Pt/Au contacts and Ti/Au metal pads. Then, the trench was formed by patterning resist using e-beam lithography. The trench was etched with a Cl$_2$:Ar inductively-coupled plasma/reactive ion-etch process and stopped in the Al$_{80}$Ga$_{20}$Sb buffer just below the channel, see cross-section in Figure 1. Thus, the reactive AlSb buffer is never exposed to air. Subsequently, the resist mask was removed *in-situ* using an NF$_3$ reactive ion etch process. Then, approximately 25 nm of SiN$_x$ was deposited *in-situ* in order to passivate and encapsulate the trenches. Mesas were defined in a similar way, but the etch was extended to the substrate. Finally a 180 nm thick SiN$_x$ layer was deposited to cover the mesa sidewalls and further encapsulate the trenches. A finalized SSD is shown in Figure 2.

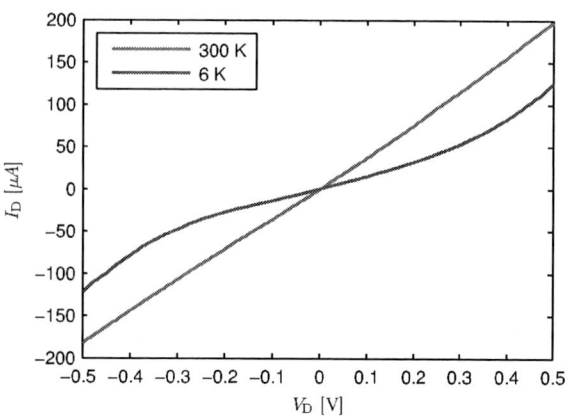

Figure 3. Measured I-V of the SSD.

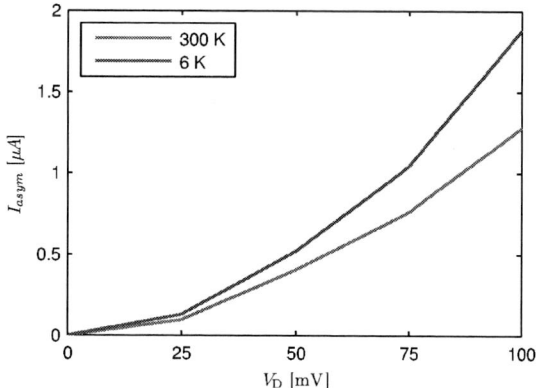

Figure 4. The extracted asymmetrical part of I_D in Figure 3.

III. MEASUREMENTS

In Figure 3, the diode current I_D versus applied voltage V_D is plotted at 300 K and 6 K. The resistance for small V_D increases upon cooling, from 2.8 kΩ to 6.9 kΩ. A possible reason is that at 6 K, charges on the trench sidewalls deplete the channel further, thus narrowing the effective channel width. For larger V_D, the resistance is similar to the 300 K case.

While I_D in Figure 3 is mainly symmetric about V_D=0, there is an asymmetric part I_{asym} that can be studied by comparing the magnitude of I_D for positive and negative V_D. Defined as

$$I_{asym} = |I_D(V_D)| - |I_D(-V_D)|, \tag{1}$$

I_{asym} is plotted in Figure 4. At 6 K, the SSD exhibits a clearly enhanced I_{asym} compared to at 300 K, indicating enhanced diode action.

From the DC-characteristics, conclusions can be drawn regarding the RF detection performance. Intrinsic voltage responsivity R_v can be calculated using

$$R_v = \Delta V_D / P = 1/2 \cdot R^2 \cdot (d^2 I_D / dV_D^2) \tag{2}$$

where R is the differential resistance (dV_D/dI_D) at the bias point, P the absorbed RF power and ΔV_D the change in V_D

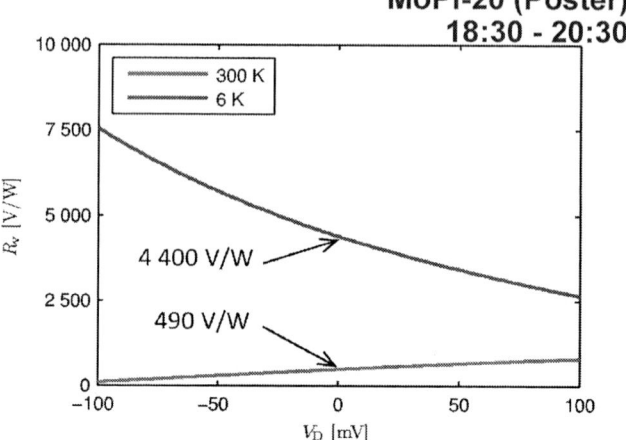

Figure 5. Estimated intrinsic voltage responsivity Rv.

induced by the absorbed RF power [4]. To find R_v, a 3rd-order polynomial was fitted to measured I-V for V_D in the range -100 mV to 100 mV, see Figure 5. At zero bias, the calculated R_v increased from 490 V/W to 4 400 V/W. The increase is both due to an increase of R and ($d^2 I_D / dV_D^2$). However, in a real system of limited system impedance, a large R may cause severe mismatch. The extrinsic responsivity $R_{v,extr}$ can be estimated with

$$R_{v,extr} = R_v \cdot (1 - \Gamma^2) \tag{3}$$

where

$$\Gamma = (R - Z_0)/(R + Z_0) \tag{4}$$

and Z_0 is the system impedance. With Z_0= 50 Ω, the zero-bias values for $R_{v,extr}$ at 300 K and 6 K are 34 V/W and 130 V/W, respectively. Thus, also the extrinsic responsivity increases upon cooling.

IV. CONCLUSION

Upon cooling from 300 K to 6 K the resistance of the SSD for small bias increased from 2.8 kΩ to 6.9 kΩ. Further, the curvature of the I-V increased, corresponding to an increase of zero-bias responsivity from 490 V/W at 300 K to 4400 V/W at 6 K.

REFERENCES

[1] I. Iniguez-de-la-Torre, H. Rodilla, J. Mateos, D. Pardo, A.M. Song, and T. Gonzalez, "Terahertz tunable detection in self-switching diodes based on high mobility semiconductors: InGaAs, InAs and InSb," *Journal of Physics.: Conf. Ser.* 193, 012082, 2009.

[2] C. Balocco, S.R. Kasjoo, X.F. Lu, L.Q. Zhang, Y. Alimi, S. Winner, and A.M. Song, "Room-temperature operation of a unipolar nanodiode at terahertz frequencies," *Applied Physics Letters*, vol. 98, 223501, May 2011.

[3] J.D. Werking, C.R. Bolognesi, L.-D. Chang, C. Nguyen, E.L. Hu, and H. Kroemer, "High-transconductance InAs/AlSb heterojunction field-effect transistors with delta -doped AlSb upper barriers," *Electron Device Letters*, vol. 13, pp. 164-166, 1992.

[4] A.M. Cowley, H. O. Sorensen, "Quantitative comparison of solid-state semiconductors" *IEEE Transactions on Microwave Theory and Techniques*, vol. MTt-14, pp 588-602, 1966

MoPI-21 (Poster)
18:30 - 20:30

Cryogenic Ultra-Low Noise Amplification
– InP PHEMT vs. GaAs MHEMT

J. Schleeh, H. Rodilla, N. Wadefalk, P. Å. Nilsson, J. Grahn

Department of Microtechnology and Nanoscience (MC2)
Chalmers University of Technology
Göteborg, Sweden
schleeh@chalmers.se

Abstract— **We present a comparative study of 130 nm high electron mobility transistors (HEMTs) fabricated on pseudomorphic InGaAs/InAlAs/InP (InP PHEMT) and InGaAs/InAlAs/GaAs (GaAs MHEMT) intended for ultra-low noise amplifiers (LNAs). The epitaxial growth, as well as the HEMT process, was performed simultaneously. When integrated in a 4-8 GHz 3-stage LNA at 300 K, the measured average noise temperature was 34 K for the GaAs MHEMT and 27 K for the InP PHEMT. When cooled down to 10 K, the InP PHEMT LNA was improved to 1.6 K, while the GaAs MHEMT LNA was only reduced to 5 K. The reason for the superior cryogenic noise performance of the InP PHEMT compared to the GaAs MHEMT in this study, was found to be a higher quality of pinch-off when cooled down.**

Keywords— Cryogenic, GaAs MHEMT, InP PHEMT, low noise

I. INTRODUCTION

The best microwave and millimeter wave noise performance is achieved with low noise amplifiers (LNAs) based on InGaAs/InAlAs/InP pseudomorphic high electron mobility transistors (InP PHEMTs) [1, 2]. However, in recent years, large progress has been achieved in high frequency and low noise optimization of InGaAs/InAlAs/GaAs metamorphic HEMTs (GaAs MHEMTs), resulting in performance comparable to the InP PHEMT. Cut-off frequencies f_T of 660 GHz have been achieved using 20 nm GaAs MHEMTs [3], as well as noise temperatures down to 3 K at 5 GHz under cryogenic conditions using 100 nm gate lengths [4].

The GaAs MHEMT is according to the International Technology Roadmap for Semiconductors (ITRS) [5] already superior to the InP PHEMT regarding noise performance at room temperature. Even though GaAs MHEMTs have proven suitable for low noise operation also at cryogenic temperature [4], no direct comparing study with InP PHEMTs has been done.

In this report we present and compare DC, RF and noise characterization of InP PHEMTs and GaAs MHEMTs at 300 K and 10 K ambient temperature.

II. DEVICE FABRICATION

To make the comparison between the InP PHEMT and the GaAs MHEMT as valid as possible, both wafers were processed side by side to minimize process variations. The epitaxial structure, previously described in [2], was except for the substrate and buffer identical for the PHEMT and MHEMT wafer. The 500 nm $In_{0.52}Al_{0.48}As$ buffer on the GaAs MHEMT wafer was grown on a 300 nm $In_{0-0.52}Al_{1-0.48}As$ metamorphic buffer on top of the GaAs substrate. For the InP PHEMT wafer, the same buffer was grown directly on the InP substrate. The gate length was 130 nm. Cross sectional STEM images of the InP PHEMT and GaAs MHEMT is shown in Fig. 1.

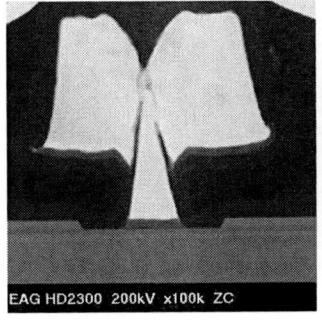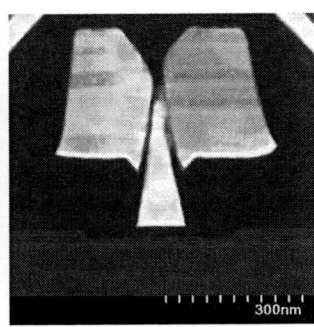

Fig. 1. Cross sectional STEM image of gate region in 130 nm gate length GaAs MHEMT (left figure) and InP PHEMT (right figure).

The ohmic contact resistance R_c slightly increased from 0.04 Ω•mm for the InP PHEMT and 0.11 Ω•mm for the GaAs MHEMT to 0.05 Ω•mm and 0.12 Ω•mm, respectively, when cooled down from 300 K to 10 K. Both technologies exhibited gate resistances of 320 Ω/mm at 300 K and 130 Ω/mm at 10 K.

III. DC, RF AND NOISE CHARACTERIZATION

DC and RF characterization was performed on both GaAs MHEMT and InP PHEMT at 300 K and 10 K in a cryogenic probe station. A strong indication for low noise performance of HEMTs is the quality of pinch-off [6]. This property has previously been investigated for state-of-the-art InP PHEMTs [7, 8].

The drain current I_d is plotted against gate voltage V_g in Fig. 2. At room temperature, the subthreshold I_d is slightly lower for the InP PHEMT compared to the GaAs MHEMT. When cooled down to 10 K, the difference is strongly increased and a superior quality of pinch-off for the InP PHEMT is evident from Fig. 2.

Another indicator for low-noise performance is the response of g_m versus I_d [6]. As the equivalent drain current temperature T_d of the Pospieszalski noise model is directly proportional to I_d, a steep response of g_m versus I_d is anticipated for superior low noise HEMTs [2, 9]. The intrinsic g_m versus I_d for the InP PHEMT and GaAs MHEMT is plotted in Fig. 3. At 300 K, the slope of g_m versus I_d was slightly higher for the InP PHEMT than the GaAs MHEMT. When cooled down, however, the difference increased substantially for I_d below 100 mA/mm.

To finally compare noise performance, both HEMT technologies were integrated and measured separately in the same benchmarking 4-8 GHz hybrid 3-stage LNA. At room temperature, see Fig. 4, the average noise temperature in the 4-8 GHz band was 34 K for the GaAs MHEMT and 27 K for the InP PHEMT. At 10 K, see Fig. 5, the average noise

MoPI-21 (Poster)
18:30 - 20:30

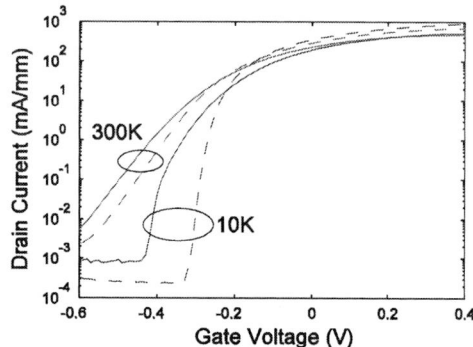

Fig. 2. Subthreshold I_d versus V_g of 2x100 μm gate width InP PHEMT (dashed) and GaAs MHEMT (solid) measured at 300 K (red) and 10 K (blue) ambient temperature and V_{ds} of 0.6 V.

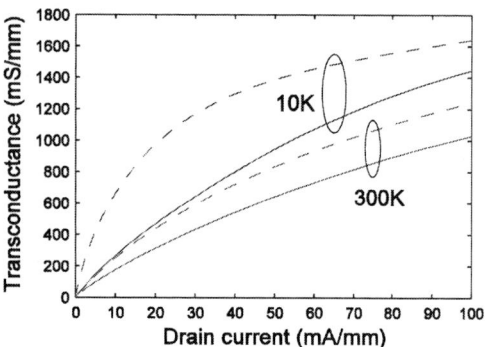

Fig. 3. Intrinsic g_m versus I_d of 2x100 μm gate width InP PHEMT (dashed) and GaAs MHEMT (solid) measured at 300 K (red) and 10 K (blue) ambient temperature and V_{ds} of 0.6 V.

temperature became 1.6 K for the InP PHEMT, and 5 K for the GaAs MHEMT.

IV. CONCLUSION

130 nm gate length InP PHEMTs and GaAs MHEMTs, containing the same active heterostructure and fabricated simultaneously, have been compared with respect to DC, RF and noise performance at 300 K and 10 K. The analysis showed superior noise improvement upon cooling of the InP PHEMT compared to the GaAs MHEMT. The reason for this is the higher quality of pinch-off for the InP PHEMT compared to the GaAs MHEMT.

ACKNOWLEDGEMENT

This research has been carried out in the GigaHertz Centre in a joint research project financed by the Swedish Governmental Agency of Innovation Systems (VINNOVA), Chalmers University of Technology, Omnisys Instruments AB, Wasa Millimeter Wave, Low-Noise Factory and SP Technical Research Institute of Sweden.

We thank IntelliEpi for assisting with epitaxial wafers for this experiment.

REFERENCES

[1] W. R. Deal, K. Leong, V. Radisic, S. Sarkozy, B. Gorospe, J. Lee, P. H. Liu, W. Yoshida, J. Zhou, M. Lange, R. Lai, and X. B. Mei, "Low Noise Amplification at 0.67 THz Using 30 nm InP HEMTs," *IEEE Microwave and Wireless Components Letters,* vol. 21, pp. 368-70, 2011.

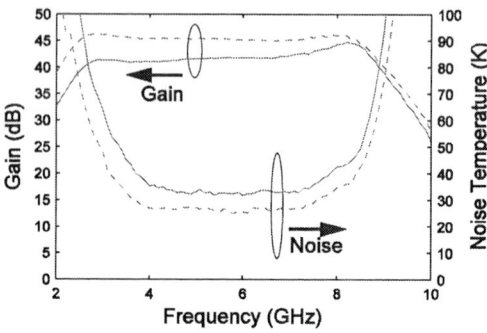

Fig. 4. Noise temperature and gain of a 3-stage 4-8 GHz LNA equipped with GaAs MHEMTs (solid) or InP PHEMTs (dashed) measured at 300 K. The amplifier was biased for best noise performance at V_{dd} = 1.25 V and I_{dd} = 45 mA for both HEMT technologies.

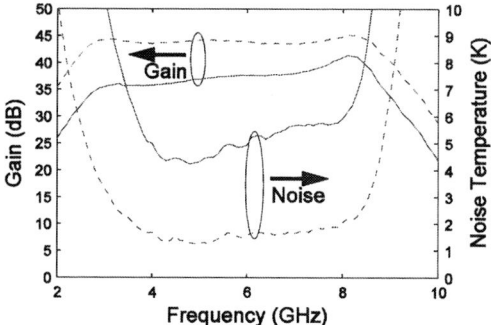

Fig. 5. Noise temperature and gain of a 3-stage 4-8 GHz LNA equipped with GaAs MHEMTs (solid) or InP PHEMTs (dashed) measured at 10 K. The amplifier was biased for best noise performance at V_{dd} = 0.57 V and I_{dd} = 24 mA when equipped with GaAs MHEMTs, and V_{dd} = 0.45 V and I_{dd} = 9.6 mA when equipped with InP PHEMTs.

[2] J. Schleeh, G. Alestig, J. Halonen, A. Malmros, B. Nilsson, P. A. Nilsson, J. P. Starski, N. Wadefalk, H. Zirath, and J. Grahn, "Ultralow-power Cryogenic InP HEMT With Minimum Noise Temperature of 1 K at 6 GHz," *IEEE Electron Device Letters,* vol. 33, pp. 664-6, 2012.

[3] A. Leuther, S. Koch, A. Tessmann, I. Kallfass, T. Merkle, H. Massler, R. Loesch, M. Schlechtweg, S. Saito, and O. Ambacher, "20 NM metamorphic HEMT with 660 GHZ FT," in *23rd International Conference on Indium Phosphide and Related Materials (IPRM),* Piscataway, NJ, USA, 2011.

[4] B. Aja Abelan, M. Seelmann-Eggebert, D. Bruch, A. Leuther, H. Massler, B. Baldischweiler, M. Schlechtweg, J. D. Gallego-Puyol, I. Lopez-Fernandez, C. Diez-Gonzalez, I. Malo-Gomez, E. Villa, and E. Artal, "4-12 and 25-34 GHz Cryogenic mHEMT MMIC Low-Noise Amplifiers," *IEEE Transactions on Microwave Theory and Techniques,* vol. PP, pp. 1-9, 2012.

[5] *ITRS - The International Technology Roadmap for Semiconductors.* Available: www.itrs.net

[6] M. W. Pospieszalski, "Extremely low-noise amplification with cryogenic FETs and HFETs: 1970-2004," *IEEE Microw. Mag.,* vol. 6, pp. 62-75, Sep. 2005.

[7] J. Shell, "The Interplanetary Network Progress Report 42-169, The Cryogenic DC Behavior of Cryo3/AZ1 InP 0.1-by-80-Micrometer-Gate High Electron Mobility Transistor Devices," May 15, 2007.

[8] J. Schleeh, H. Rodilla, N. Wadefalk, P. A. Nilsson, and J. Grahn, "Characterization and Modeling of Cryogenic Ultra-Low Noise InP HEMT," *To be published in Transactions on Electron Devices,* 2013.

[9] M. W. Pospieszalski, "Modeling of noise parameters of MESFETs and MODFETs and their frequency and temperature dependence," *IEEE Trans. Microwave Theory Tech.,* vol. 37, pp. 1340-1350, 1989.

978-1-4673-6130-9/13 $31.00 © 2013 IEEE

MoPI-22 (Poster)
18:30 - 20:30

High Performacne InAs/AlSb HEMT with Refractory Iridium Schottky Gate Metal

Wen-Yu Lin, [1] Chao-Hung Chen, [1] Hsien-Chin Chiu, [1] Fan-Hsiu Huang, [1]
W. J. Hsueh, [2] Yue-Ming Hsin,[2] and Jen-Inn Chyi[2]

[1]Dept. of Electronics Engineering, Chang Gung University, Taoyuan, Taiwan, R.O.C

[2]Dept. of Electrical Engineering, National Central University, Jhongli 32001, Taiwan
TEL:+886-3-2118800 FAX:+886-3-2118507 Email: hcchiu@mail.cgu.edu.tw

Abstract—In the work, a novel approach in fabricating high-performance of InAs/AlSb high electron mobility transistors using iridium (Ir) gate technology was proposed and investigated. The Ir-gate exhibited a superior metal work function which was beneficial for increasing Schottky barrier height (Φ_B) of InAs/AlSb heterostructure from 0.54 to 0.58 eV. The Ir-gate InAs/AlSb HEMT exhibited a V_{th} of -0.9 V, a maximum drain current of 270 mA/mm, and a peak transconductance of 420 mS/mm. In contrast, the V_{th} of Ti-gate InAs/AlSb HEMT was -1.5 V, a maximum drain current of 257mA/mm, and a peak transconductance of 280mS /mm, respectively. It was suggested that Ir interface presented a high potential for high power transistor applications.

Keywords—AlSb, iridium, Titanium Schottky barrier height.

I. INTRODUCTION

The Sb-based devices have intrinsic advantages of high-speed and low power consumption that can provide the technology required for these applications. Recently, many groups have also reported progress in InAs-based and InSb-based channel HEMTs. Because of there has been interest in the potential of III–V FETs for advanced logic applications, which can enhance digital circuit functionality [1],[2]. Therefore, complementary circuits in the Sb-based material system are highly desirable. An excellent electron mobility as high as 20,000~30,000 cm²/V-s in the InAs/AlSb quantum wells and low sheet resistance could be achieved.

In this work, we demonstrated for the first time a low leakage current and low noise performace of InAs/AlSb HEMT by using Ir-gate structure, which is melting point is 2450°C. Therefore, the Ir-gate devices exhibited a high thermal stability, which achieves a maximum drain current of 270 mA/mm, a peak transconductance of 420 mS/mm, a V_{th} was -0.9 V. Ir-gate exhibited a higher work function than titanium (Ti) due to the Schottky barrier height was improved effectively and also reduced the gate leakage current, which is especially important for power transistors to avoid extra power consumption and linearity degradation at high power operation. Therefore, Ir-gate technology for the InAs/AlSb HEMT has great potential for high PA applications.

II. DEVICE STRUCTURE AND FABRICATION

Figure 1.showed the brief epitaxy structure were grown molecular beam epitaxy (MBE) on semi-insulating GaAs substrates. Two AlSb planar layers sandwiched the InAs undoped channel layer This InAs/AlSb HEMT demonstrated a sheet charge density of 8×10^{12} cm^{-2} together with a Hall mobility of 19,000 cm²/V-s at 300K [3]. Devices were processed by conventional optical lithography and lift-off process. Ohmic contacts were realized by using Pd/Ti/Pt/Au (20nm/40nm/40nm/50nm) alloy followed by a 300 °C, 10 sec RTA annealing in N$_2$ ambient. To define an active region, BCl$_3$ gas was used for dry etching by Reactive Ion Etching (RIE) system, which active region depth of 100 nm. An Ir/Ti/Au (10nm/20nm/300nm) composited metal was deposited by electron-beam evaporator for gate electrodes. A 300nm Ti/Au was deposited for interconnection and probe pads. Finally, a 300nm SiN$_x$ was deposited using plasma enhance chemical vapor deposition (PECVD) chamber at 200°C for 20 seconds for device passivation layer.

III. EXPERIMENTAL RESULTS

The Schottky barrier height (Φ_B) is an important parameter to determine the signal dynamic range of InAs/AlSb HEMT. It is expectable that the Schottky barrier height improvement is effective to reduce the gate leakage current, which is especially important for power transistors to avoid extra power consumption and linearity degradation at high power operation. For E/D-mode digital IC design, it is also beneficial to have a large Schottky barrier height to increase the logic noise margin. The current-voltage (*I-V*) measurement is the most popular and direct method to determine the Schottky barrier height. The Schottky barrier height can be calculated from Eq (1) and given by:

$$\Phi_B = \frac{kT}{q} ln(\frac{AA^*T^2}{I_s}) \tag{1}$$

where A is the diode area, k is the Boltzmann constant and A^*, Φ_B, η represents the effective Richardson's constant, the effective barrier height, the ideality factor at the measurement temperature T, respectively. The Φ_B was determined by the I_g-V_g characteristics of fabricated InAs/Al Sb HEMT. The Φ_B is 0.58 eV and 0.54 eV for Ir-gate and Ti-gate, respectively. Ir exhibited the highest work function in metals to obtain a higher Φ_B of InAs/AlSb HEMT. The 1µm gate length Ti-gate and Ir-gate InAs/AlSb HEMTs were characterized on-wafer for dc performance at room temperature. The drain-to-source current and transconductance (g_m) versus gate-to-source voltage (V_{gs}) characteristic biased at V_{ds} = 0.4 V of two devices shown in Figure 2. The threshold voltage (V_{th}) of the device is defined as the V_{gs} when the I_{ds} reaches 1mA/mm. The measured V_{th} were -1.5V and -0.9 V for Ti-gate and Ir-gate InAs/AlSb HEMT, respectively. Form figure 2, the Ti-gate device have leakage current than Ir-gate device due to Ir-gate technology have a higher work function than Ti-gate.

The most important aspect of the device described in the present paper is their low gate leakage characteristics. Ir-gate with low gate leakage allow us to probe the generation of holes with impact ionization directly. As described the holes that are generated are also attracted to negatively charged gate. These holes, if collected by the gate , will provide an excess gate leakage current over and above the nominal gate leakage observed in Ti-gate. Form figure 3, The Ir-gate leakage current is lower than Pt-gate. Because Ir-gate exhibited a superior metal work function can improved to Schottky barrier. Therefore, the Ir-gate can effectively reduce the impact ionization effect. To analyze the trapping/detrapping phenomena in devices of different isolation technology $1/f$ noise spectra with various gate bias voltages were measured. Figure 4 presented the slope extracted at 100Hz was −1.38 for Ir-gate HEMTs and −1.12 for Pt-gate HEMTS. Based on the experimental results, the fluctuation in Ir-gate and Pt-gate HEMTs were mobility fluctuation because its curve slopes of the V_{GS}-V_{th} as a function of $S_{ID}/I_D{}^2$ density were closed to -1.

IV. CONCLUSION

In summary, we adopted the Ir/Ti/Au composited gate metals to replace traditional Ti/ Pt /Au gate on the InAs/AaSb HEMT due to its high work function and high thermal stability. The demonstrated Ir-gate InAs/AlSb HEMT achieved positive shift of V_{th} value and reduce leakage current which is important for industry applications, for high power amplifier applications.

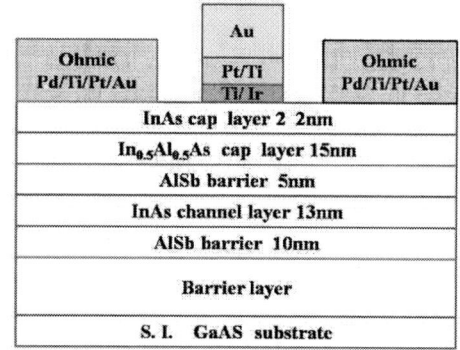

Figure 1. shows the InAs/AlSb for Ir and Ti gate strcture.

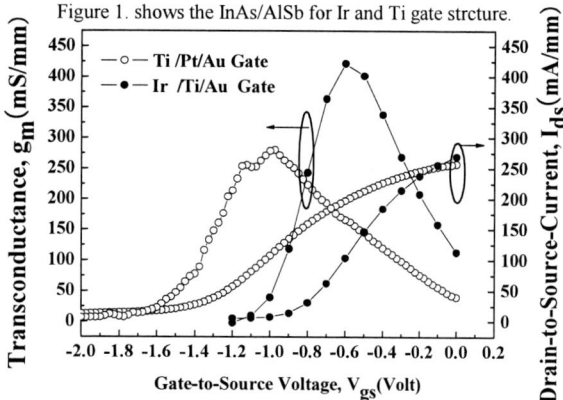

Figure 2. Transfer curves of a 1μm InAs/AlSb HEMTs.

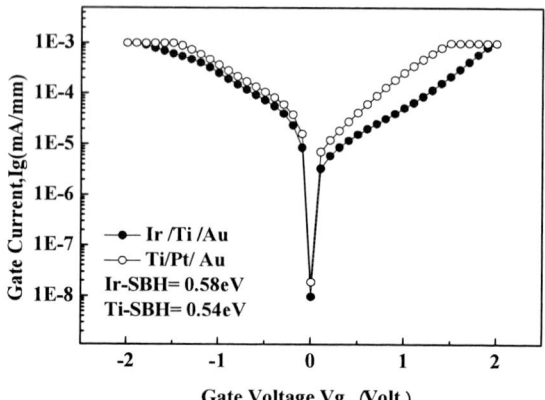

Fig.3 The Schottky characteristics for two devices.

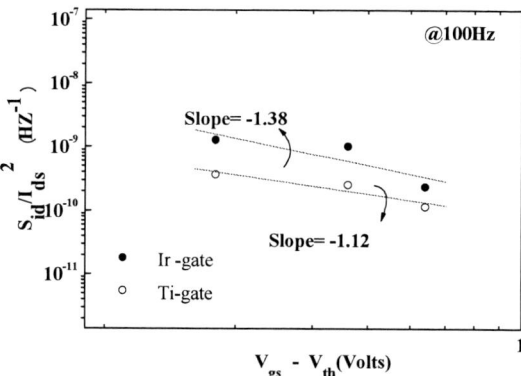

Fig.4 The flicker noise spectra characteristics of $S_{ID}/I^2{}_D$ versus $V_{GS} - V_{th}$ at a fixed frequency of 100Hz

V. REFERENCES

[1] R. Chau, S. Datta, M. Doczy, B. Doyle, B. Jin, J. Kavalieros, A. Majumdar, M. Metz, and M. Radosavljevic, "Benchmarkingnanotechnology for high-performance and low-power logictransistor applications," IEEE Trans. Nanotechnol., vol. 4, no. 2, pp. 153–158, Mar. 2005.

[2] D.-H. Kim, J. A. del Alamo, J.-H. Lee, and K.-S. Seo, "Logic suitability of 50-nm In0.7Ga0.3As HEMTs for beyond-CMOS applications," IEEE Trans. Electron Devices, vol. 54, no. 10, pp. 2606–2613, Oct. 2007.

[3] B. R. Bennett, M. G. Ancona, J. B. Boos, and B. V. Shanabrook, "Mobility enhancement in strained p-InGaSb quantum wells," Appl. Phys. Lett., vol. 91, no. 4, p. 042 104, Jul. 2007.

MoPI-23 (Poster)
18:30 - 20:30

Influence of gate-channel distance
in low-noise InP HEMTs

P. Å. Nilsson, H. Rodilla, J. Schleeh, N. Wadefalk, J. Grahn

Department of Microtechnology and Nanoscience (MC2)
Chalmers University of Technology
Göteborg, Sweden
per-ake.nilsson@chalmers.se

Abstract—**The effect on the electrical properties, relevant to noise, from the gate-channel distance (barrier layer thickness) in 130 nm gate-length InP HEMTs was investigated. An increased quality of pinch-off was seen in HEMTs with an 8 nm barrier layer thickness compared to an 11 nm barrier. For the 8 nm barrier material the gate leakage increased from 1 μA/mm to 7 μA/mm at -1 V gate bias.**

Keywords—InP HEMT; LNA; noise

I. Introduction

InGaAs/InAlAs/InP High Electron Mobility Transistors (InP HEMTs) are considered to be the best transistors for cryogenic low noise amplifiers (LNAs) at microwave frequencies. Such LNAs are used e.g. in radio astronomy and in deep space communication where a reduced noise figure in the LNA means a smaller antenna area for the same sensitivity.

To achieve low noise in a HEMT, the device is biased at a low drain current to keep the drain temperature, T_d, low. At this bias point, the transconductance, g_m, should be as high as possible, i.e. the quality of pinch-off must be good [1]. Simultaneously, the gate leakage current must be kept low in order to keep the noise down. A higher quality of pinch-off can be achieved by reducing the gate-channel distance, but this will sooner or later increase the gate leakage. Hence a trade-off between these parameters must be made.

We have previously fabricated state-of-the art InP HEMTs, with a noise temperature of 1 K at 6 GHz [2]. In this work, we investigate similar InP HEMTs with a reduced gate-channel distance.

II. Device Design and Fabrication

The details of our standard HEMT device structure have been published previously [2]. In short, we use a 500 nm $In_{0.52}Al_{0.48}As$ buffer grown on an InP substrate. This is followed by a 15 nm $In_{0.65}Ga_{0.35}As$ channel, a 3 nm $In_{0.52}Al_{0.48}As$ spacer, a planar Si δ-doping layer (5×10^{12} cm^{-2}), an 11 nm $In_{0.52}Al_{0.48}As$ barrier, and a 20 nm thick $In_{0.53}Ga_{0.47}As$ heavily doped cap layer (fig. 1.) We use Ge-based ohmic contacts with a typical resistivity of 50 mΩmm. The gate area is formed by e-beam lithography. A recess in the cap layer is etched using succinic acid which is selective to the barrier

---

This research has been carried out in the GigaHertz Centre in a joint research project financed by the Swedish Governmental Agency of Innovation Systems (VINNOVA), Chalmers University of Technology, Omnisys Instruments AB, Wasa Millimeter Wave, Low-Noise Factory and SP Technical Research Institute of Sweden.

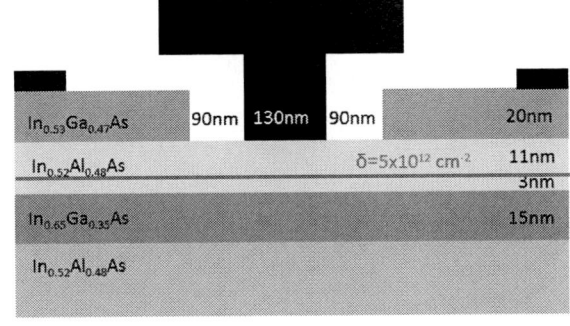

Figure 1. HEMT device structure with the standard barrier thickness of 11 nm. In this work, HEMTs with 8 nm barrier thickness were also made.

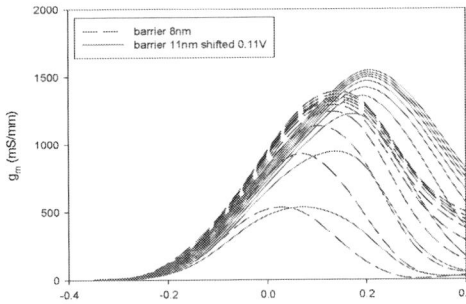

Figure 2. Simulated transconductance for 11 nm (red solid line) and 8 nm (blue dashed line) barrier materials versus V_{GS}. V_{DS} is varied from 0.1V to 1.0V in steps of 0.1V. An improved quality of pinch-off for the thinner barrier can be seen. The g_m curves for the 11 nm barrier is shifted 0.11V in the figure for the comparison.

material. For the mushroom gates, we use a Ti/Pt/Au stack, and the devices are passivated with silicon nitride.

To vary the gate-channel distance and thus improve the quality of pinch off, the thickness of the barrier layer was reduced. We used Monte Carlo (MC) simulations, and selected a barrier thickness of 8 nm for the experiments. All other device parameters were kept constant. The MC simulations showed an improved quality of pinch off for the material with a thinner barrier (fig. 2.) The maximum g_m was rather similar for the 11 nm barrier material and the 8 nm barrier material. One

978-1-4673-6130-9/13 $31.00 © 2013 IEEE

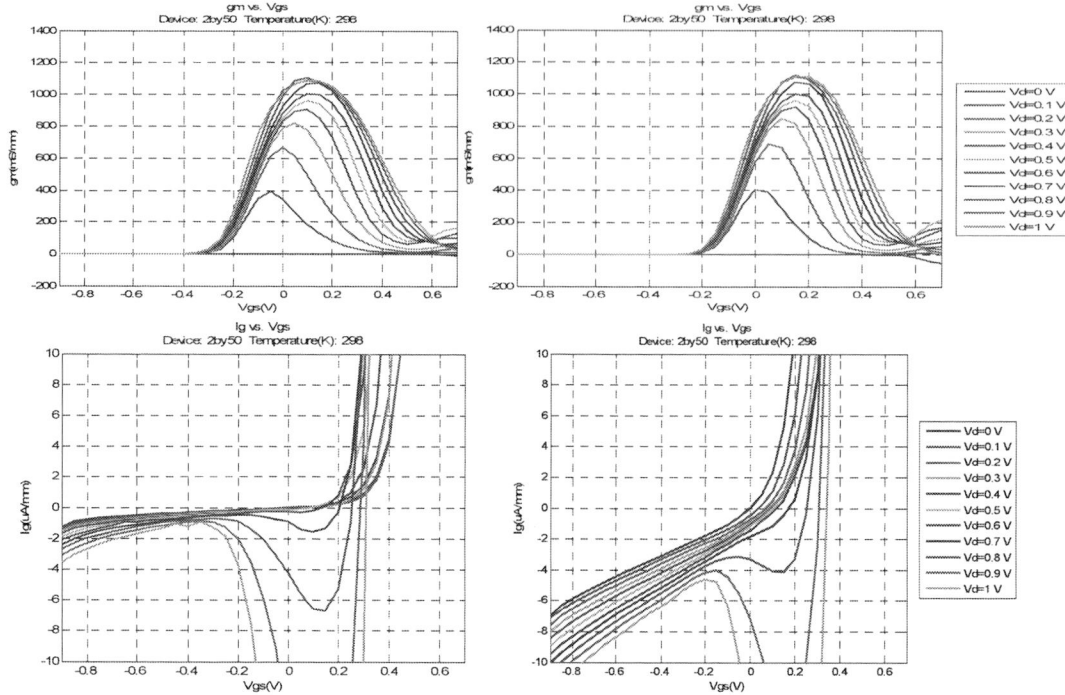

Figure 3. Measured transconductance for devices with 11 nm barrier thickness (upper left) and 8 nm barrier thickness (upper right) and gate leakage for the same devices (lower left and right).

might expect the maximum g_m to increase when the gate-channel distance becomes low. On the other hand, when the channel is in close proximity to the surface, the number of carriers is affected. An increase of the gate-source capacitance C_{gs} was observed in MC simulations when decreasing the barrier thickness from 11 nm to 8 nm (fig. 4) while the difference in the gate-drain capacitance C_{gd} was negligible.

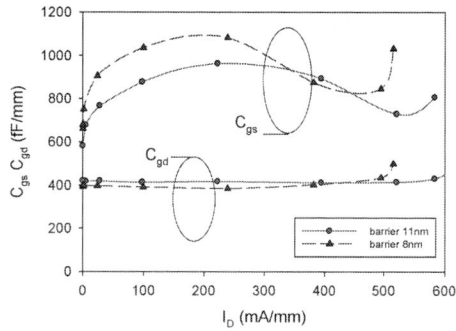

Figure 4. Simulated C_{gs} and C_{gd} versus I_D at V_{DS} =0.6V for 11 nm (red solid line) and 8 nm (blue dashed line) barrier thickness.

The measured $g_m(V_{GS})$ on fabricated devices (upper part Fig. 3.) adhered rather well to the MC simulation results. The improved quality of pinch-off was clearly seen for the thinner barrier with an increase in measured C_{gs} from 790 fF/mm to 850 fF/mm at V_{DS}=1V and I_D=225mA/mm. The gate leakage increased from 1 μA/mm to 7 μA/mm at V_{GS}= -1 V, but is still very low for the 8 nm barrier material (lower part Fig. 3.) at typical low-noise bias conditions V_{DS}= 0.4 V – 0.6 V [2].

III. CONCLUSIONS

A study of the electrical properties relevant for noise in InP HEMTs was performed for various barrier thicknesses. The measured results agreed well with MC simulations, and an increased quality of pinch-off was seen for the thinner barrier material while the gate leakage was kept low and C_{gs} increased.

REFERENCES

[1] M. W. Pospieszalski, "Extremely low-noise amplification with cryogenic FETs and HFETs: 1970-2004," *IEEE Microw. Mag.*, vol. 6, pp. 62-75, Sep. 2005.

[2] J. Schleeh, G. Alestig, J. Halonen, A. Malmros, B. Nilsson, P. A. Nilsson, J. P. Starski, N. Wadefalk, H. Zirath, and J. Grahn, "Ultralow-power Cryogenic InP HEMT With Minimum Noise Temperature of 1 K at 6 GHz," *IEEE Electron Device Letters*, vol. 33, pp. 664-6, 2012.

MoPI-24 (Poster)
18:30 - 20:30

Terahertz Oscillators using Resonant Tunneling Diodes with InAlGaAs/InP Composite Collector

R. Sogabe[1], K. Shizuno[1], H. Kanaya[1], S. Suzuki[1], M. Asada[1], H. Sugiyama[2], and H. Yokoyama[2]

[1] Graduate School of Interdisciplinary Science and Engineering, Tokyo Institute of Technology,
2-12-1-S9-3 Ookayama, Meguro, Tokyo 152-8552, Japan
[2] NTT Photonics Laboratories, NTT Corporation
*E-mail address: safumi@quantum.pe.titech.ac.jp

Abstract—**We proposed a resonant tunneling diode (RTD) with InAlGaAs/InP composite collector for reduction in transit delay caused by the gamma to L valley transition at the collector depletion region. Terahertz oscillators fabricated with this RTD show room-temperature fundamental oscillations of 680-770 GHz with the RTD areas of 1-1.5 square microns. Higher frequency will be possible by reducing the RTD area.**

Keywords—resonant tunneling diode; terahertz oscillator; slot antenna; InAlGaAs/InP composite collector

I. INTRODUCTION

Recently, the terahertz (THz) range has received considerable attention because it can be employed in various applications [1]. In particular, high-capacity, short-distance wireless communication is an important application of this range. Demonstrations of THz communication have been intensively conducted [2-4]. Compact, coherent, and high-power solid-state sources are considered as key components for such applications. Toward THz frequency operation, HBTs [5, 6], HEMTs [5, 7], and Si-CMOS [8, 9] are being intensively studied. Resonant tunneling diodes (RTDs) [10-14], which have the highest oscillation frequency among electronic single oscillators, are also one of the candidates of THz sources.

A reduction in transit delay at the collector depletion region is effective for high frequency oscillation in RTDs having low capacitances with thick collector spacer layers. A fundamental oscillation at 1.04 THz was achieved with the graded emitter structure that can reduce the electric field and Γ-L transition responsible for a long transit time [11]. However, the suppression of the Γ-L transition was insufficient in the graded emitter. In this report, we propose a novel RTD structure with InAlGaAs/InP composite collector layer which has a large Γ-L valley separation for complete suppression of the Γ-L transition, and measured oscillation characteristics of the THz oscillators using this RTD.

II. DEVICE STRUCTURE

The THz oscillator was fabricated by integrating an RTD with a slot antenna. The structure, fabrication process, and

This work was supported by Scientific Grants-in-Aid from the Ministry of Education, Culture, Sports, Science and Technology, Japan, the Industry–Academia Collaborative R&D Program from the Japan Science and Technology Agency, Japan, and the Strategic Information and Communications R&D Promotion Programme from the Ministry of Internal Affairs and Communications.

measurement system for the oscillation characteristics are described in [10]. The length of the slot antenna is 20 μm. The oscillation is obtained if the absolute value of the negative differential conductance (NDC) of the RTD (G_{RTD}) compensates for the radiation loss of the antenna. The oscillation frequency is determined by the resonance frequency of the resonance circuit constructed mainly by the inductance of the antenna and the capacitances of the RTD and antenna.

G_{RTD} degrades with increasing frequency as a result of the intrinsic delay time τ, which consists of the dwell time τ_{RTD} in the resonance region and the transit time τ_{dep} across the collector depletion region, as $\tau = \tau_{RTD} + \tau_{dep}/2$. G_{RTD} is approximately expressed as $G_{RTD} = G_{RTD0} \times \cos\omega\tau = (3/2)\cos\omega\tau \times \Delta I/\Delta V$, where G_{RTD0} denotes the absolute value of the NDC in the DC characteristics, ΔI and ΔV are the current and voltage widths of the NDC region, and ω is the angular frequency. The capacitance of RTD is expressed as the sum of the depletion layer capacitance C_{dep} and the additional capacitance associated with the delay time τG_{RTD0} [10]. A small value of τ is necessary to achieve high-frequency oscillation.

A thick undoped InGaAs was used to collector spacer layer for reduction of the depletion layer capacitance in conventional RTDs [11]. Because the Γ-L valley separation of InGaAs is relatively small, τ_{dep} was increased due to Γ-L transition. To employ materials with large Γ-L separation such as InP at the collector spacer layer is effective in suppressing the Γ-L transition. However, by the complete replacement of InGaAs with InP at the collector spacer, the resonance level in the quantum well becomes lower than the conduction band edge of InP, and the tunneling through the barrier becomes difficult. To solve this problem, stepwise InAlGaAs layers were inserted between the collector-side tunneling barrier and the InP collector spacer, as shown in Fig. 1. The thicknesses of the stepwise InAlGaAs region and the InP layer were 10 and 15 nm, respectively. A GaInAs graded emitter structure was also employed to reduce the electric field. The first resonance level in the well and the peak voltage were reduced by an indium (In)-rich strained well structure. A cap layer with an In-rich strained composition and high-doping concentration was used to reduce the contact resistance to the electrode. The peak current density was 15.5 mA/μm^2, and the widths of current density and voltage of the NDC region were $\Delta J = 8.5$ mA/μm^2 and $\Delta V = 0.38$ V. The peak current density was comparable to that of a conventional RTD having an InGaAs uniform collector spacer (12 mA/μm^2) [11].

MoPI-24 (Poster)
18:30 - 20:30

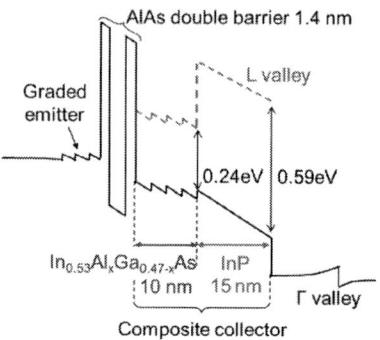

Figure 1. Band diagram of the RTD with InAlGaAs/InP composite collector for suppression of Γ-L transition. 4-step graded $In_{0.53}Al_xGa_{0.47-x}As$ (x = 0, 0.05, 0.1, 0.15 from the emitter side) was inserted to ensure tunneling through the barrier into the condcution band of the collector and smooth connection to InP.

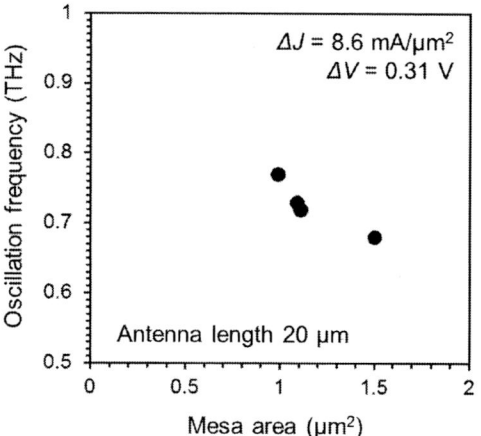

Figure 2. Experimental oscillation frequency as a function of mesa area for the RTDs with InAlGaAs/InP composite collector.

III. OSCILLATION CHARACTERISTICS

The output power of oscillation was detected by a Schottky barrier diode integrated with a broad band bow-tie antenna with lock-in technique, and the oscillation frequency was measured using a simple Fably-perot interferometer constructed with two parallel Si plates.

The dependence of oscillation frequency on mesa area is shown in Fig. 2. We obtained fundamental oscillation frequencies of 680-770 GHz for the oscillators with RTD areas of 1 to 1.5 μm^2. The oscillation frequency increased with decreasing mesa area due to reduction in the capacitance of RTD. Although smaller mesa areas of less than 1 μm^2 were also fabricated, the oscillation frequency was not able to be measured with this measurement system, because the output power was lower than the noise level of the detector due to the small mesa area. The oscillation frequency was slightly higher (~50 GHz) than that of the conventional RTD [11] with the same RTD area, probably due to the reduction in additional

capacitance τG_{RTD0} through the reduced τ. This is the expected effect of the composite collector structure. In the present large-area samples, however, C_{dep} is dominant in the RTD capacitance, and thus, a significant increase in oscillation frequency is not expected, even if τ is successfully reduced by the composite collector. The effect of τ will be clearly obtained as an extension of the highest limit of oscillation frequency in the small-area samples. Therefore, it is necessary in future work to measure the frequencies of small-area devices with low output powers using a liquid He-cooled bolometer and to find the frequency limit as well as the precise values of the transit time τ_{dep}, in order to clearly show the effect of the composite collector.

REFERENCES

[1] M. Tonouchi, "Cutting-edge terahertz technology," *Nat. Photonics*, vol. 1, pp. 97-105, 2007.

[2] I. Kallfass, J. Antes, T. Schneider, F. Kurz, D. Lopez-Diaz, S. Diebold, H. Massler, A. Leuther, and A. Tessmann, "All Active MMIC-Based Wireless Communication at 220 GHz," *IEEE Trans. Terahertz Sci. Technol.* vol. 1, no. 2, pp. 477-487, 2011.

[3] K. Ishigaki, M. Shiraishi, S. Suzuki, M. Asada, N. Nishiyama, and S. Arai, "Direct intensity modulation and wireless data transmission characteristics of terahertz-oscillating resonant tunnelling diodes," *Electron. Lett.*, vol. 48, no. 10, pp. 582-583, 2012..

[4] H.-J. Song, K. Ajito, Y. Muramoto, A. Wakatsuki, T. Nagatsuma and N. Kukutsu, "24 Gbit/s data transmission in 300 GHz band for future terahertz communications," *Electron. Lett.*, vol. 48, no. 15, pp. 953-954, 2012.

[5] L. A. Samoska, "An Overview of Solid-State Integrated Circuit Amplifiers in the Submillimeter-Wave and THz Regime," *IEEE Trans. Terahertz Sci. Technol.*, vol. 1, no. 1, pp. 9-24, 2011.

[6] M. Seo, M. Urteaga, J. Hacker, A. Young, Z. Griffith, V. Jain, R. Pierson, P. Rowell, A. Skalare, A. Peralta, R. Lin, D. Pukala, and M. Rodwell, "InP HBT IC Technology for Terahertz Frequencies: Fundamental Oscillators Up to 0.57 THz," *IEEE J. Solid-State Circuits*, vol. 46, no. 10, pp. 2203-2214, 2011.

[7] K. Leong, G. Mei, V. Radisic, S. Sarkozy, W. Deal, "THz Integrated Circuits using InP HEMT Transistors," presented at *Int. Conf. Indium Phosphide and Related Materials*, Santa Barbara, USA, Plenary I, Aug. 2012.

[8] Q. J. Gu, Z. Xu, H.-Y. Jian, X. Xu, M.-C. F. Chang, W. Liu, and H. Fetterman, "Generating Terahertz Signals in 65nm CMOS with Negative-Resistance Resonator Boosting and Selective Harmonic Suppression," *Symp. VLSI Circuits Dig.*, pp.109-110, 2010.

[9] O. Momeni, E. Afshari, "High Power Terahertz and Millimeter-Wave Oscillator Design: A Systematic Approach," *IEEE J. Solid-State Circuits*, vol. 46, no. 3, pp. 583-597, 2011.

[10] M. Asada, S. Suzuki, and N. Kishimoto, "Resonant Tunneling Diodes for Sub-Terahertz and Terahertz Oscillators," *Jpn. J. Appl. Phys.*, vol. 47, no.6, pp.4375-4384, 2008.

[11] S. Suzuki, M. Asada, A. Teranishi, H. Sugiyama, and H. Yokoyama, "Fundamental oscillation of resonant tunneling diodes above 1 THz at room temperature", *Appl. Phys. Lett.*, vol. 97, 242102, 2010.

[12] M. Feiginov, C. Sydlo, O. Cojocari, and P. Meissner, "Resonant-tunnelling-diode oscillators operating at frequencies above 1.1 THz," *Appl. Phys. Lett.*, vol. 99, 233506, 2011.

[13] H. Kanaya, H. Shibayama, S. Suzuki, and M. Asada, "Fundamental oscillation up to 1.31 THz in resonant tunneling diodes with thin well and barriers," *Appl. Phys. Express*, vol. 5, 124101, 2012.

[14] Y. Koyama R. Sekiguchi,, and T. Ouchi, "Above 1 THz oscillations from RTD oscillators with integrated patch antennas," presented at *Int. Symp. Frontiers in Terahertz Technology*, Nara, Japan, P1.8, Nov. 2012.

978-1-4673-6130-9/13 $31.00 © 2013 IEEE

120nm AlSb/InAs HEMT without gate recess : 290GHz f_T and 335GHz f_{max}

Gardès C., Bagumako S.M., Desplanque L., Wichmann N., Bollaert S., Danneville F., Wallart X., Roelens Y.

Institut d'Électronique de Microélectronique et de Nanotechnologie (IEMN), UMR CNRS 8520, Université Lille I
BP 60069, 59652 Villeneuve d'Ascq Cedex, FRANCE
cyrille.gardes@iemn.univ-lille1.fr

Abstract—In this paper, we report on high frequency performances of AlSb/InAs high electron mobility transistor (HEMT) with 120nm gate length at room temperature. The excellent combined cut-off frequencies f_T/f_{max} of 290/335 GHz simultaneously obtained at drain bias of 0.36V is another demonstration of the ability of AlSb/InAs HEMT for high frequency operation with low-power consumption. Small-signal equivalent circuit parameters have been extracted.

Keywords—AlSb/InAs HEMT; antimonide-based compound semiconductor; high frequency performances.

I. INTRODUCTION

Though the best high frequency performances are obtained for InAlAs/InGaAs HEMT technology which is more mature [1, 2], AlSb/InAs HEMTs are potentially excellent candidates for low-voltage, low-power consumption operation in the case of high-speed analog and digital applications [3]. The AlSb/InAs heterostructure is a 6.1Å material system that combines very high conduction band discontinuity with the high peak velocity and high electron mobility of the low bandgap InAs channel. While AlSb/InAs heterostructures are grown since the 1980s [4, 5], InAs-channel HEMT with interesting RF figures-of-merit have been obtained since the last ten years [6, 7].

Up to now, the highest f_T/f_{max} reported are 260/280 GHz obtained simultaneously for a 100-nm HEMT at V_{ds}=0.4V [8]. The best extrinsic f_T of 303GHz has been reached for a transistor with 120nm gate length at drain bias of 0.44V [9]. In this paper, we present the highest combination of cut-off frequencies reported for AlSb/InAs HEMTs. The main modifications regarding our previous work [9, 10] lie in an optimization of heterostructure growth conditions [11], no need for gate recess and the use of alternative metallic gate stack [12].

II. EPITAXIAL STRUCTURE AND DEVICE FABRICATION

A. Epitaxial structure

The AlSb/InAs heterostructure was grown by molecular beam epitaxy on 3-inches semi-insulating GaAs substrate. A thick AlSb buffer is used to accommodate the large lattice mismatched between 6.1Å materials and GaAs substrate. Then, the structure consists of a 120Å InAs channel, a 65Å AlSb spacer, a Te δ-doping plane and a composite Schottky barrier with a 25Å $Al_{0.8}Ga_{0.2}Sb$ layer and a 50Å $Al_{0.5}In_{0.5}As$ layer (fig. 1). The $Al_{0.5}In_{0.5}As$ layer in the composite Schottky barrier avoid oxidation of $Al_{0.8}Ga_{0.2}Sb$ with air exposure and acts as a hole barrier [13]. Hall measurements at room temperature exhibit a sheet carrier density of 1.5×10^{12} cm^{-2} and electron mobility of 26000cm²/(Vs), giving sheet resistance of 160Ω/□.

Protection layer	$Al_{0.5}In_{0.5}As$	50 Å
Barrier layer	$Al_{0.8}Ga_{0.2}Sb$	25 Å
δ-doping plane	Te 4.5×10^{12} cm^{-2}	
Spacer layer	AlSb	65 Å
Channel layer	InAs	120 Å
Barrier layer	AlSb	500 Å
Buffer layer	$Al_{0.8}Ga_{0.2}Sb$	2500 Å
buffer	AlSb	15000 Å
S.I. Substrate	**GaAs**	

Figure 1. AlSb/InAs heterostructure

B. Device fabrication

HEMTs fabrication starts with ohmic contact evaporation of Pd/Pt/Au after e-beam lithography, followed by rapid thermal annealing at 275°C. Despite the absence of heavily-doped cap layer in the heterostructure, contact resistance, obtained by transmission-line model measurements, is still below 0.05Ω.mm. Schottky T-gate is realized using bilayer resist e-beam lithography process and Mo/Pt/Au metallization. Then, Ti/Au bonding pads are evaporated. Finally, the active area is defined by chemical deep mesa isolation using HF/H₂O₂ solution to completely remove the AlSb buffer, leading to air-bridge gate. Device features are a two-finger 120-nm long gate with 2x15μm transistor width (fig. 2). Source-drain spacing is 1.2μm.

Figure 2. 120-nm AlSb/InAs HEMT.

This work is supported by the national research agency under projects Low IQ (N° ANR-08-NANO-022) and SMIC (N° ANR-11-ASTR-031-03)

MoPl-25 (Poster)
18:30 - 20:30

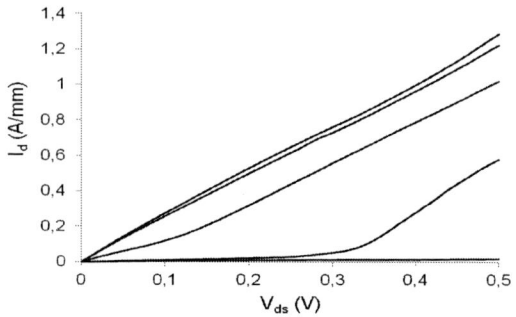

Figure 3. Drain current-voltage characteristic of 120-nm AlSb/InAs HEMT. Vgs is varying from 0V to -1.2V with -0.3V step.

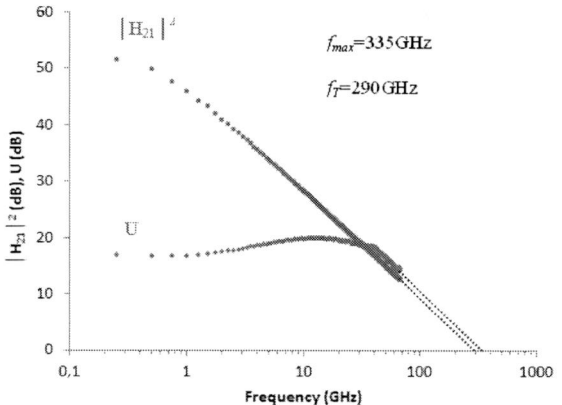

Figure 4. f_{max} and f_T extrapolated from Mason's unilateral gain U and current gain $|H_{21}|^2$ for Vds=360mV and Vgs=-850mV

III. STATIC AND DYNAMIC MEASUREMENTS

Drain current-voltage characteristics are plotted in fig. 3. Pinch-off voltage is -1.1V. Maximum drain currents are 0.65A/mm and 1.00A/mm for drain bias of 250mV and 400mV respectively. These are similar to our previous results [9, 10] despite the higher sheet resistance of the heterostructure and the higher source-drain spacing in the current device. This is confirmed by the following extraction of access resistances R_s and R_d.

HF measurement setup consists in a 67 GHz Agilent PNA for S-parameters on-wafer measurements and an Agilent HP4142 generator for DC biasing. Extrinsic current gain $|H_{21}|^2$ and unilateral power gain U for V_{ds}=0.36V and V_{gs}=-0.85V are presented in fig. 4. Cut-off frequencies are the highest obtained simultaneously for AlSb/InAs HEMT: (f_T, f_{max}) extrapolated from wideband measurements are (290GHz, 335GHz). It should be quoted that in these biasing conditions, DC power consumption is 110mW/mm. In order to explain these excellent figures-of-merit, intrinsic parameters have been extracted from small signal equivalent circuit model. They are presented in table 1. R_g corresponds to the lower values reported for 120nm AlSb/InAs HEMT [10]. Moreover, C_{gs} and C_{gd} are reduced (in

the order of 30%) regarding previously published values [8, 10] while intrinsic transconductance g_m is about 20% higher than previously reported results for such drain bias [9]. This is related to better aspect ratio of the heterostructure which also does not require any recess (no ohmic cap layer to etch) before deposit of the gate and finally the change of our gate metal scheme.

IV. CONCLUSION

In this study, we reported on microwave performances of AlSb/InAs HEMTs without gate recess. Combined (f_T, f_{max}) of (290GHz, 335GHz) for 120nm HEMT have simultaneously been obtained with V_{ds}=0.36V and V_{gs}=0.85V. Extraction of parameters from the small-signal equivalent circuit confirmed improvement of C_{gs}, C_{gd} and intrinsic transconductance.

TABLE I. SMALL SIGNAL EQUIVALENT CIRCUIT PARAMETERS AT VDS=360MV AND VGS=-850MV

Vds (mV)	Id (A/mm)	Rg (Ω/mm)	Rd (Ω.mm)	Rs (Ω.mm)	gm (S/mm)	gd (S/mm)	Cgs (fF/mm)	Cgd (fF/mm)
360	0.3	140	0.13	0.15	1.95	0.59	435	237

REFERENCES

[1] Y.Yamashita et al., "Pseudomorphic $In_{0.52}Al_{0.48}As/In_{0.7}Ga_{0.3}As$ HEMTs with an ultrahigh f_T of 562GHz", Electron Dev. Lett., vol.23, n°10, pp.573-575, 2002.

[2] W. Deal et al., "THz monolithic integrated circuits using InP high electron mobility transistors" Terahertz Sci. & Technol., IEEE Trans. on, vol.1, n°1, pp.25-32, 2011.

[3] B.R. Bennett, R. Magno, J.B. Boos, W. Kruppa and M.G. Ancona "Antimonide-based compound semiconductors for electronic devices : A review" Solid-State Electronics, vol.49, n°12, pp.1875-1895, 2005.

[4] C. Chang, L.L. Chang, E.E. Mendez, M.S. Christie and L. Esaki "Electron densities in InAs-AlSb quantum wells" J. of Vac. Sci. & Technol. B, vol.2, n°2, pp.214-216, 1984.

[5] G. Tuttle and H. Kroemer "An AlSb/InAs/AlSb quantum well HFT" Electron Dev., IEEE Trans. on, vol.34, n°11,p.2358, 1987.

[6] W.R. Deal et al., "A low power/low noise MMIC amplifier for phased-array applications using InAs/AlSb HEMT" MTT-S, Microwave Symp. Dig., pp.2051-2054, 2006.

[7] B.Y. Ma et al., "InAs/AlSb HEMT and its application to ultra-low-power wideband high-gain low-noise amplifiers" MTT, IEEE Trans. on, vol.54, pp.4448-4455, 2006.

[8] R. Tsai et al., "260GHz f_T, 280GHz f_{max} AlSb/InAs HEMT technology" Conf. Digest of 2005 Device Research Conference, pp.257-258, 2005.

[9] Y. Roelens et al., "Tellurium δ-doped 120nm AlSb/InAs HEMTs : towards sub-100mV electronics" Conf. Digest of 2010 Device Research Conference, pp.53-54, 2010.

[10] A. Olivier et al., "High frequency performance of tellurium δ-doped AlSb/InAs HEMTs at low power supply" Proc. of the 5th EUMIC Conf., pp.162-165, 2010.

[11] L. Desplanque et al., "AlSb nucleation induced anisotropic electron mobility in AlSb/InAs HEMTs heterostructures on GaAs" Appl. Phys. Lett., vol.100, n°26, p.262103-4, 2012.

[12] Y.C. Chou et al., "The effect of gate metals on manufacturability of 0.1μm metamorphic AlSb/InAs HEMTs for ultra low-power applications" Proc. of IPRM Conf. Dig., 2008.

[13] J. Boos et al., "AlSb/InAs HEMTs for low-voltage, high-speed applications", Electron Dev., IEEE Trans. on, vol.45, pp. 1869-1875, 1998.

978-1-4673-6130-9/13 $31.00 © 2013 IEEE

MoPI-26 (Poster)
18:30 - 20:30

Monte Carlo Simulation of InAlAs/InGaAs HEMTs with Buried Gate

Akira Endoh[1,2], Issei Watanabe[1], Akifumi Kasamatsu[1], and Takashi Mimura[1,2]

[1]National Institute of Information and Communications Technology, 4-2-1 Nukui-kitamachi, Koganei, Tokyo 184-8795, Japan
[2]Fujitsu Laboratories Ltd., 10-1 Morinosato-Wakamiya, Atsugi, Kanagawa 243-0197, Japan
E-mail: aendoh@nict.go.jp

Abstract—We carried out Monte Carlo (MC) simulation of InAlAs/InGaAs high electron mobility transistors (HEMTs) with buried gate. We employed a T-shaped structure as a gate electrode. The maximum transconductance g_{m_max} and gate capacitance C_g increase with increasing the buried depth d. The extent of increase in the g_{m_max} is more than that in C_g. As a result, the cutoff frequency f_T increases with increase the buried depth d. These phenomena agree with our previous experimental results.

Keywords—HEMTs; InAlAs; InGaAs; Monte Carlo simulation; transconductance; gate capacitance; cutoff frequency

I. INTRODUCTION

InP-based $In_{0.52}Al_{0.48}As/In_xGa_{1-x}As$ ($x \geq 0.53$) high electron mobility transistors (HEMTs) are one of the most promising devices for future ultrahigh-speed applications. To achieve higher-speed operations, reducing gate length L_g is a straightforward method. Besides reducing the L_g, the gate-channel distance must be reduced to suppress the short-channel effects [1]. To reduce the gate-channel distance, there are two methods: one is recessed-gate technology [2] and another is gate metal sinking process [3]. In our previous work [4, 5], we examined the effect of gate buried depth d on the maximum transconductance g_{m_max} and cutoff frequency f_T using the recessed-gate technology. We found that the g_{m_max} and f_T increases with increasing d. Recently, the world record level f_T values of HEMTs are achieved by the buried gate technologies such as recessed-gate [6] and gate metal sinking [7]. In this work, we carried out Monte Carlo (MC) simulation of $In_{0.52}Al_{0.48}As/In_{0.53}Ga_{0.47}As$ HEMTs with buried gate.

II. MONTE CARLO SIMULATION

MC simulations were carried out by using the program, "COSMOS," developed by Mizuho Information & Research Institute, Inc. [8] Fig. 1 shows a model structure of the HEMT with buried gate. We used a three-valley model (Γ, L, X) with nonparabolicity for the conduction band structures of $In_{0.53}Ga_{0.47}As$, $In_{0.52}Al_{0.48}As$, and InP. The electron scattering mechanisms considered were polar optical phonon scattering, non-polar optical phonon scattering, acoustic phonon scattering, inter-valley phonon scattering, and ionized impurity scattering. Dirichlet boundary conditions were applied to all metal-semiconductor interfaces, and Neumann boundary conditions (the zero normal derivative of the potential) were applied to other surfaces. We used a T-shaped gate as shown in Fig. 1. We used SiO_2 between gate head and semiconductor layers. All

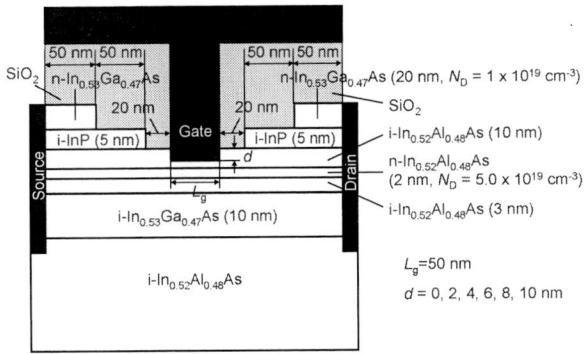

Figure 1.   Model structure of InAlAs/InGaAs HEMT with buried gate.

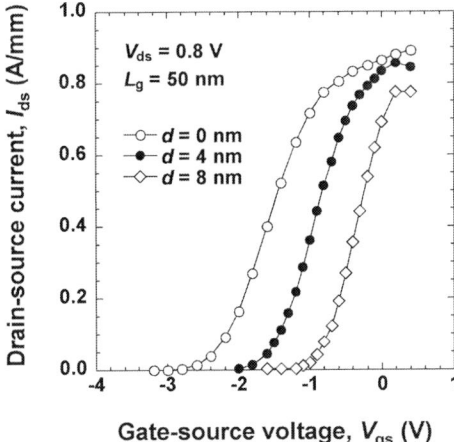

Figure 2.   Drain-source current vs. gate-source voltage (I_{ds}-V_{gs}) characteristics of 50-nm-gate HEMTs with gate buried depth d of 0, 4, and 8 nm. The drain-source voltage V_{ds} is 0.8 V.

simulations were carried out at a lattice temperature of 300 K. Potential was calculated by the finite difference method. The time step was set to 0.5 fs. The electron population was represented by approximately 100000 superparticles.

III. RESULTS AND DISCUSSION

Fig. 2 shows the drain-source current vs. gate-source voltage (I_{ds}-V_{gs}) characteristics of 50-nm-gate HEMTs with

978-1-4673-6130-9/13 $31.00 © 2013 IEEE

MoPI-26 (Poster)
18:30 - 20:30

$$f_T = \frac{g_{m\_max}}{2\pi C_g}. \tag{2}$$

Fig. 4 shows the gate buried depth d dependence of f_T under a V_{ds} of 0.8 V. f_T increases with increasing d. This trend agrees with our experiment [4, 5]. Note that the increase of f_T with increase d was not observed in the MC simulations of the HEMTs with rectangular gate. Therefore, the head of the T-shaped gate plays a very important role in determining the trend of f_T.

IV. SUMMARY

In summary, we carried out MC simulation of InAlAs/InGaAs HEMTs with buried gate. The maximum transconductance g_{m_max} and gate capacitance C_g increase with increasing the buried depth d. The extent of increase in the g_{m_max} is more than that in C_g. Therefore, the cutoff frequency f_T increases with increase the buried depth d. These phenomena agree with our previous experiments. We found that the head of the T-shaped gate plays a very important role in determining the trend of f_T.

ACKNOWLEDGMENT

This work was supported in part by "The research and development project for the expansion of radio spectrum resources" of the Ministry of Internal Affairs and Communications, Japan.

REFERENCES

[1] A. Endoh, Y. Yamashita, K. Shinohara, M. Higashiwaki, K. Hikosaka, T. Mimura, S. Hiyamizu, and T. Matsui, "Fabrication Technology and Device Performance of Sub-50-nm-Gate InP-Based High Electron Mobility Transistors," Jpn. J. Appl. Phys., vol. 41, pp. 1094–1098, February 2002.

[2] Y. Yamashita, A. Endoh, K. Shinohara, K. Hikosaka, T. Matsui, S. Hiyamizu, and T. Mimura, "Pseudomorphic $In_{0.52}Al_{0.48}As/In_{0.7}Ga_{0.3}As$ HEMTs With an Ultrahigh f_T of 562 GHz," IEEE Electron Device Lett., vol. 23, pp. 573–575, October 2002.

[3] K. Shinohara, W. Ha, M. J. W. Rodwell, and B. Brar, "Extremely High g_m > 2.2 S/mm and f_T > 550 GHz in 30-nm Enhancement-Mode InP-HEMTs with Pt/Mo/Ti/Pt/Au Buried Gate," Proc. 19th IPRM, TuA2-2, pp. 18–21, May 2007.

[4] A. Endoh, Y. Yamashita, K. Shinohara, K. Hikosaka, T. Matsui, S. Hiyamizu, and T. Mimura, "InP-Based High Electron Mobility Transistors with a Very Short Gate-Channel Distance," Jpn. J. Appl. Phys., vol. 42, pp. 2214–2218, April 2003.

[5] A. Endoh, Y. Yamashita, K. Shinohara, M. Higashiwaki, K. Hikosaka, T. Matsui, S. Hiyamizu, and T. Mimura, "InP HEMTs: Physics, Applications, and Future," Conf. Dig. 61st DRC, pp. 5–8, June 2003.

[6] D.-H. Kim and J. A. del Alamo, "30-nm InAs PHEMTs With f_T = 644 GHz and f_{max} = 681 GHz," IEEE Electron Device Lett., vol. 31, pp. 806–808, August 2010.

[7] D.-H. Kim, B. Brar, and J. A. del Alamo, "f_T = 688 GHz and f_{max} = 800 GHz in L_g = 40 nm $In_{0.7}Ga_{0.3}As$ MHEMTs with g_{m_max} > 2.7 mS/μm," IEDM Tech. Dig., pp. 319–322, December 2011.

[8] http://www.mizuho-ir.co.jp/solution/research/semiconductor/devicemeister/montecarlo/index.html

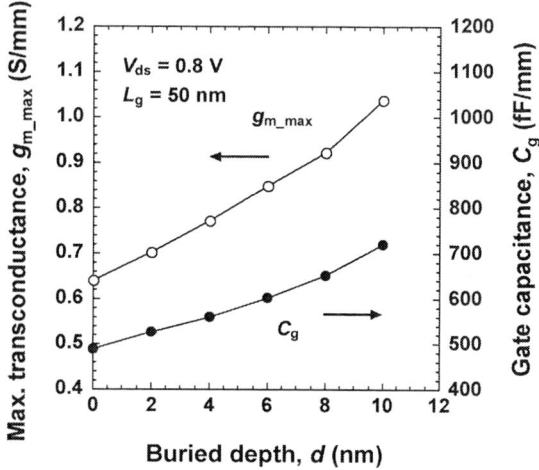

Figure 3. Gate buried depth d dependence of maximum transconductance g_{m_max} and gate capacitance C_g. The drain-source voltage V_{ds} is 0.8 V.

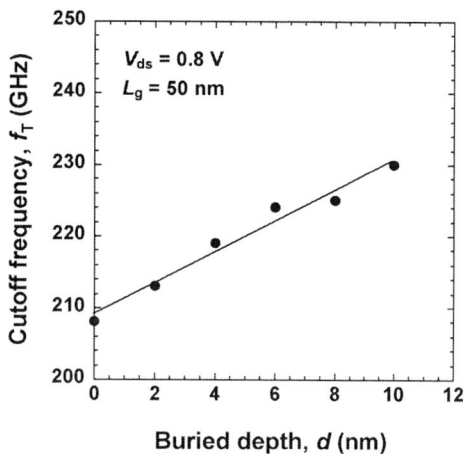

Figure 4. Gate buried depth d dependence of cutoff frequency f_T. The drain-source voltage V_{ds} is 0.8 V.

buried depth d of 0, 4, and 8 nm under a drain-source voltage V_{ds} of 0.8 V. The positive threshold voltage shift occurs with increasing d. Furthermore, the maximum transconductance g_{m_max} increases with increasing d. The g_{m_max} values are 0.639, 0.772, and 1.037 S/mm for d = 0, 4, and 8 nm, respectively. Fig. 3 shows the gate buried depth d dependence of the maximum transconductance g_{m_max} and gate capacitance C_g. C_g was calculated by

$$C_g = \frac{dQ}{dV_{gs}} \tag{1}$$

where Q is the gate charge. C_g values were calculated under a fixed V_{ds} of 0.8 V and a V_{gs} with g_{m_max}. Both of g_{m_max} and C_g increase with increasing d. The extent of the increase in g_{m_max} is more than that in C_g. This trend agrees with our experiments [4]. Using g_{m_max} and C_g values, we obtained the cutoff frequency f_T by

978-1-4673-6130-9/13 $31.00 © 2013 IEEE

MoPI-27 (Poster)
18:30 - 20:30

Terahertz GaAs Schottky diode mixer and multiplier MIC's based on e-beam technology

V. Drakinskiy[1a], P. Sobis[2], H. Zhao[1], T. Bryllert[1,3], and J. Stake[1]

[1]Terahertz and Millimetre Wave Laboratory, Department of Microtechnology and Nanoscience,
Chalmers University of Technology, SE-412 96 Gothenburg, Sweden
[2]Omnisys Instruments AB, SE-43132 V. Frölunda, Sweden
[3]Wasa Millimeter Wave AB, Gothenburg, Sweden
[a]vladimir.drakinskiy@chalmers.se

Abstract— We present the progress of the technological development of a full e-beam based monolithically integrated Schottky diode process applicable for sub-millimetre wave multipliers and mixers. Evaluation of the process has been done in a number of demonstrators showing state-of-the-art performance, including various multiplier circuits up to 200 GHz with a measured flange efficiency of above 35%, as well as heterodyne receiver front-end modules operating at 340 GHz and 557 GHz with a measured receiver DSB noise temperature of below 700 K and 1300 K respectively.

Keywords— Schottky diodes, passive circuits, membrane, submillimeter wave mixers, multipliers

I. INTRODUCTION

There is a need for efficient and reliable heterodyne receivers operating in the sub-millimetre wave band above 300 GHz for future space science missions and earth observation instruments. The sub-millimetre wave regime allows the study of different meteorological phenomena such as water vapour, the ice and water content in clouds, and ice particle sizes and distribution, which are important parameters for the hydrological cycle of the climate system and the energy budget of the atmosphere.

Today Schottky diode mixers and multipliers are the key elements for millimetre and sub-millimetre wave room-temperature heterodyne receiver systems. At frequencies up to around 400 GHz discrete diode technology can be applied. The planar Schottky diode topology has proven reliable and is today used in most commercial mixer and multiplier circuits.

At higher frequencies (> 400 GHz) monolithic integration is needed due to better fabrication and alignment tolerances as well as to enable more advanced circuit integration. Moreover, the performance and functionality of discrete diode circuit designs, is limited by the shape and thickness of the supporting substrate. One of the solutions is the fabrication of monolithic integrated circuits (MICs) supported by a thin membrane. Numerous results based on this technique have been reported [1]. In this paper we present a full e-beam based Terahertz MIC GaAs Schottky membrane process enabling advanced circuit integration well up in the THz range.

II. DIODE FABRICATION

A. Schottky diode mixer on membrane

The Chalmers diode process is based on electron beam lithography, with a beam spot of less than 5 nm, allowing precise and repeatable anode and air bridge formation. Hence, this process module can also be utilized for submicron size anodes and terahertz monolithic integrated circuits (TMICs). Scanning Electron Microscope images of a mixer diode is shown in Figure 1.

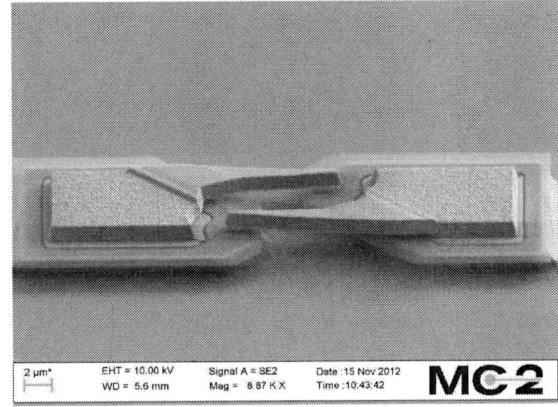

Figure 1. SEM image of an antiparallel diode with an anode area of 0.1 μm² designed for operating at 1.2 THz and fabricated at MC2 Chalmers.

For the diodes on membrane, the starting structure is a semi-insulating GaAs substrate supporting a 3 μm thick GaAs layer sandwiched in between two AlGaAs etch stop layers and a buffer and an active layers. The standard diode fabrication process is as follows:

- Deposition of a stress-balanced PECVD SiO_2 layer.

- Patterning of the ohmic contacts, wet etching through the SiO_2 layer and the active layer of GaAs and deposition of the ohmic contacts metallization with following lift off process.

- Annealing of the ohmic contacts.

- Patterning of the Schottky contacts, wet etching through the SiO_2 layer and deposition of the Schottky contacts metallization with following lift off process.

- Patterning of the air bridge and deposition of metallization with following lift off process.

- Isolation of the diode by wet etching.

- Patterning of the membrane shape, wet etching of 3 μm GaAs using a selective etchant, which stops etching on the bottom AlGaAs layer.

- Patterning of passive circuitry e.g. beamleads, waveguide probes and filter structures and deposition of metallization with following lift off process.

978-1-4673-6130-9/13 $31.00 © 2013 IEEE

- Thinning down the sample from the backside to the AlGaAs layer which is then etched away to release the devices.

Figure 2. SEM image of a released monolithically integrated Schottky membrane mixer designed for operating at 557 GHz and fabricated at MC2 Chalmers.

In Figure 2, a 557 GHz membrane mixer MIC developed under the TeraComp FP7 EU project is shown. With this particular mixer design an optimum receiver noise level of less then 1300 K DSB including all losses has been reached with several assembled mixer modules, using an external IF LNA with a T_{min} of 30 K. The result by itself is a redefinition of state of the art performance for room temperature receivers at these frequencies, but also an important indication for the device quality of our process [2].

B. Schottky doublers on membrane

The device technology described above can also be used for GaAs Schottky multipliers. Results for our narrowband GaAs Schottky varactor multipliers show a very good agreement between model simulations and measurements indicating good process control.

In Figure 3 a Schottky membrane doubler mounted in a waveguide block module is shown. The design is optimized for a high efficiency and low input power. The measurements show state-of-the-art results that are presented in Figure 4.

Figure 3. Photograph of an assembled Schottky membrane doubler, designed for an operating output frequency of 170 GHz and fabricated at MC2 Chalmers.

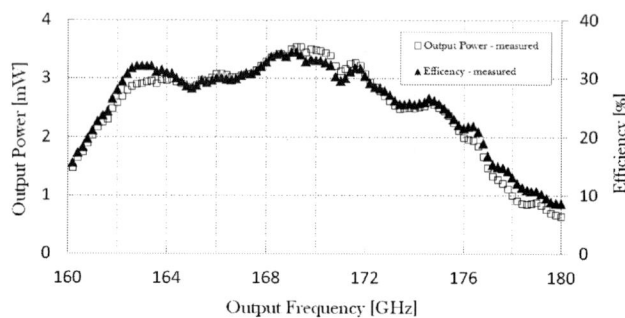

Figure 4. Measured results for a low power 170 GHz high efficiency Schottky membrane doubler, with a measured bandwidth of nearly 10% and with a 3.5 mW of output power running at 35% efficiency.

Figure 5 shows the measured results for a broad band Schottky doubler, which covers a full waveguide band with more than 10% conversion efficiency and good return-loss.

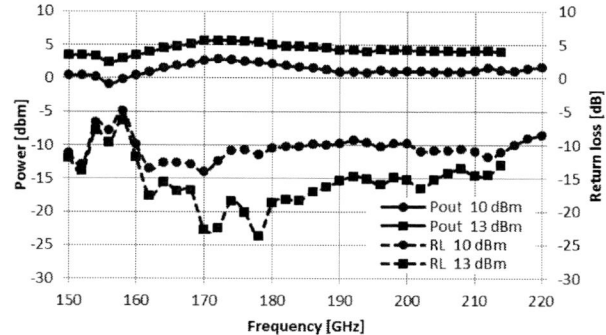

Figure 5. Measured results for a broadband Schottky membrane doubler covering a full waveguide band.

III. CONCLUSIONS

The fabrication process of monolithically integrated Schottky diode mixers and multipliers for THz applications has been developed. The agreement between the results and simulations indicates good control and stability of the process.

ACKNOWLEDGMENT

The work was supported by the EU FP7 project "TERACOMP" under grant № 242424. The work was also carried out in the GigaHertz Centre, Gothenburg, Sweden.

REFERENCES

[1] P.H. Siegel, R.P. Smith, S. Martin, and M. Gaidis, "2.5 THz GaAs monolithic membrane-diode mixer," IEEE Trans. Microw. Theory Tech., vol.47, no. 5, pp. 596-604, May 1999.

[2] H. Zhao, V. Drakinskiy, P. Sobis, J. Hanning, T. Bryllert, A.-Y. Tang, and J. Stake, "Development of a 557 GHz GaAs monolitic membrane-diode mixer", 24th International Conference on Indium Phosphide and Related Materials, Audust 2012.

MoPI-28 (Poster)
18:30 - 20:30

Simulation and Fabrication of InGaAs Planar Gunn Diode on InP Substrate

Vasileios Papageorgiou, Ata Khalid, Chong Li and David R.S. Cumming

School of Engineering
University of Glasgow
UK

Abstract—**This paper describes the simulation and fabrication of the first planar Gunn diode based on InGaAs on InP substrate. Gunn devices were simulated using the Sentaurus Device software. The fabricated planar Gunn diodes are 1.3 µm long and 120 micron wide and the measured and simulated results are in excellent agreement.**

Keywords—Gunn diode; simulation; mm-wave devices

I. INTRODUCTION

The first millimetre-wave planar Gunn diodes on GaAs, operating at 108 GHz have recently been presented [1]. It has also been demonstrated that the inclusion of a pseudomorphically grown $In_{0.20}Ga_{0.80}As$ layer in the structure increases the frequency of oscillation to 118 GHz [2]. However, on a GaAs substrate the indium content is limited to 23% since a further increase would strain the structure. The alternative is to use an $In_{0.53}Ga_{0.47}As$ layer, lattice-matched to an InP substrate, to tap into the superior electrical properties of this material system.

An investigation into $In_{0.53}Ga_{0.47}As$ as a viable material for planar Gunn diodes was carried out in our laboratory. Consequently this work has demonstrated the operation of $In_{0.53}Ga_{0.47}As$ channel planar Gunn diodes at 164GHz [3].

In this paper we initially describe the simulation of the device performance using the Sentaurus Device modeling tool and then the fabrication process of $In_{0.53}Ga_{0.47}As$ based planar Gunn diode. The measured DC charaterics of the devices are successfully predicted by the simulation model, indicating that the model provides a reliable guideline for the fabrication process.

II. SIMULATION & FABRICATION

A. Sentaurus Device simulator

The Sentaurus Device simulation software solves transport equations, Poisson's equation and the electron and hole continuity equations. Excellent agreement between the simulated and the measured data has been reported for 90 nm CMOS technology using small-signal 3D simulations [5] as the software provides more accurate solutions of the Boltzmann's equation using a Monte Carlo approach. For the purpose of this work, a doping-dependent model is used for the calculation of the mobility that degrades due to the

scattering of the carriers by the impurity ions. As the Gunn diode operates at high biasing voltages, the high-field saturation model is used at the same time. The latter takes into account the Transferred-Electron effect for the prediction of the negative differential resistance (NDR). The high-field mobility μ_{hf} is calculated by

$$\mu_{hf} = \frac{\mu_{low} + \left(\dfrac{v_{sat}}{E}\right)\left(\dfrac{E}{E_0}\right)^4}{1 + \left(\dfrac{E}{E_0}\right)^4} \quad (1)$$

where μ_{low} is the low field mobility, v_{sat} the saturation velocity, E the driving field and E_0 the reference field strength. The parameters that have been used for the simulations are detailed in the table below.

TABLE I. DEVICE PARAMETERS

Parameter	InP	$In_{0.53}Ga_{0.47}As$
Permittivity	12.4	13.9
Bandgap (eV)	1.336	0.857
Affinity (eV)	4.4	4.55
Low field mobility ($cm^2V^{-1}s^{-1}$)	4500	16000
Electron saturation velocity ($cm \times s^{-1}$)	1×10^{17}	0.92×10^{17}
Reference field strength ($V \times cm^{-1}$)	4000	4000

B. Device Fabrication

InGaAs layers were grown using molecular beam epitaxy (MBE) on a 600 µm thick semi-insulating InP substrate. A 300 nm thick layer of $In_{0.53}Ga_{0.47}As$ was grown directly on to the substrate, without any buffer layer and with a doping density of 8×10^{16} cm^{-3}, to form the active channel layer. This was followed by a 200 nm thick cap layer of $In_{0.53}Ga_{0.47}As$ with a doping density of 2×10^{18} cm^{-3}. The channel layers are designed to keep the $n \times L_{ac}$ product of the devices above 10^{12} cm^{-2}, where n is the free carrier density and L_{ac} is the separation distance between the anode and cathode [4]. Fig.1 shows the layer structure of the $In_{0.53}Ga_{0.47}As$ planar Gunn diode and the device used for the simulations.

This work was supported by UK EPSRC and e2v Technologies (UK) Ltd.

MoPI-28 (Poster)
18:30 - 20:30

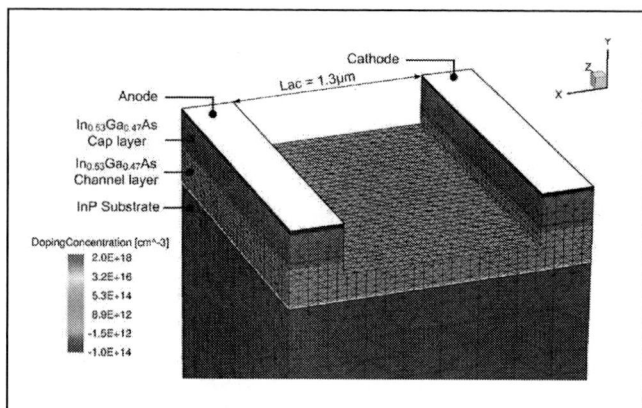

Figure 1.   Device structure.

The cap layer helps to achieve lower Ohmic contact resistance to the devices. For the device terminals, anode and cathode contact regions were defined by electron beam lithography using polymethylmethacrylate (PMMA) resist. The metal alloy Au/Ge/Au/Ni/Au was deposited by e-beam evaporation followed by lift-off. The contacts were not annealed and contact resistances of the order of $0.12\ \Omega$ mm were achieved. The mesa was etched using a $1{:}1{:}10{::}H_2SO_4{:}H_2O_2{:}H_2O$ etching solution at an etch rate of 60 nm/s. A 50 Ω coplanar waveguide (CPW) feed structure was then deposited using 800 nm of gold to form the probe pads for subsequent on-wafer RF and DC measurements. Finally, the unwanted $In_{0.53}Ga_{0.47}As$ contact layer between the anode and cathode was completely etched away using a $1{:}1{:}8\ H_3PO_4{:}H_2O_2{:}H_2O$ etching solution. The etch end-point is determined by successive etch and electrical measurement steps so that the etch depth is accurate to within 5 nm. Fig. 2 shows the results of the stepped-etching up to the point that the optimum depth has achieved. In this case a 120 nm channel layer is also removed in order to avoid the burn out of the device that can be caused by the very high electric field appearing at the contacts. The fabricated device yield (for L_{ac}=1.3 μm to 15 μm) was as high as 90% due to the simple device layer structure and the fabrication process where only four fabrication steps are needed.

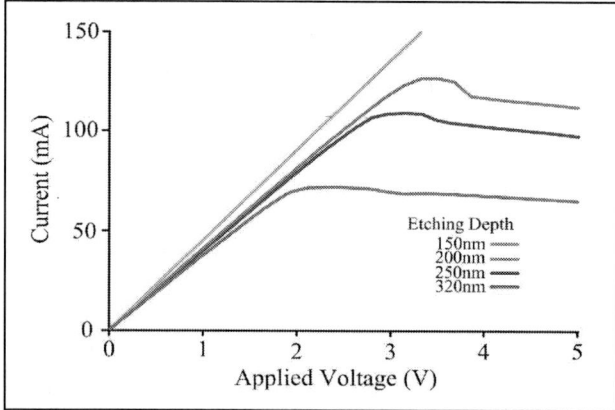

Figure 2.   Simulated data of device etching process for the achievement of the optimum device current.

DC characteristics of the devices were measured with a semiconductor device analyzer (Agilent Technologies B1500A) on a semi-automated probe station (Cascade Microtech Summit 12K). It can be clearly seen in Fig. 3 that the negative differential region (NDR), the peak voltage V_{pk} of 2.5 V and the peak current, I_{pk}, of ~70.0 mA are well predicted by the simulation. The drop in the measured current is most likely caused by the combined effect of the bias-associated heating effect and the NDR. This very good matching between the simulated and the measured performance can help in the design of the next generation $In_{0.53}Ga_{0.47}As$ based planar Gunn devices for mm-wave and THz applications.

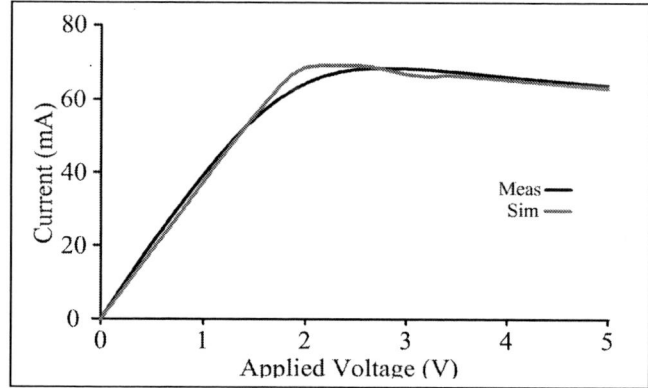

Figure 3.   Simulated and measured output current of planar Gunn diode.

III.   CONCOLUSIONS

We have described the simulation and the fabrication of the first planar Gunn diode based on $In_{0.53}Ga_{0.47}As$ on InP substrate. The Sentaurus Device simulation tool was used for the modelling of the devices, providing guidance for the optimum current level that has to be achieved by the fabrication process. The simulated and the measured DC characteristic are in excellent agreement. This would prove very useful in designing the next generation of $In_{0.53}Ga_{0.47}As$ based planar Gunn devices.

REFERENCES

[1] A. Khalid, N. J. Pilgrim, G. M. Dunn, M. C. Holland, C. R. Stanley, I. G. Thayne, and D. R. S. Cumming, "A planar Gunn diode operating above 100 GHz," *IEEE Electron Device Lett.*, vol. 28, no. 7, pp. 849-851, Oct. 2007.

[2] C. Li, A. Khalid, S. H. Paluchowski Caldwell, M. C. Holland, G. M. Dunn, I. G. Thayne, and D. R. S. Cumming, "Design, fabrication and characterization of $In_{0.23}Ga_{0.77}As$-channel planar Gunn diodes for millimeter wave applications," *Solid-State Electron.*, vol. 64, no.1, pp.67–72, Oct. 2011.

[3] A. Khalid, C. Li, V. Papageorgiou, G. M. Dunn, M. J. Steer, I. G. Thayne, M. Kuball, C. H. Oxley, M. Montes Bajo, A. Stephen, J. Glover and D. R. S. Cumming, "$In_{0.53}Ga_{0.47}As$ Planar Gunn Diodes Operating at a Fundamental Frequency of 164 GHz," *IEEE Electron Device Lett.*, vol. 34, no. 1, pp. 39-41, Jan. 2013.

[4] W. Kowalsky and A. Schlachetzki, "InGaAs Gunn Oscillators," *IET Electronics Letters*, vol. 20, no.12, pp. 502-503, Jun. 1984.

[5] T. Tatsumi, "Geometry Optimization of Sub-100nm Node RF CMOS Utilizing Three Dimensional TCAD Simulation," in *Proc. ESSDERC*, 2006, pp. 319-322

978-1-4673-6130-9/13 $31.00 © 2013 IEEE

MoPI-29 (Poster)
18:30 - 20:30

5 GHz Low-Power RTD-Based Amplifier MMIC With a High Figure-Of-Merit of 24.5 dB/mW

Jongwon Lee, Jooseok Lee, Jaehong Park, and Kyounghoon Yang

Department of Electrical Engineering, Korea Advanced Institute of Science and Technology (KAIST)
373-1, Guseong-Dong, Yuseong-Gu, Daejeon, Republic of Korea
temuchin80@kaist.ac.kr

Abstract—A low dc power RTD microwave amplifier utilizing a hybrid-coupled reflection-type topology is presented. The low-power microwave amplifier, which consists of a quadrature hybrid coupler and two RTDs, is implemented using an InP-based MMIC technology. The fabricated RTD amplifier shows low dc power consumption of 470 μW with high gain of 11.5 dB at 5 GHz, resulting in high FOM of 24.5 dB/mW. The amplifier IC is the first demonstration of a low-power microwave amplifier based on the RTD device technology.

Keywords—amplifier, InP, monolithic microwave integrated circuit, negative differential resistance, resonant tunneling diode

I. Introduction

An ultra-low power MMIC (monolithic microwave integrated circuit) amplifier is one of the critical building blocks in high data rate WLANs (wireless local area networks) at the 5 GHz frequency band, which is widely used for portable electronic devices. In order to reduce the power dissipation, a variety of design technologies have been employed in the conventional transistor-based MMIC amplifiers [1]-[3]. The power consumption of the conventional amplifiers at the 5 GHz band is still a few milli-watts or more due to the inherently limited trans-conductance of MOS transistors at a dc power below the milli-watt scale [3].

By utilizing a hybrid-coupled reflection-type topology that exploits the intrinsic NDR (negative differential resistance) characteristics of the quantum-effect diodes, RTD (resonant tunneling diode)-based amplifiers can achieve ultra-low power consumption of sub-milliwatts while maintaining a high power gain characteristic. A HITD-based amplifier exhibiting a low dc power characteristic of 400 μW with a gain of less than 5 dB at 5.8 GHz has been reported previously [4]. In this work, we demonstrate an RTD-based microwave amplifier with low dc power and high gain characteristics. The low-power microwave amplifier based on the RTD device technology is implemented for the first time by using a BCB -based MMIC technology.

II. Amplifier Design and Device Technology

Fig. 1 shows the circuit configuration of the RTD-based low-power amplifier, which incorporates a quadrature hybrid coupler and two identical RTDs. The RTDs are separately connected to ports 3 and 4 of the hybrid coupler. The other two ports (ports 1 and 2) of the coupler serve as the input/output (I/O) ports of the amplifier. When the hybrid coupler is lossless and reciprocal under the ideal condition, the power gain of the amplifier is equivalent to the square of the magnitude of the

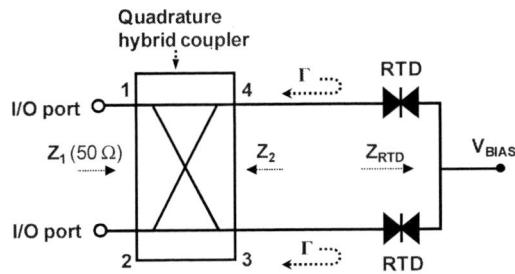

Fig. 1. Circuit configuration of the RTD-based low-power amplifier.

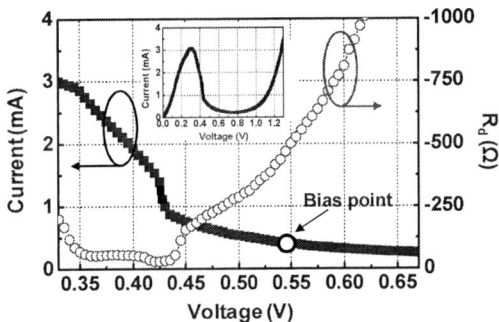

Fig. 2. Measured dc I-V and negative differential resistance (R_d) of the fabricated RTD with an emitter mesa area of 1.8×2.6 μm^2 (inset: the full I-V characteristic).

voltage reflection coefficient (Γ) between an output characteristic impedance of the coupler (Z_2) and an impedance of the RTD (Z_{RTD}) as shown in Fig. 1. The Γ is given by [5]

$$\Gamma = \frac{Z_2 - Z_{RTD}}{Z_2 + Z_{RTD}} \qquad (1)$$

A bias voltage (V_{bias}) of the RTD is selected as 0.54 V to achieve the low dc power consumption as shown in Fig. 2. In order to have a high power gain characteristic, the Z_2 of the quadrature hybrid coupler is designed to be about 400 Ω at a center frequency of 5 GHz, which is close to the NDR magnitude, $|R_d|$, of the selected RTD bias point.

The 5 GHz low-power microwave amplifier based on the RTD device technology is implemented for the first time by using a BCB-based MMIC technology as shown in Fig 3. The

978-1-4673-6130-9/13 $31.00 © 2013 IEEE

MoPI-29 (Poster)
18:30 - 20:30

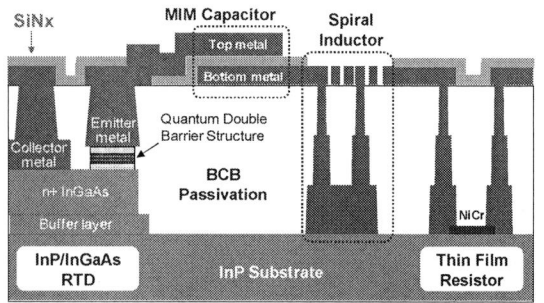

Fig. 3. Schematic cross-sectional view of the monolithically integrated RTD-based amplifier.

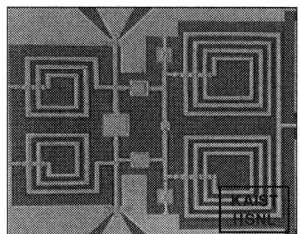

Fig. 4. Microphotograph of the fabricated RTD-based amplifier.

InP-based RTD, MIM capacitor, spiral inductor and thin film resistor were monolithically integrated in the stacked RTD layer structure. The fabrication sequence of the RTD, spiral inductor and thin film resistor has been described previously [6]. The MIM capacitor has been fabricated using a low-k BCB layer prepared by a CF_4/O_2 reactive ion etching process, bottom and top metal layers prepared by a Ti/Au thermal evaporation process and a PECVD-deposited SiNx dielectric layer. The measured dc I-V characteristic and negative differential resistance (R_d) of ($\partial V/\partial I$) of the fabricated RTD with an emitter mesa area of 1.8×2.6 μm^2 are shown in Fig. 2. The RTD exhibits a peak voltage (V_P) of 0.31 V, a peak current of 3.1 mA and a PVCR of 13.5 at room temperature. The measured return loss of the quadrature hybrid coupler, which consists of four MIM capacitors and four spiral inductors, was less than -15 dB at the design frequency of 5 GHz. The capacitance per area of the integrated MIM capacitors was measured to be 160 pF/mm^2. The inductance of the used spiral inductors was in a range of 0.9 -2.5 nH for proper operation of the hybrid coupler.

III. MEASUREMENT RESULTS

The microphotograph of the fabricated 5 GHz RTD amplifier is shown in Fig. 4. The performance of the MMIC amplifier was characterized by on-wafer probing using a vector network analyzer. The RTD bias voltage and current were 0.54 V and 435 μA, respectively. All scattering parameters of S12 and S21, each representing a transmission gain, and S11 and S22, each expressing a return loss, were measured to confirm the operational performances of the hybrid-coupled reflection-type amplifier with the inherent bi-directional property [5] as shown in Fig. 5. It was observed that the RTD-based amplifier provides the transmission gain over the frequency range from 4.8 to 5.3 GHz. At 5 GHz, the RTD-based amplifier exhibited

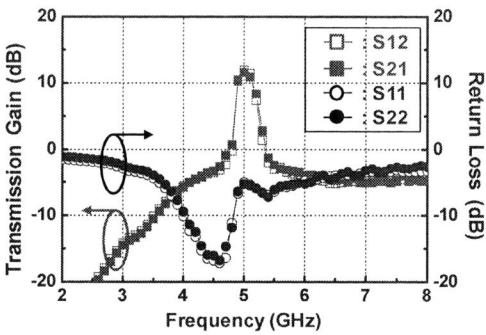

Fig. 5. Measured transmission gain and return loss of the low-power RTD-based amplifier.

transmission gains of S12 and S21 with a maximum value more than 11.5 dB and low dc power consumption of 470 μW. The corresponding return losses of S11 and S22 were measured to be about -5 dB at the maximum-gain frequency of 5 GHz. The high return loss characteristic of the RTD-based amplifier will be improved by a further optimization in circuit design and fabrication process. The widely used FOM (figure-of-merit) for low-power RF applications, which is the ratio of the gain to the dc power consumption, is calculated to be as high as 24.5 dB/mW.

IV. CONCLUSTION

In order to achieve the low dc power of sub-milliwatts, an RTD-based amplifier utilizing the hybrid-coupled reflection-type topology is designed and fabricated. The RTD-based amplifier, which is implemented by using a BCB-based MMIC technology, has shown low dc power consumption of 470 μW with high gain of 11.5 dB at 5 GHz, resulting in high FOM of 24.5 dB/mW. The implemented IC is the first demonstration of a low-power MMIC amplifier based on the RTD device technology.

REFERENCES

[1] I. Song, H. Jhon, H, Jong, M. Koo, and H. Shin, "A low power low noise amplifier with subthreshold operation in 130 nm CMOS technology," *Microwave Opt. Technol. Lett.* vol. 50, no. 11, pp. 2762-2764, Nov. 2008.

[2] D. Wu, R. Huang, W. Wong, and Y. Wang, "A 0.4-V low noise amplifier using forward body bias technology for 5 GHz application," *IEEE Microw. Wireless Compon. Lett.*, vol. 17, no. 7, pp. 543-545, July 2007.

[3] H. Hsieh and L. Lu, "A CMOS 5-GHz micro-power LNA," *in IEEE RFIC Symp. Dig.*, 2005, pp. 31-34.

[4] A. Cidronali, V. Nair, G. Collodi, J. Lewis, M. Camprini, et. al., "MMIC applications of heterostructure interband tunnel devices," *IEEE Trans. Microw. Theory Tech*, vol. 51, no. 4, pp. 1351-1367, April 2003.

[5] S. Chung, S. Chen, and Y. Lee, "A novel bi-directioanl amplifier with applications in active van atta retrodirective arrays," *IEEE Trans. Microw. Theory Tech*, vol. 51, no. 2, pp. 542-547, Feb. 2003.

[6] Y. Jeong, S. Choi, and K. Yang, "Novel antiphase-coupled RTD microwave oscillator operating at extremely low DC-power consumption," *IEEE Trans. on Nanotech.*, vol. 9, No. 3, pp. 338–341, May 2010.

TuD1-2 (Invited)
9:00 - 9:30

IPRM2013, May 19 - 23, 2013, Kobe, Japan
The 25th International Conference on Indium Phosphide and Related Materials

Light emission between 2 and 4 μm:
Innovative active region designs for
InP- and GaSb-based devices

Gerhard Boehm, Stephan Sprengel, Kristijonas Vizbaras, Christian Grasse, Tobias Gruendl, Ralf Meyer,
and Markus-Christian Amann
Walter Schottky Institut, Technische Universität München
Am Coulombwall 4, 85748 Garching, Germany
boehm@wsi.tum.de

Abstract—**This work shows different approaches to cover the spectral range from 2 to 4 μm with active regions based on InP and GaSb for devices suitable as light sources for gas-sensing applications. For shorter wavelengths up to 2.3 μm type-I InP-based quantum wells with rectangular and triangular shape are the preferred choice, beyond that GaSb-based type-I active regions were studied to cover the wavelengths up to 4 μm. An alternative method is implementing type-II structures on InP to exploit the advantages of this well-known material system for device fabrication. For the different methods device designs, growth issues and applications will be discussed.**

Keywords—*mid-infrared lasers, GaSb, InP, Type-II, VCSEL, gas-sensing*

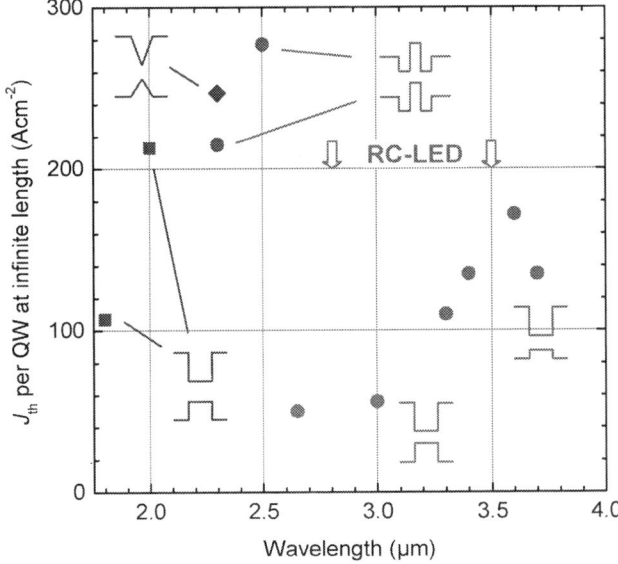

Figure 1. Overview of threshold current densities per QW at infinite length $J_{th,\,inf,\,QW}$ of edge emitting lasers with different types of active regions: Rectangular (■) and triangular (◆) shaped type-I structures on InP measured in cw-mode, type-II on InP (●) and type-I on GaSb (●) measured in pulsed mode, all at room temperature. Resonant cavity LEDs at 2.8 and 3.5 μm realized on InP are indicated.

I. INTRODUCTION

For Tunable Diode Laser Absorption Spectroscopy light sources in the wavelength range from 2 to 4 μm are highly desirable, because many interesting and environmentally relevant gases like CO, NH_4 and CO_2 exhibit strong absorption lines in this region. Apart from single-mode emission at room temperature with a small beam divergence, also low power consumption is essential for compact gas sensing applications. All these requirements can be fulfilled with a vertical-cavity surface-emitting laser (VCSEL) realized already on InP and GaSb for wavelength up to 2.6 μm. Historical approaches to fabricate light emitters for the infrared wavelength range were based on type-I active regions to realize VCSELs on GaAs and InP up to 2.0 μm, or made use of a quantum cascade scheme (QCLs) to cover the range from beyond 10 μm down to 4 μm. Also QCLs using second harmonic generation and intraband cascade lasers (ICLs) [1] were successfully fabricated to enter the region below 4 μm. However, the necessary electrical power of these devices is above hundreds of milliwatts which is not suitable for compact measurement solutions. This work wants to exemplify the impact of various active region designs on the performance of lasers using data of realized edge emitters (Fig. 1) – first on InP and then on GaSb substrates.

II. ACTIVE REGION DESIGN

A. Type-I Active Regions on InP with rectangular wells

Around the year 2000, the first VCSEL with rectangular shaped wells emitting at 2004 nm for CO_2 detection was fabricated and is nowadays commercially available [2]. The wavelength was limited due to the accumulated strain in the high compressively strained GaInAs QWs. To prevent relaxation it was necessary to reduce the growth temperature significantly compared to active regions emitting at shorter wavelength. The reduced optical quality of the active region resulted in a doubling of the threshold current density at infinity length per QW $J_{th,\,inf,\,QW}$ of fabricated edge emitters (Fig. 1).

B. Type-I Active Regions on InP with triangular wells

Extending the wavelength further required a redistribution of the Indium content towards the center of the QW to exploit a

978-1-4673-6130-9/13 $31.00 © 2013 IEEE

TuD1-2 (Invited)
9:00 - 9:30

better overlap of the charge carrier wave function with lower-bandgap and, hence, more Indium-containing material. With these triangular shaped wells it was possible to reduce the average strain compared to rectangular ones. Additionally a high tensile strain had to be applied in the barriers to reduce the overall strain in the active region. All these optimizations combined with an even lower growth temperature led to an active region containing five QWs and emitting at 2.3 μm. As Fig. 1 reveals, however, an increased $J_{th, inf, QW}$ compared to the 2 μm active region occurred and also the fabricated VCSEL devices showed a significantly increased threshold current. In spite of these difficulties, VCSEL devices could be realized out of this material and a gas detection experiment with CO at 2360 nm was successfully carried out [3]. Concerning VCSEL devices, obviously the maximum achievable wavelength on InP cannot be further increased using type-I active regions.

C. *Type-II Active Regions on InP*

The most promising approach to extend the accessible wavelengths on InP beyond 2.3 μm is the application of type-II heterostructures as the optical active part of the laser structure. In these structures the transition energies become smaller than the bandgap energy of the used materials. GaInAs and GaAsSb that can both be grown lattice-matched on InP exhibit such a type-II band alignment. With the combination of these two materials light emission up to a wavelength around 4 μm has theoretically been predicted [4]. In type-II structures electrons and holes are spatially separated. Here the electrons are located in the GaInAs, the holes in the GaAsSb layer. Important for the active region design is the optimization of the wave function overlap, because spontaneous and stimulated emission rates depend on the square of it. On the other hand the thicknesses and the strain of the layers are also influencing the transition energy. An optimization of the device performance must include both, a large wave function overlap and low transition energy at the same time. In Fig. 1 $J_{th, inf, QW}$ of two edge emitting lasers working at 2.3 and 2.5 μm are shown [5], which have similar values compared with the type-I lasers. For both types the tendency to higher $J_{th, inf, QW}$ for longer wavelength is visible, even when the type-II structures are measured in pulsed mode and the type-I lasers in continuous wave (cw) mode. In spite of this behavior it is promising to fabricate VCSEL structures with these type-II active regions, because for the well-known fabrication on InP substrate simply the active region is substituted and the thicknesses of all other layers only have to be adjusted to the new wavelength. The expected low output power of some hundred microwatts in cw-mode at room temperature will be sufficient for gas-sensing experiments. To demonstrate the potential of the type-II active regions on InP, two electrically pumped resonant cavity LED-structures were fabricated emitting around 2.8 and 3.5 μm [6].

D. *Type-I Active Regions on GaSb*

Type-I GaSb-based active regions have the potential to cover the whole desired wavelength range from 2 μm to about 4 μm with type-I active regions. Fig. 1 reveals the superior material quality of the GaSb-based material system compared with InP-based material for wavelengths beyond 2 μm. However, because of some uncertainties on the material parameters for quaternary and quinternary material combinations as well as the challenges for growth and device fabrication, the GaSb-based material is usually used only for a wavelength range not accessible with InP that means beyond 2.3 μm. In the range between 2.3 and 2.6 μm electrically pumped VCSEL structures have already been successfully realized. For longer wavelength not only the non-radiative losses, such as Auger, and free-carrier absorption, but also intrinsic material problems of the (Al)GaInAsSb material system strongly increase. These materials suffer from a miscibility gap and are thermally unstable, which is a great issue especially for VCSEL fabrication. For wavelength above 3 μm the hole-confinement becomes critical and, in order to suppress leakage, complex quinternary barriers have to be introduced. Optimized edge emitters have been successfully realized up to a wavelength of 3.7 μm (Fig. 1) [7]. The devices show laser performance in pulsed mode up to 20°C, but due to the weak hole-confinement cw-performance is difficulty to achieve.

III. SUMMARY

In conclusion, we reviewed the state-of-the-art potential of active regions for VCSEL devices in the range between 2 and 4 μm with edge emitter results. On InP substrate the urging strain issues of the active material requires a change of the quantum well structure, first the shape of the well and then the type of the transition to achieve longer wavelengths. For GaSb based structures the excellent gain of the material competes with intrinsic material problems and the weak confinement of the holes. From the present point of view, it is still unclear whether the concepts on InP or GaSb will lead to an easier fabrication or to more reliable VCSEL devices in the wavelength range beyond 2.5 μm.

REFERENCES

[1] M. Kim, C.L. Canedy, W.W. Bewely, C.S. Kim, J.R Lindle, J. Abell, I. Vurgaftman and J.R. Meyer, "Interband cascade lasers emitting at 3.75 μm in continous wave above room temperature", Appl. Phys. Lett, **92**, 191110 (2008)

[2] Vertilas GmbH, www.vertilas.com

[3] G. Boehm, A. Bachmann, J. Rosskopf, M. Ortsiefer, J. Chen, A. Hangauer, R. Meyer, R. Strzoda and M.-C. Amann, "Comparison of InP- and GaSb-based VCSELs emitting at 2.3 μm suitable for carbon monoxide detection", J. Cryst. Growth, **323**, pp. 442-445, (2011)

[4] J.Y.T. Huang, L.J. Mawst, T.F. Kuech, X. Song, S.E. Babcock, C.S. Kim, I. Vurgaftman, J.R. Meyer and A.L. Holmes Jr, "Design and characterization of strained InGaAs/GaAsSb typ-II 'W' quantum wells on InP substrates for mid-IR emission", J. Phys. D: Appl. Phys., **42**, 025108 (8pp) (2009)

[5] S. Sprengel, A. Andrejew, K. Vizbaras, T. Gruendl, K. Geiger, G. Boehm, C. Grasse and M.-C. Amann, "Type-II InP-based lasers emitting at 2.55 μm", Appl. Phys. Lett, **100**, p. 041109, (2012)

[6] C. Grasse, T. Gruendl, S. Sprengel, P. Wiecha, K. Vizbaras, R. Meyer and M.-C. Amann, "GaInAs/GaAsSb-based type-II micro-cavity LED with 2-3 μm light emission grown on InP", J. Cryst. Growth (2012), http://dx.doi.org/10.1016/j.jcrysgro.2012.07.001

[7] Kristijonas V. and M.-C. Amann, "Room-temperature 3.73 μm GaSb-based type-I quantum-well lasers with quinternary barriers", Semicond. Sci. Technol., **27**, 032001 (4pp), (2012)

TuD1-3 (Oral)
9:30 - 9:45

MOCVD Growth and Device Characterization of InP/GaAsSb/InP DHBTs with a GaAs Spacer

Takuya HOSHI, Hiroki SUGIYAMA, Haruki YOKOYAMA, Norihide KASHIO,

Kenji KURISHIMA, Minoru IDA, and Hideaki MATSUZAKI

NTT Photonics Laboratories, NTT Corporation

3-1 Morinosato Wakamiya, Atsugi, Kanagawa, 243-0198 Japan

E-mail: hoshi.takuya@lab.ntt.co.jp

Abstract—This paper describes the impact of a GaAs spacer inserted between the InP emitter and GaAsSb base of InP/GaAsSb/InP double-heterojunction bipolar transistors (DHBTs) grown by metalorganic chemical vapor deposition. The GaAs spacer suppresses the incorporation of excess Sb atoms from the GaAsSb base into the InP emitter, which reduces the recombination current at the junction. The fabricated DHBTs with a 0.5-μm-wide emitter show dc current gain of over 90 at current density (J_C) of 10 mA/μm². The 0.25-μm-emitter DHBTs yield current-gain cut-off frequency (f_t) of 384 GHz and maximum oscillation frequency (f_{max}) of 264 GHz at J_C of 14 mA/μm².

Keywords—*DHBT; GaAsSb; GaAs spacer; MOCVD.*

I. INTRODUCTION

InP/GaAsSb/InP double-heterojunction bipolar transistors (DHBTs) are very attractive for application to terahertz devices because they can offer high cut-off frequency and sufficiently high breakdown voltage [1]. However, achieving high-current-gain DHBTs is still challenging because it is difficult to form a high-quality InP-emitter/GaAsSb-base interface. Moreover, their type-II heterojunction (or negative conduction-band offset) causes electron pile-up in the emitter, which enhances the recombination current at the junction.

In order to achieve a high-quality heterojunction, it is important to suppress the incorporation of Sb from the GaAsSb base into the InP emitter during growth. For this purpose, exposing the GaAsSb surface to AsH₃ atmosphere during the growth interruption (before the growth of the InP emitter) in metalorganic chemical vapor deposition (MOCVD) has been proposed [2]. However, the appropriate growth-interval time and precursor-supply sequences greatly depend on the growth apparatus. Thus, the optimization of growth conditions becomes very important. In this paper, we propose a thin GaAs spacer inserted between the InP and GaAsSb in order to improve the quality of the emitter/base junction. The structural properties of epitaxial layers and results of fabricated DHBTs are described.

II. EXPERIMENTS

DHBT layers were grown on 3-inch Fe-doped (001) InP substrates in a low-pressure vertical MOCVD reactor. The switching sequence to form the GaAs spacer is shown in Fig. 1. After the growth of the GaAsSb base, the GaAs spacer is grown immediately without any growth interruption by turning off the supply of TMSb and CBr₄. Then, TEGa and AsH₃ are turned off, and InP is grown by supplying TMIn and PH₃. This simple sequence is expected to reduce the incorporation of Sb into the InP emitter without any onerous growth optimization.

We fabricated large-area DHBTs to investigate the effect of the GaAs spacer. They consist of an InGaAs/InP subcollector, a 200-nm-thick InP collector, a 32-nm-thick GaAs$_{0.5}$Sb$_{0.5}$ base doped at $p = 4 \times 10^{19}$ cm⁻³ with C, a GaAs spacer, a 70-nm-thick unintentionally doped InP emitter, and an InGaAs cap layer. The thickness of the GaAs spacer ranges between 0 and 5 nm. The conduction-band offset is still negative at the heterojunction between the InP and the strained GaAs. The thick unintentionally doped emitter is useful for reducing the tunneling-recombination current at the junction and thus appropriate for evaluating the heterojunction quality from the dc current gain of large-area devices.

We also prepared small-size DHBTs for high-frequency characterization. They consist of a 100-nm-thick InP collector doped at $n = 6 \times 10^{16}$ cm⁻³ with Si, a 30-nm-thick GaAs$_{0.6}$Sb$_{0.4}$ base doped at $p = 4 \times 10^{19}$ cm⁻³, a 2-nm-thick GaAs spacer, and a 20-nm-thick unintentionally doped InP emitter. Thinning the emitter and collector is a prerequisite for high-current-density operation. Details of fabrication are described elsewhere [3].

III. RESULTS AND DISCUSSION

Fig. 2 shows the x-ray diffraction (XRD) profiles for (004) reflection for the DHBTs with various GaAs spacer thickness. The profiles show good agreement between the experimental and simulation results, which suggests that the incorporation of Sb into the GaAs is sufficiently small, and that the designed thickness is obtained for the GaAs. Fig. 3 shows the AFM images for the surface of the GaAs(Sb), which was exposed by removing the InP emitter with HCl/H₃PO₄ solution. The epi-wafer without the spacer exhibits bumpy surface morphology with RMS roughness of approximately 1 nm. In contrast, step-flow like surface morphology is observed for the epi-wafer with the spacer. The RMS roughness decreases to 0.17~0.36 nm, which is almost the same value as those of the GaAsSb films grown for reference. SIMS measurements also revealed that the Sb incorporation into the InP emitter was small when the GaAs spacer was employed. Taken together, these results indicate that the GaAs spacer plays an effective role in suppressing the interface mixing due to the Sb migration and in achieving an abrupt heterointerface.

Fig. 4 shows Gummel plots for fabricated large-area devices. When the GaAs spacer thickness was less than 2 nm, the ideality factors of collector and base current were ~1.0 and ~1.1, respectively. The dc current gain (β), measured at a current density (J_C) of 6 μA/μm^2, increases from 25 to 30 as the thickness of the spacer increases from 0 to 2 nm. These results support that the improvement of the crystal quality at the heterojunction is due to the presence of the GaAs spacer. However, a marked decrease of the current gain was observed for the devices with a 3- and 5-nm-thick spacer. For instance, the device with the 5-nm-thick spacer exhibits the base ideality factor of ~1.3 and current gain of only 10. Such degradation implies the generation of defects associated with strain relaxation. Therefore, a 2-nm-thick GaAs spacer is preferable for the DHBTs.

To investigate the influence of the GaAs spacer on high-current-density operation, we fabricated high-frequency DHBTs with a thinner emitter and collector. Fig. 5 shows the common-emitter I-V characteristics of the DHBT with an emitter size of 0.5×4 μm^2. The DHBT exhibits dc current gain of over 90 at $J_C = 10$ mA/μm^2, which is reasonable for the base sheet resistance of 1300 Ω/sq. For the 0.25-μm-emitter DHBTs, a peak current-gain cut-off frequency (f_t) of 384 GHz and a peak maximum oscillation frequency (f_{max}) of 264 GHz are obtained at $J_C = 14$ mA/μm^2. Consequently, the presence of the GaAs spacer does not pose any penalty on the capability of high-current injection. The proposed GaAs spacer is useful for application to very high-speed DHBTs.

IV. SUMMARY

We have demonstrated InP/GaAsSb/InP DHBTs with a GaAs spacer inserted between the InP emitter and the GaAsSb base. The XRD, AFM and SIMS measurements have revealed effective suppression of the incorporation of Sb from the GaAsSb base into InP emitter. The current gain of a large-area device increases with increasing thickness of the GaAs spacer from 0 to 2 nm, but decreases significantly for the thickness of 5 nm. High-frequency devices show high-current-density operation and reasonably high current gain.

REFERENCES

[1] H. G. Liu, O. Ostinelli, Y. Zeng, and C. R. Bolognesi, Proc. IEDM 2007, pp. 667-670, Dec. 2007.

[2] F. Brunner, S. Weeke, M. Zorna, and M. Weyers, J. Crystal Growth 272, 111 (2004).

[3] N. Kashio, K. Kurishima, Y. K. Fukai, M. Ida, and S. Yamahata, IEEE Trans. Electron. Dev. 57, 373 (2010).

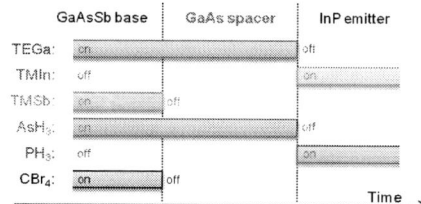

Fig.1. Switching sequence to form the GaAs spacer.

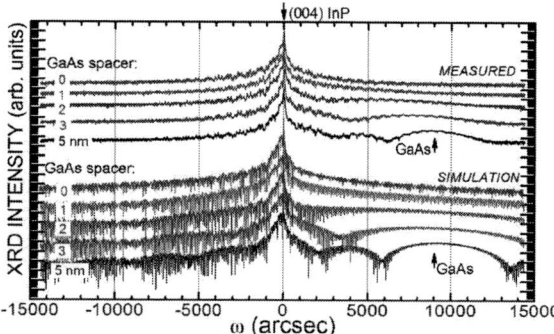

Fig.2. XRD profiles of DHBTs.

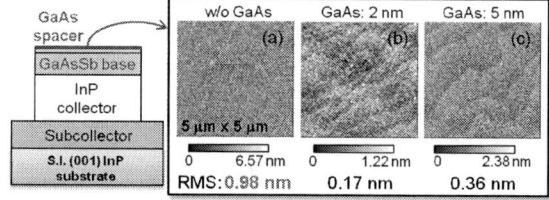

Fig.3. AFM images of the surface of GaAs(Sb) exposed by removing the InP emitter with HCl/H$_3$PO$_4$ solution.

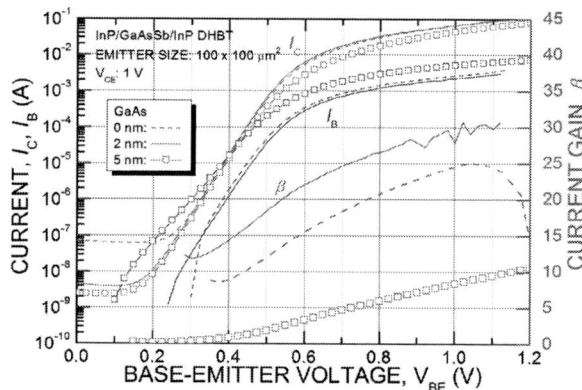

Fig. 4. Gummel plots of the large-area devices. The collector-emitter voltage, V_{CE}, is 1 V.

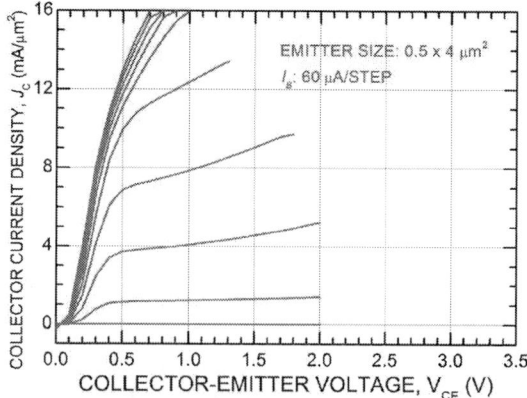

Fig. 5. Common-emitter collector I-V characteristics of the DHBT with an emitter size of 0.5×4 μm^2.

TuD1-4 (Oral)
9:45 - 10:00

High Growth Rate Gallium Phosphide for Red LEDs

Stephen Farrell, Chris Ebert, Devon Dyer

Veeco Instruments, Inc.
Somerset, NJ 08873 U.S.A.
sfarrell@veeco.com

Abstract— **Growth rates of up to 17 microns per hour for Gallium Phosphide grown by MOCVD are achieved and material doping properties are analyzed.**

Keywords—gallium, phospide, light-emitting diode, MOCVD

I. INTRODUCTION

Red LEDs have long been used as indicator lights due to their relatively low operating voltage, low cost, and ease of manufacturability. Recently, red LEDs have emerged as a popular instrument for color rendering when employed alongside blue-emitting GaN-based LEDs in lamps for general lighting applications. The advantages of using red LEDs for this purpose are particularly evident for warm color temperatures where the alternative, phosphor coatings, become less efficient. The market for "warm" LED lighting includes residential and other traditionally incandescent-lit areas, and is thus potentially very large. At the time of this writing, the cost for a 60W incandescent replacement is $15-20, which is still prohibitively high when compared to an incandescent bulb (less than $0.25) or even a similar brightness compact fluorescent lamp ($2-3). While the ratio of red-to-blue chips contained in an LED bulb will obviously vary by end-use application and manufacturer, it is typical for each lamp to contain several red LEDs. As such, a measureable fraction of the LED bulb cost is associated with these red LED components. Most red LED epi structures contain a very thick GaP layer on the p-side of the device which serves to promote current spreading. This encourages photon excitation over a larger area and increases the LED's total output power. The GaP thickness required to achieve sufficient current spreading is approximately 8-10μm. At nominal MOCVD growth rates of 3-5μm/hr, this represents a significant portion of the overall growth time for a typical red LED, on the order of 50-60%. It is therefore evident that an increase GaP-layer growth rate will significantly reduce the growth time and per-die cost of red LEDs. In this paper, we explore several different growth rates of GaP up to 17μm per hour. These high growth-rate conditions are examined in both bulk GaP test run layers as well as in red LEDs emitting in the 630nm range. The doping properties of these structures, including magnesium (Mg) interfaces and intrinsic carbon (C) levels, are investigated, as well as the effects of growth rate and V/III ratio on doping concentrations.

II. GAP IN LIGHT EMITTING DIODES

The epi structure for the LED devices grown in this study is shown in Table I. A si-doped 18-period AlGaAs DBR was grown on a 350μm thick GaAs substrate. A 20 period InGaAlP/InGaP MQW structure, cladded by InAlP, was then grown, followed by transition layers to accommodate the

TABLE I.   BASELINE RED LED EPITAXIAL STRUCTURE

LED Structure	Layer	Thickness (Å)	Dopant
Current Spreading	GaP2	45000	Mg
	GaP1	45000	Mg
Transition Layers	InGaP	32	Mg
	$In_{0.5}(Ga_{0.3}Al_{0.2})P$	400	Mg
p-clad	InAlP	6400	Mg
20 periods	$In_{0.5}(Ga_{0.25}Al_{0.25})P$	1850	
	$In_{0.5}(Ga_{0.25}Al_{0.25})P$	133	
	InGaP	66	
n-clad	$In_{0.5}(Ga_{0.25}Al_{0.25})P$	870	
	InAlP	3700	Si
18 periods	$Al_{0.84}Ga_{0.16}As$	518	Si
	$Al_{0.40}Ga_{0.60}As$	420	Si
Substrate	GaAs	350μm	n+

lattice-mismatched (3.7%), fully relaxed GaP layer. The current spreading layer was grown in two stages at 730°C. The first stage was grown at a rate of 4.3μm/hr for all devices, while the second stage was grown at varying growth rates. The baseline had 4.5μm of 4.3μm/hr GaP followed by 4.5μm of 9μm/hr GaP. The details of the GaP growth conditions are shown in Table II. The total GaP thickness for all devices was 9μm. All growths in this study were performed on a Veeco TurboDisc K475 reactor using 15° offcut GaAs substrates.

The C background profiles for the four LED structures, obtained by secondary ion mass spectroscopy (SIMS), are shown in Figure 1. One can see from the figure that Samples A and B, with similar GaP2 growth rates, also had similar GaP2 carbon doping ($1.6\times10^{18}cm^{-3}$ and $1.7\times10^{18}cm^{-3}$, respectively). Sample C, with the GaP2 grown at 13.6μm/hr, had a C background of $1.3\times10^{18}cm^{-3}$, while Sample D, with the fast GaP2 layer at 17.4μm/hr, had the lowest C background at slightly less than $1.1\times10^{18}cm^{-3}$. This is a positive and perhaps surprising result, as it is often assumed that growing faster leads to a higher incorporation of impurities. For the incorporation of C in GaP, it seems the opposite is true. For all samples the oxygen background levels were below $5\times10^{15}cm^{-3}$. Another consideration in the growth of GaP in red LEDs is the

TABLE II.   GAP GROWTH CONDITIONS FOR FOUR LED STRUCTURES

Two-Step GaP Growth Conditions		Sample			
		A	B	C	D
GaP1	Thickness (μm)	4.5	0.5	4.5	4.5
	Growth Rate (μm/hr)	4.3	4.3	4.3	4.3
GaP2	Thickness (μm)	4.5	8.5	4.5	4.5
	Growth Rate (μm/hr)	9	8.7	13.6	17.4

TuD1-4 (Oral)
9:45 - 10:00

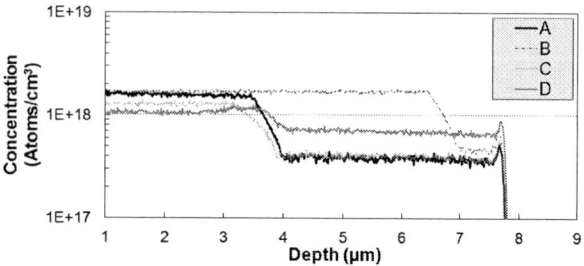

Figure 1. Carbon levels for LEDs with different GaP growth conditions.

effect of growth conditions on the diffusion of the p-type dopant Mg. This is because a primary influence on device performance is the proximity of the Mg doping front to the device's active region. Ideally, one would prefer to have a perfectly abrupt Mg interface that terminates just before the first MQW barrier. In practice, of course, this is difficult to obtain. Here we looked at the effect of our varying GaP growth rate on the extent of this Mg front.

Figure 2 shows the Mg doping profiles for our four devices. The target Mg doping level was the same for all GaP2 layers, however one can observe a small variation for the four samples. All samples show a local Mg peak at the interface of the GaP and the p-side InAlP cladding. Figure 2 includes an overlay of the Al profile so that we can see where the Mg front terminates with respect to the MQW. We can see that in samples A and C the Mg drops off sharply in the InAlP and is below $1 \times 10^{17} cm^{-3}$ in the wells. By contrast, samples B and D have a second Mg peak in the MQW region. Sample B has a thicker, high-doped GaP2 while sample D has the fastest Gap2 growth rate. It appears as though, in both cases, the higher total amount of Mg leads to an extended doping front. As mentioned earlier, this is undesirable as it is potentially harmful to device performance. Sample A wafers were fabricated into chips and probed. For a die size of 200μm, the forward voltage was 2.01V while the output intensity was 46mcd at 20mA with a dominant wavelength of 632nm. LED device data from the faster growth rate GaP wafers will be obtained as part of our ongoing investigation into the effects of GaP growth conditions on LED performance.

III. GAP TEST RUN EXPERIMENTS

As a follow-up to our experiments with growing GaP at high growths rates in LEDs, we performed two test run growths. The first test run was a V-III ratio "staircase" in

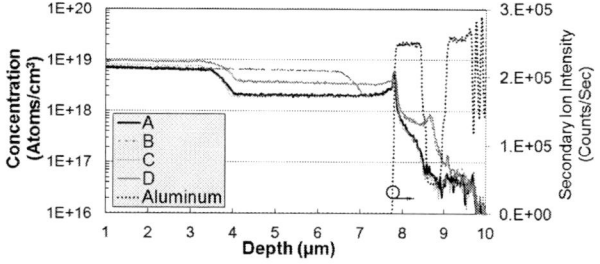

Figure 2. Mg doping for LEDs with different GaP growth conditions.

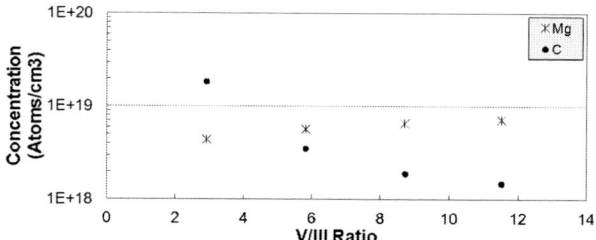

Figure 3. Dependence of Mg and C incorporation on V-III ratio for GaP.

which the GaP growth rate was fixed but the phosphine (PH₃) flow was varied. PH₃ flows of 800, 600, 400, and 200sccm were used, corresponding to V-III ratios of 11.5, 8.7, 5.8, and 2.9. A plot of the Mg and C levels as a function of V-III ratio is shown in Figure 3. While the Mg shows a weak linear dependence, increasing with high V-III ratios, the C decreases for higher V-III ratios, with a $1/x^{1.86}$ dependence and $R^2=0.97$.

The second GaP test run that was grown was a growth rate staircase. We grew four layers of GaP with each layer having a progressively slower growth rate. Growth rates were measured, using Veeco's in-situ Real Temp™ reflectivity system, to be 16.9, 12.7, 8.6, and 4.3μm/hr. Again using SIMS, the Mg and C levels were obtained and are plotted as a function of growth rate in Figure 4. It is clear from the figure that the Mg shows little or no variation with growth rate, suggesting the Mg incorporation is kinetically limited.

Consistent with what we saw in our LEDs previously, the C incorporation actually decreases with increasing growth rate. As with V-III ratio, the C doping dependence on growth rate can be fitted to a $1/x^n$ function. In this case, for $n=1.24$, R^2 is greater than 0.99. The carbon doping data from our LED samples has been added to Figure 4 as well.

IV. CONCLUSION

We have used a Veeco TurboDisc K475 reactor to achieve GaP growth rates as high as 17μm/hr. It was found that the background carbon concentration decreases with increasing growth rate, while the Mg concentration is constant for growth rates higher than 4μm/hr. SIMS was used to obtain doping profiles and the data suggests that Mg diffusion for GaP grown at 13.6μm/hr is equivalent to GaP grown at 8.7μm/hr. Lastly, we have shown that reducing the V-III ratio at a given growth rate will result in higher carbon incorporation.

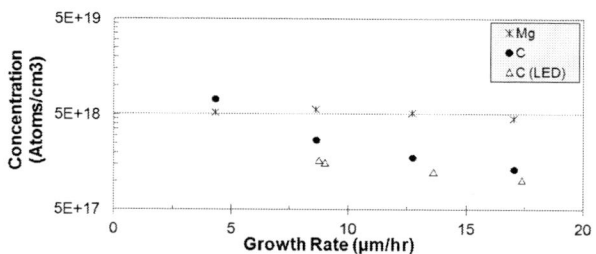

Figure 4. Dependence of Mg and C incorporation on GaP growth rate

TuD1-5 (Oral)
10:00 - 10:15

MOCVD growth of carbon-doped InGaAs layers using ethyl-base metal organic materials

Hideo Yokohama[1,2], Kenji Shiojima[1], and Gako Araki [2]

[1] Graduate School of Electrical and Electronics Engineering, University of Fukui, 3-9-1 Bunkyo, Fukui-shi, Fukui
910-8507, Japan

[2] OPTRANS Corporation, 241 Noborito, Tama-ku, Kawasaki-shi, Kanagawa 214-0014, Japan
Tel: +81-44-932-6491, Fax: +81-44-932-8281, E-mail: yokohama@optrans.com

Abstract—**Ethyl-base metal organic materials and pulse-doping technique were employed in C-doped p-InGaAs epitaxial growth. By using triethylindium, a linear relationship between In supply ratio and In content was observed with less growth-temperature variation, comparing with trimethylindium. Pulse-doping for C with CBr_4 was significantly suppressed metal organic materials consumption. As a result of that, good controllability of C doping by CBr_4 flow and a large carrier density of 1.4×10^{19} cm^{-3} were achieved.**

Keywords—*MOCVD; Ethyl-based material; C-doped InGaAs; Pulse-doping*

I. INTRODUCTION

InP/InGaAs Heterojunction bipolar transistors (HBTs) are expected for over 40 Gbit/s high-speed data communication systems. Heavy doping in p-type base layer is an important issue to improve both device characteristics and reliability. Carbon is a good candidate for a p-type dopant because of its low diffusion coefficient and extremely high-level doping comparing with Zn. These features make C used for the base layer of InP/InGaAs HBTs. When C-doped p-InGaAs layers are grown by metalorganic chemical vapor deposition (MOCVD), Hydrogen atoms are incorporated and bond with C. As a result of that, C acceptors become inactive. It has been reported that thermal treatment in inert gases is effective to diffuse out hydrogen atoms [1-3]. However, excessive thermal treatment could result in the degradation of device characteristics.

We have demonstrated MOCVD growth of C-doped p-InGaAs layers by using ethyl-base gases instead of methyl base; i.e., triethylindium (TEIn), triethylgallium (TEGa) and triethylarsine (TEAs), and carbon tetrabromide (CBr_4) . As ethyl-based materials are easy to be decomposed, the hydrogen incorporation in InGaAs layers was reduced. Since the activation efficiency of C-acceptors in the base layers increased, good HBT characteristics were obtained as a first report. Additionally, low-temperature growth would be expected by the ethyl-base growth, which is required for high C doping.

In this paper, we report differences of the growth mechanism between ethyl- and methyl-base growth from the aspects of (i) In incorporation ratios as a function of the growth temperature, (ii) effects of CBr_4 doping on material consumption and carrier controllability, and (iii) doping

sequence. Finally, the advantages of the ethyl-base growth for C-doped InGaAs will be discussed.

II. EXPERIMENTAL PROCEDURE

AIXTRON CCS system (4" x10), which has a vertical reaction chamber with a closed coupled showerhead, was used for the epitaxial growth. The InGaAs layers were grown on (100) oriented Fe-doped semi-insulating InP substrates. Either TEIn or TMIn for an In source, TEGa for Ga, AsH_3 for As were used, respectively. For C-doping, H_2-diluted CBr_4 was used. Two doping sequences on the CBr_4 supply were carried out. One is a conventional way, continuous CBr_4 supply (bulk doping). The other is alternative supply between In source material (TEIn or TMIn) and CBr_4 (pulse doping).

The In composition was determined by X-ray diffraction analysis. The carrier concentration was measured by the Van der Pauw method at room temperature.

III. RESULTS AND DISCUSSION

We experimented several temperatures to growth un-InGaAs by using $TMIn+TEGa+AsH_3$ and $TEIn+TEGa+AsH_3$. The growth temperature ranged from 450 to 635°C. Figure 1 shows calculated consumption of metal organic materials depend on growth temperature of un-InGaAs. It has been similar consumption of TMIn, TEGa and TEIn, when the growth temperature is high at 635°C . However, as the growth temperature decreases, in the case of $TMIn+TEGa+AsH_3$ with consumption of TEGa increased suddenly. It can be considered that in the lower temperature intermediate react with methyl radical supply from methyl-base materials (TMIn), and incorporation of Ga atoms decreased. On the other hands in the case $TEIn+TEGa+AsH_3$, growth temperature dependence was not observed 635°C and 475°C. Growth temperature below the 500°C is necessary in carbon doping to InGaAs layers, so ethyl-base materials ($TEIn+TEGa+AsH_3$) is better growth efficiency in the lower growth temperature.

Next, we focus on the CBr_4 doping characteristics. Figure 2 shows metalorganic material consumption as a function of CBr_4 flow. In the bulk doping, significantly large consumption is observed due to a high activity of CBr_4. However, by the pulse doping, the consumptions decrease to be almost constant values over the wide CBr_4 flow range for both methyl- and ethyl-base materials. It is clearly shown that the pulse doping is effective way to improve material efficiency.

Figure 3 shows the relationship between carrier concentration and CBr_4 flow in the pulse-doped p-InGaAs. The carrier concentration is proportional to the CBr_4 flow, and as high as 1.4×10^{19} cm^{-3} when the flow is 32 µmol/min. The pulse doping is also effective for high C-doping.

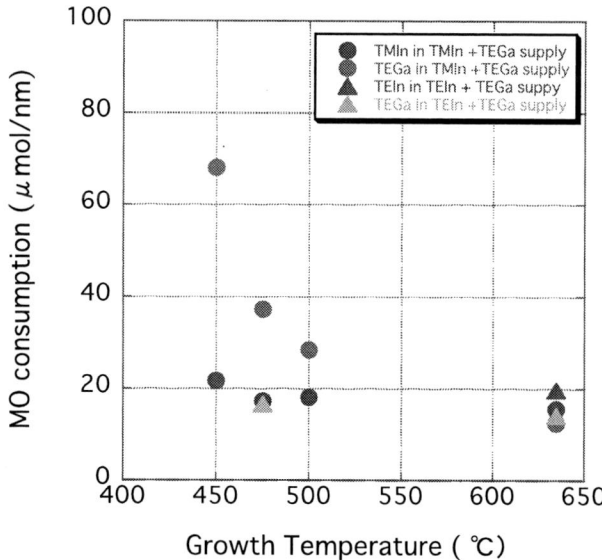

Figure 1. Calculated consumption of metal organic materials dependence of un-InGaAs growth temperature.

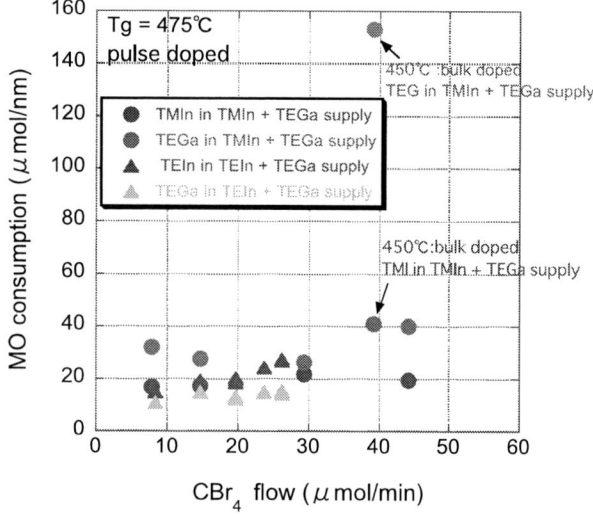

Figure 2. Metalorganic material consumption as a function of CBr_4 flow.

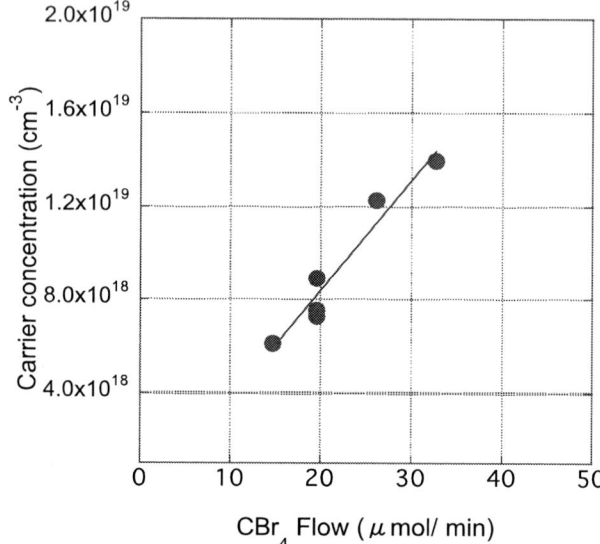

Figure 3. Relationship between carrier concentration and CBr4 flow in the pulse-doped p-InGaAs.

IV. SAMMMARY

We have investigated differences of the growth mechanism between ethyl- and methyl-base growth from the aspects of In incorporation ratio, effects of CBr_4 doping, and doping sequence in the C-dope InGaAs growth.

It was found that Good In composition controllability against the In supply ratio was obtained by the TEIn use instead of TMIn. The pulse doping of CBr_4 was effectively improved material efficiency or both methl- and ethyl-base materials. Finally, carrier concentration as high as 1.4×10^{19} cm^{-3} was obtained.

These results indicate that the use of ethyl-base materials and pulse doping are promising ways to provide high performance InP/InGaAs HBTs.

ACKNOWLEDGMENT

A part of this research is supported by NEDO (New Energy and Industrial Technology Development Organization).

REFERENCES

[1] Hiroshi Ito, Jpn. J. Appl. Vol. 35 (1996) pp. L1155-L1157, Part2, No. 9B, 15 September 1996.

[2] N.Watanabe, S.Yamahata, and T. Kobayashi, Journal of Crystal Growth, Vol. 200, Issue 3-4, pp. 599-602, December 1999.

[3] N. Watanabe, M. Uchida, H. Yokohama, G. Araki, : Ext. Abtr. (11th GAAG Symposium - Munich 2003); European Microwave week 2003.

TuD2-1 (Invited)
11:00 - 11:30

High-speed directly modulated laser for applications beyond 100GbE

Wataru Kobayashi, Takeshi Fujisawa, Toshio Ito, Takayuki Yamanaka,
Yasuo Shibata, Takashi Tadokoro[*], and Hiroaki Sanjoh
NTT Photonics Laboratories, NTT Corporation, Kanagawa, Japan

Abstract—Recent results are reported for a 1.3-μm InGaAlAs directly modulated laser (DML) with a ridge waveguide structure. We realized a 3-dB-down electrical-to-optical (E/O) response of 34 GHz for a 150-μm DML. Clear eye openings were obtained at 43 Gb/s up to 60°C and at 50 Gb/s at 25°C.

Keywords—Direct modulation; ridge waveguide; multiple quantum-well distributed feedback laser

I. INTRODUCTION

The rapid increase in data traffic is continuing to accelerate as Ethernet technology evolves on the Internet. This continuous growth in data traffic has made it necessary to increase the Ethernet data rate thus generating the need for higher speed light sources for optical network systems. To meet this demand, 100Gbit Ethernet (100GbE) was standardized in 2010 [1] and 25 Gb/s x 4 channels was employed in the 1.3-μm wavelength band. However, the continuous growth of data traffic will inevitably require a higher data rate with a bandwidth wider than 100GbE for future Ethernet systems.

A directly modulated laser (DML) is an attractive candidate that can be operated at a bit rate of more than 25 Gb/s. The advantages of the DML are that it is cost-effective, has a low power consumption, and is simple and easy to drive. Figure 1 summarizes previous reports on 1.3-μm DMLs designed for high-speed operation [2-9]. As shown in the figure, DML performance was improved in two ways, namely both high temperature and high speed operation were achieved. The purpose of the increased operating temperature is to reduce the power consumption needed for a thermo-electric cooler by realizing uncooled or semi-cooled operation. However, it is difficult to realize a high operating temperature and high-speed operation simultaneously. Generally, frequency relaxation oscillation (*fr*) limits the DML speed, and shortening the laser cavity is considered an effective way of improving *fr*. But when the cavity is short, the chip resistance increases, and this leads to the likelihood of self-heating. For this reason, when the cavity is short, *fr* levels off at a high injection current and high operating temperature.

In this study, we employed an InGaAlAs multiple-quantum-well (MQW) material system as used in previous work [2-9]. We employed a ridge waveguide structure, which can be realized with a simple fabrication process and that is suitable for multi-lane integration. This paper summarizes the

results we obtained for our 1.3-μm DML operating at over 25 Gb/s.

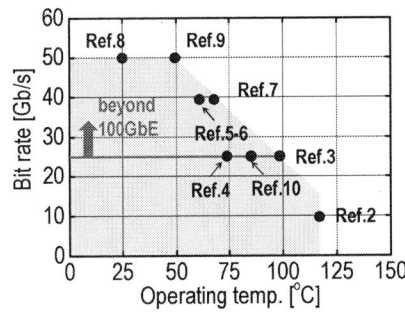

Fig. 1. Recent reports on high-speed 1.3-μm DML.

II. DEVICE STRUCTURE AND STATIC CHARACTERISTICS

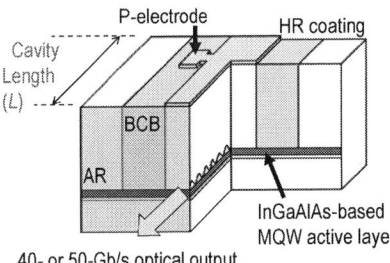

Fig. 2. Schematic of fabricated DML.

Figure 2 is a schematic of the fabricated InGaAlAs DML. The MQW consists of 8 QWs and 9 barriers with thicknesses of 6 and 10 nm, respectively. The cavity length (*L*) was set at 200 μm for 25-Gb/s operation [10]. To achieve operation at over 25 Gb/s, we shortened the cavity to increase *fr*. But too short a cavity cannot provide high temperature operation as described above. We carefully designed the cavity to achieve high-speed operation at high temperature. For these reasons, we set the cavity length at 150 μm. The coupling constant of the grating (κ) for this cavity was designed to be 220 cm^{-1}. In addition, a ridge waveguide structure was buried in benzocyclobutene (BCB) to reduce the capacitance of the metal pad region. The front and rear facets of the DML were coated with AR and HR films.

Figure 3 shows typical light power current (L-I) characteristics at 25°C for 150- and 200-μm DMLs. The

---

[*] Currently with the Department of Electrical and Electronic Engineering, Tokyo Denki University, Tokyo, Japan.

978-1-4673-6130-9/13 $31.00 © 2013 IEEE

TuD2-1 (Invited)
11:00 - 11:30

threshold currents (I_{th}) are 6.6 and 7.8 mA, respectively. As indicated in Fig. 3, when L is 150 μm, I_{th} is reduced and the linearity of the L-I curve is maintained at an LD (I_{LD}) injection current of 60 mA.

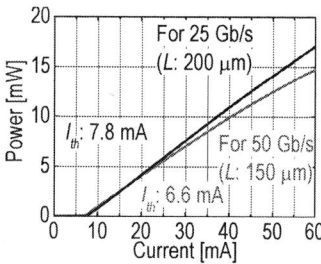

Fig. 3. L-I characteristics of 150- and 200-μm DMLs.

Figure 4 shows the L dependence of fr versus the square root of (I-I_{th}). The figure shows that as L become shorter, the fr slope becomes larger. From the L-I linearity, which is shown in Fig. 3, we estimated a maximum fr of about 26 GHz for a 150-μm DML [8].

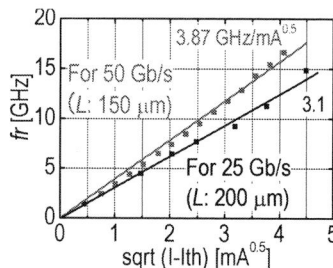

Fig. 4. L dependence of fr versus (I-Ith)$^{0.5}$.

III. MODULATION PERFORMANCE

To perform a high-speed experiment using these DMLs, we assembled butterfly modules whose RF input was a K connector.

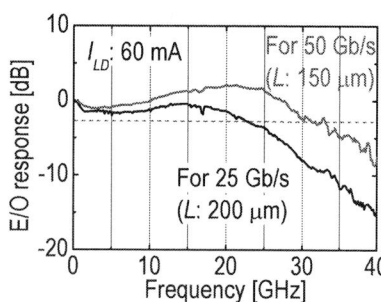

Fig. 5. E/O response for 150- and 200-μm DMLs.

Figure 5 shows the measured small signal electrical to optical (E/O) response of 150- and 200-μm DMLs. The red line indicates that the 3-dB-down frequency bandwidth (f_{3dB}). I_{LD} was 60 mA. As the cavity becomes shorter, the f_{3dB} value increases. But too short a cavity cannot realize an increased fr due to self-heating. We also measured the f_{3dB} of a 100-μm DML and the value was 28 GHz [11]. The f_{3dB} value for the 150-μm DML was 34 GHz, and we used this DML to perform high-speed modulation experiments.

43 Gb/s, NRZ, PRBS: 2^{31}-1, V_{pp}=2.2 V, I_{LD}: 55 mA (60°C)

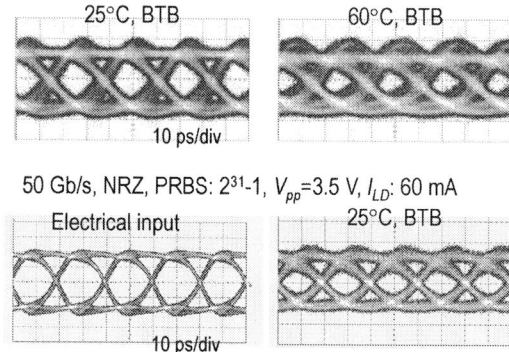

Fig. 6. 43-Gb/s and 50-Gb/s eye diagrams.

Figure 6 shows 43- and 50-Gb/s eye diagrams. The upper diagrams are for 43 Gb/s at 25 and 60°C [6]. The experimental conditions are shown in the figure. Clear eye opening was obtained up to 60°C. The lower eye diagrams were obtained for 50-Gb/s operation. A 50-Gb/s electrical input signal and optical waveform are depicted. We obtained a dynamic extinction ratio (DER) of 4.5 dB and a modulated output power of 5 dBm [8]. These results show that our DML is an attractive candidate for applications to 100GbE light sources and beyond.

IV. CONCLUSION

We have reported our recent results for a 1.3-μm InGaAlAs DML with a ridge waveguide structure designed to operate at over 25 Gb/s. By shortening the cavity from 200 μm for 25 Gb/s operation to 150 μm, we achieved an f_{3dB} value of 34 GHz and clear eye openings at 43 and 50 Gb/s. These results indicate that an InGaAlAs DML with a ridge waveguide structure is a promising candidate for applications beyond 100GbE.

REFERENCES

[1]   http://www.ieee802.org/3/ba
[2]   K. Takagi et al., Photon. Technol. Lett, 16, 2415-2417, (2004).
[3]   K. Adachi et al., CLEO'10, CME4, (2010).
[4]   Y. Yamasaki et al., ISLC'12, TuB2, (2012).
[5]   K. Nakahara et al., Photon. Technol. Lett, 19, 1436-1438, (2007).
[6]   T. Tadokoro et al., J. Lightwave Technol, 30, 2520-2524, (2012).
[7]   T. Simoyama et al., OFC' 2011, OWD3, (2011).
[8]   W. Kobayashi et al., ISLC'2012, TuB1, (2012).
[9]   T. Simoyama et al., ECOC'2012, P2.11, (2012).
[10] T. Tadokoro et al., Photon. Technol. Lett, 21, 1154-1156, (2009).
[11] W. Kobayashi et al., J. Sel. Topics Quantum Electron, accepted for publication.

978-1-4673-6130-9/13 $31.00 © 2013 IEEE

TuD2-2 (Oral)
11:30 - 11:45

IPRM2013, May 19 - 23, 2013, Kobe, Japan
The 25th International Conference on Indium Phosphide and Related Materials

Simultaneous 40-Gbps Direct Modulation of 1.3-µm Wavelength AlGaInAs Distributed-Reflector Laser Arrays on Semi-Insulating InP Substrate

Manabu Matsuda, Ayahito Uetake, Takasi Simoyama, Shigekazu Okumura,
Kazumasa Takabayashi, Mitsuru Ekawa, and Tsuyoshi Yamamoto

Fujitsu Laboratories Ltd.
10–1 Morinosato-Wakamiya, Atsugi, Japan
matsuda.manabu@jp.fujitsu.com

Abstract—40-Gbps direct modulation of 1.3-µm wavelength AlGaInAs distributed-reflector laser arrays on semi-insulating InP substrate are investigated. Clear eye-opening is demonstrated under simultaneous operation of two lasers.

Keywords—DR laser; laser array; direct modulation

I. INTRODUCTION

The explosive increase in data traffic requires next-generation high-speed data transmissions. Since directly modulated lasers (DMLs) have advantages in cost effectiveness and low power consumption, increase of their modulation speed have been actively studied [1-8]. High-speed direct modulation up to 50 Gbps have been reported [3, 7]. To operate a DML at high-speed under low driving current, it is essential to make the cavity length short. It is, however, difficult to fabricate a laser with a short active region of less than 150 µm only by cleaving. To suppress the increase in threshold gain is also important in the short cavity laser. For these purposes, we introduced the concept of distributed-reflector (DR) laser [9] and have demonstrated low-driving-current high-speed direct modulation in 25, 40, and 50 Gbps [2, 6-8].

From 100-Gbps Ethernet, multiple lasers are used in a transmitter. So, making integrated laser arrays is also highly required to realize compact modules. In the previous work, we have reported 25-Gbps direct modulation of 4-wavelength AlGaInAs DR laser arrays with 4 wavelengths on LAN-WDM grid on semi-insulating (SI) InP substrates [8]. In this work, we investigated 40-Gbps direct modulation of 1.3-µm wavelength DR laser arrays for future expansion of network capacity by higher modulation speed per channel.

II. DEVICE STRUCTURE

Figure 1 shows the schematic view of the fabricated DR laser array on SI-InP substrate. The gratings with different periods were formed and the coupling coefficient of the grating is set at 210 cm^{-1} to obtain sufficient optical feedback in the short active region. The MQW active layers consist of compressively strained 12 AlGaInAs wells. The peak wavelength of photoluminescence was 1304 nm. The DBR mirror regions have lattice-matched AlGaInAs passive

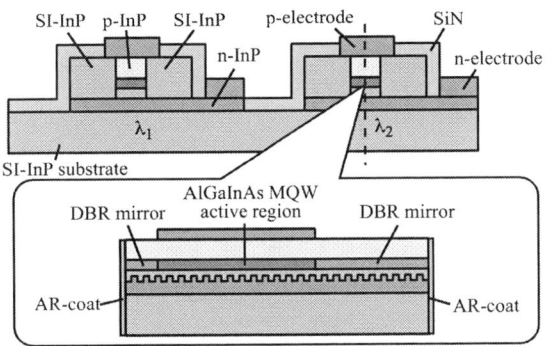

Figure 1. Schematic view of fabricated DR laser array.

waveguides. We set the length of the DFB active region as short as 125 µm. The 100-µm-long rear DBR mirror acts as high-reflective mirror. The 25-µm-long front DBR mirror introduces moderate reflectivity to increase the optical feedback in the short-cavity laser. The mesa stripe including the active region and the DBR mirrors was defined by dry-etching, and was buried with Fe-doped InP to form the semi-insulating buried-heterostructure (SI-BH). The width of the active layers was around 1.4 µm. To separate each laser electrically, coplanar electrodes were formed. The spacing between the lasers was set to 250 µm. The both facets of the laser arrays were covered with AR coatings.

III. DEVICE CHARACTERISTICS

Figure 2 shows the CW light-current characteristics of two different wavelength lasers in a fabricated laser array at various temperatures under simultaneous operation conditions. During the characterization of one laser in the array, the other laser was biased with the injection current of 50 mA. The threshold currents of the two lasers were 5.4 and 5.1 mA at 25°C, respectively. Figure 3 shows the lasing spectra at 25°C under simultaneous operation, and both lasers were biased at 50 mA. Lasing wavelengths were 1303.1 and 1307.7 nm, respectively. The spacing between two wavelengths was about 400 GHz, that is the spacing of LAN-WDM grid. These lasers have almost the same spectral shape, which indicates the high single-mode yield of the DR laser structure. The −3dB bandwidths of small-signal EO-response of these lasers biased

978-1-4673-6130-9/13 $31.00 © 2013 IEEE

TuD2-2 (Oral)
11:30 - 11:45

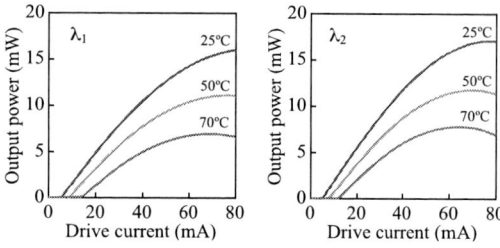

Figure 2.   Light-current characteristics of two different wavelength lasers under simultaneous operation.

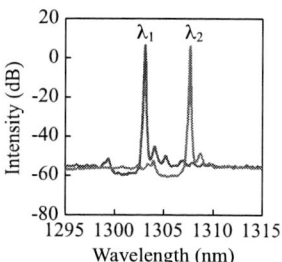

Figure 3.   Lasing spectra of two different wavelength lasers at 25°C, 50mA under simultaneous operation.

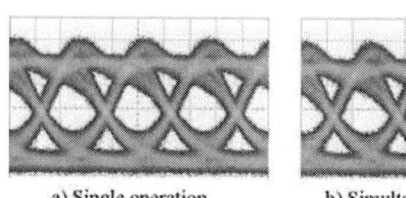

a) Single operation   b) Simultaneous operation

Figure 4.   40-Gbps eye diagrams of λ_1 laser under single and simultaneous operation at 25°C.

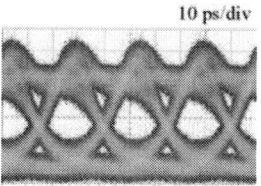

Figure 5.   40-Gbps eye diagrams of λ_1 laser after 10-km SMF transmission under simultaneous operation at 25°C.

at 100 mA were about 30 GHz at 25°C. This shows these lasers have a potential of high-speed operation.

IV.   MODULATION CHARACTERISTICS

We carried out 40-Gbps direct modulation using the laser array. We used non-return-to-zero (NRZ) signals and set the word length of the pseudo random bit sequence (PRBS) to $2^{31}-1$. We applied modulation signals with the swing voltages of 3 V peak to peak. The center bias currents were set to 56.8 mA to achieve the dynamic extinction ratio of 5 dB. Figure 4 shows 40-Gbps eye-diagrams of λ_1 laser in the array at 25°C under single operation (left) of λ_1 laser and simultaneous operation (right) of both lasers. In the case of simultaneous operation, both lasers were applied the same bias current and modulation voltage. We introduced a different signal phase between the electrical signals applied to λ_1 laser and λ_2 laser. We obtained clearly opened eye patterns even if both lasers were modulated simultaneously. This shows the influence of crosstalk between these two lasers is negligible. We also performed SMF transmission experiments under the same driving condition. Figure 5 shows eye-diagrams of after 10-km transmission of λ_1 laser in the array at 25°C under simultaneous operation of both lasers. We confirmed eye-openings after 10-km transmission.

V.   CONCLUSION

We demonstrated 40 Gbps direct modulation of 1.3-μm wavelength AlGaInAs DR laser arrays. Clear eye-opening was observed when two lasers were modulated simultaneously. We also showed eye-opening after 10-km transmission under 40-Gbps direct modulation of two lasers simultaneously. These results indicate that the 1.3-μm AlGaInAs DR laser arrays have a potential as light sources for over-100-Gbps Ethernet. High-

speed performance of these DR laser arrays will be improved by optimization of the laser structure.

REFERENCES

[1]   K. Adachi, K. Shinoda, T. Kitatani, D. Kawamura, T. Sugawara, and S. Tsuji, " 40-Gb/s/ch operation of 1.3-μm four-wavelength lens-integrated surface-emitting DFB laser array," International Semiconductor Laser Conference 2012 (ISLC 2012), Paper TuB5, 2012.

[2]   T. Simoyama, M. Matsuda, S. Okumura, A. Uetake, M. Ekawa, abd T. Yamamoto, "40-Gbps transmission using direct modulation of 1.3-μm AlGaInAs MQW distributed-reflector lasers up to 70°C," OFC/NFOEC 2011, Paper OWD3, 2011.

[3]   W. Kobayashi, T. Tadokoro, T. Ito, T. Fujisawa, T. Yamanaka, Y. Shibata, and M. Kohtoku, " High-speed operation at 50 Gb/s and 60-km SMF transmission with 1.3-mm InGaAlAs-based DML," International Semiconductor Laser Conference 2012 (ISLC 2012), Paper TuB1, 2012.

[4]   G. Sakiano, T. Takiguchi, Y. Hokama, T. Nagira, H. Yamaguchi, E. Ishimura, A.Sugitatsu and T. Shimura, "25.8Gbps direct modulation AlGaInAs DFB lasers with Ru-doped InP buried heterostructure for 70°C operation," OFC/NFOEC 2012, Paper OTh3F.3, 2012.

[5]   Y. Yamasaki, N. Kaida, T. Takeuchi, T. Hasegawa, N. Okada, K. Akiyama, G. Chifune, Y. Onishi, K. Uesaka, N. Ikoma, T. Fujii, and T. Nakabayashi, "High reliability 1.3-μm buried heterostructure AlGaInAs-MQW DFB laser operated at 28-Gbit/s direct modulation," International Semiconductor Laser Conference 2012 (ISLC 2012), Paper TuB2, 2012.

[6]   M. Matsuda, T. Simoyama, A. Uetake, S. Okumura, M. Ekawa and T. Yamamoto, "Uncooled, low-driving-current 25.8 Gbit/s direct modulation using 1.3 mm AlGaInAs MQW distributed-reflector lasers," Electron. Lett., vol. 48, no. 8, pp. 450-453, April 2012.

[7]   T. Simoyama, M. Matsuda, S. Okumura, A.Uetake, M. Ekawa, and T. Yamamoto, "50-Gbps direct modulation using 1.3-μm AlGaInAs MQW distribute-reflector lasers," European Conference of Optoelectronics and Communications 2012 (ECOC 2012), P2.11, 2012.

[8]   T. Simoyama, M. Matsuda, S. Okumura, A. Uetake, M. Ekawa, and T. Yamamoto, "4-Wavelength 25.8-Gbps directly modulated laser array of 1.3-μm AlGaInAs distributed-reflector lasers," International Semiconductor Laser Conference 2012 (ISLC 2012), Paper TuB3, 2012.

[9]   J.-I. Shim, K. Komori, S. Arai, I. Arima, Y. Suematsu, and R. Somchai, "Lasing characteristics of 1.5 μm GaInAsP-InP SCH-BIG-DR lasers," IEEE J. Quantum Electron., vol 27, pp. 1736–1745, 1991.

978-1-4673-6130-9/13 $31.00 © 2013 IEEE

TuD2-3 (Oral)
11:45 - 12:00

Static and dynamic characteristics of InAs/AlGaInAs /InP quantum dot lasers operating at 1550 nm

Johann Peter Reithmaier[1], Vitalii Ivanov[1], Vitalii Sichkovskyi[1], Christian Gilfert[1], Anna Rippien[1], Florian Schnabel[1], David Gready[2] and Gadi Eisenstein[2]

[1]Institute of Nanostructure Technologies and Analytics (INA), CINSaT, University of Kassel, Germany
[2]Department of Electrical Engineering, Technion, Haifa 32000, Israel
jpreith@ina.uni-kassel.de

Abstract—With high modal gain InAs/AlGaInAs/InP quantum dot laser material short cavity ridge waveguide lasers were fabricated with cavity lengths down to 275 µm. These devices show new record values in direct digital signal modulation at 22 GBit/s. In addition also strong improvement are expected from this laser material in the static properties, in particular on the linewidth due to the suppression of the linewidth enhancement factor. First distributed feedback lasers on similar quantum dot laser material were processed and preliminary linewidth measurements indicate a significant linewidth reduction. Quantitative investigations are under way and will be presented at the conference.

Keywords—quantum dot laser, optical communication, high frequency modulation, narrow linewidth, distributed feedback laser

I. INTRODUCTION

Recent advances in epitaxial growth of InP based quantum dot (QD) laser material led to significant advances in laser performance including record modal gain values of more than 10 cm^{-1} per dot layer [1]. These superb characteristics are due to a high degree of confinement in the slightly elongated dots and due to low residual losses. Using a spatially resolved model [2], we optimized the epitaxial layer structure for the fabrication of short ridge waveguide lasers. On first 350 µm long devices a 3 dB bandwidth of about 8 GHz and up to 20 GBit/s digital modulation could be obtained [3]. In this paper we report on shorter devices, which show new record modulation speeds up to about 23.5 GBit/s with a clear open eye at 22 GBit/s.

With QD lasers it is theoretically possible to fully suppress the linewidth enhancement factor, which would have a big advantage for the suppression of the wavelength chirp for direct modulation or for ultra-narrow linewidth lasers as local oscillators in coherent communication. Recently narrow linewidth emission below 200 kHz could be shown [4]. For this purpose distributed feedback lasers were fabricated on very similar QD laser material to investigate these properties on our more symmetric and high modal gain QD material. To allow the fabrication of lateral gratings with well-defined vertical geometry, GaInAsP etch stop layers were integrated into the layer structure. First high-resolution linewidth measurements indicate already very narrow emission linewidths. A quantitative analysis will be worked out in the next months and will be discussed at the conference.

II. DEVICE STRUCTURE AND FABRICATION

The laser structures used are schematically shown in Fig. 1. The growth parameters are similar to previous work [5] but the waveguide layers were reduced from 200 to 100 nm to reduce significantly the carrier transport time. Details about the epitaxy are described in the literature [2]. For distributed feedback (DFB) lasers the layer design was modified by integration of GaInAsP etch stop layers, which allow the fabrication of well-defined lateral gratings after ridge waveguide processing. The slight asymmetry in the waveguide was chosen to get an improved overlap of the guided wave with the lateral grating. In the inset of Fig. 4, an image of a DFB laser is shown with a first order grating before planarization and contact metallization. The lasers for high-speed modulation were bonded onto a microwave test fixture, which is fed by a microwave probe. The DFB lasers were mounted on similar sub-mounts but are only contacted for continuous wave operation.

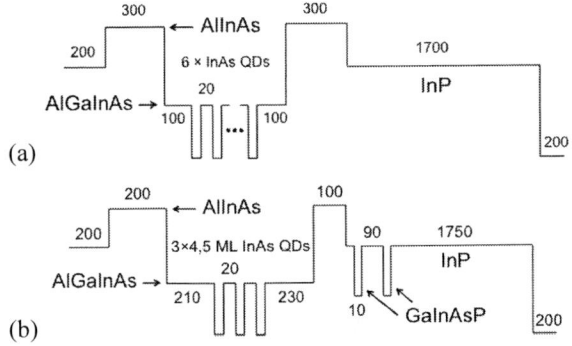

Fig. 1. Schematic layer designs for high speed (a) and DFB lasers (b).

III. RESULTS

The lasers were tested for both their static and dynamic characteristics. All the investigated lasers show stable ground state lasing at 1.55 µm or above. The static characteristics are shown in Fig. 1, which describes temperature dependent P-I curves. The CW threshold current changes from 26 mA at 16°C where the maximum power is around 7 mW to 45 mA at 50°C where the power drops to 1.5 mW.

Dynamical properties were obtained in a series of modulation experiments at various temperatures and for different operating conditions. Small signal bias dependent small signal responses at 16 °C are shown in Fig. 3. The maximum observable bandwidth is 8.3 GHz (16 °C) and drops to 6.5 GHz for 45°C.

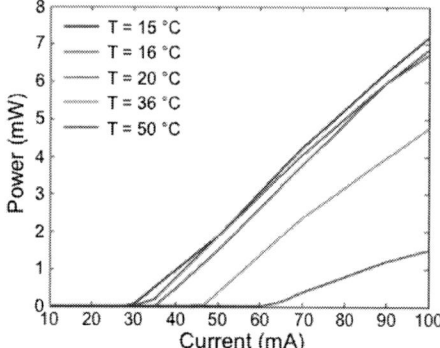

Fig. 2. Temperature dependent L-I curves for a 275 µm long RWG QD laser.

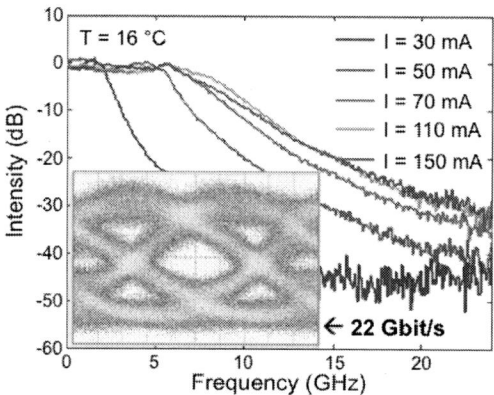

Fig. 3. Small signal response of 275-µm long RWG QD laser for different bias current. The inset shows the eye diagram for 22 GBit/s digital modulation.

These bit rates seem inconsistent with the relatively low small signal response. This is a rather common observation in QD lasers also at 1300 nm [6]. It is observable in lasers where the gain is large but the small signal response is limited by a nonlinear gain compression factor. This is indeed the case for the responses shown in Fig. 2. This effect can occur in principle in any laser but the conditions needed are possible only in some QD lasers [7].

In comparison to previously characterized 350 µm long the devices, the thermal effects are more pronounced in the short laser and therefore the shorter photon lifetime does not yield a wider bandwidth. Nevertheless, the large signal response is significantly improved and is now limited to 23.5 GBit/s with a clear open eye at 22 GBit/s as shown in the inset of Fig. 3.

DFB lasers of different grating periods were characterized in cw operation at 20 °C. Lasers with 830 µm cavity lengths without facet coatings exhibit threshold currents of about 45 mA with an output power of 6 mW per facet.

The spectral emission spectra of two lasers with different grating periods are shown in Fig. 4. A side mode suppression of more than > 50 dB can be observed. First high-resolution linewidth measurements based on a heterodyne technique indicates significant linewidth narrowing in comparison to

conventional DFB lasers. Quantitative investigations are under way and will be discussed at the conference.

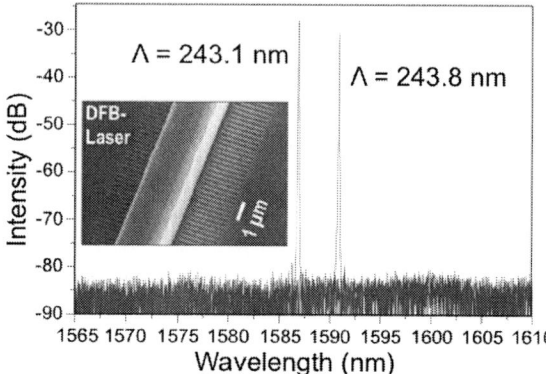

Fig. 4. Spectra of two DFB lasers with two different grating periods as labeled showing SMSRs of > 50 dB. The inset shows an SEM picture of the device after etching of first order lateral gratings.

IV. CONCLUSIONS

To conclude, we have described static and dynamic properties of InAs/InP QD lasers operating at 1550 nm. Due to the high modal gain of 70 cm^{-1}. 275 µm long devices are lasing in ground state up to 50 °C. They emit 6.5 mW under CW conditions at room temperature and more than 1 mW at 50 °C. The lasers exhibit a small signal bandwidth around 8 GHz and record digital modulation speeds of up to 22 GBit/s. These fast responses are the result of good carrier confinement and a well-designed layer structure based on an advanced spatially resolved dynamical model.

Based on a modified layer structure with integrated etch stop layers, QD DFB lasers were realized showing SMSRs > 50 dB and well controlled emission wavelengths. High-resolution linewidth measurements are in work and will be presented at the conference. The work was partially supported by the Israeli Science Foundation. The financial support by the BMBF through project *Monolop* is gratefully acknowledged.

REFERENCES

[1] C. Gilfert, V. Ivanov, N. Oehl, M. Yacob, and J. Reithmaier,
Applied Physics Letters 98, 201102 (2011).

[2] D. Gready and G. Eisenstein, to be published IEEE JSTQE, special issue on semiconductor lasers, 2013.

[3] V. Ivanov, C. Gilfert, F. Schnabel, A. Rippien, J.P. Reithmaier, D. Gready, G. Eisenstein, C. Bornholdt, ISLC, San Diego, 2012, postdeadline paper, DOI: 10.1109/ISLC.2012.6348317

[4] Z.G. Lu, P.J. Poole, J.R. Liu, P.J. Barrios, Z.J. Jiao, G. Pakulski, D. Poitras, D. Goodchild, B. Rioux andA.J. SpringThorpe, Electron. Lett. 47 (2011), doi: 10.1049/el.2011.0946.

[5] D. Gready, G. Eisenstein, C. Gilfert, V. Ivanov, J.P. Reithmaier, IEEE Photon. Technol. Lett. 24, 809 (2012).

[6] Y. Tanaka, M. Ishida, K. Takada, Y. Maeda, T. Akiyama,
T. Yamamoto, H.-Z. Song, M. Yamaguchi, Y. Nakata, K. Nishi,
M. Sugawara, and Y. Arakawa, LEOS Annual Meeting, 668 (2009)

[7] D. Gready and G. Eisenstein, Photonic Society annual meeting 2012.

TuD2-4 (Oral)
12:00 - 12:15

Frequency-resolved optical gating measurements of sub-ps pulses from InAs/InP quantum dash based mode-locked lasers

C. Calò[1], H. Schmeckebier[2], K. Merghem[1], R. Rosales[1,2], F. Lelarge[3], A. Martinez[1], D. Bimberg[2] & A. Ramdane[1]

[1]CNRS Laboratory for Photonics and Nanostructures, 91460 Marcoussis, France
[2]Institut für Festkörperphysik, Technische Universität Berlin, Hardenbergstr. 36, 10623 Berlin, Germany
[3]III-V Lab, a joint Laboratory of "Alcatel Lucent Bell Labs", "Thales Research & Technology" and "CEA-LETI",
Route de Nozay, 91460 Marcoussis, France

Corresponding author: cosimo.calo@lpn.cnrs.fr

Abstract—**Mode-locking of single-section Fabry-Pérot lasers based on InAs/InP quantum dashes is studied by second-harmonic generation frequency-resolved optical gating (SHG-FROG). The devices take advantage of an optimized epitaxial structure with modal gain of 50 cm^{-1} showing a broad and flat emission spectrum of width in excess of 14 nm. Self-starting pulses of a width down to 430 fs are observed after intracavity dispersion compensation using single-mode fiber (SMF).**

I. INTRODUCTION

Monolithic semiconductor mode-locked lasers (MLLs) are becoming very attractive for a wide range of applications, such as short pulse generation for optical time division multiplexing (OTDM), all-optical clock recovery, optical frequency comb generation for coherent orthogonal frequency division multiplexing (OFDM) or optical sampling [1]. Standard semiconductor passive MLLs include a reverse biased absorber section which locks the phases of the longitudinal modes by periodically creating a window of net gain upon absorber saturation by the optical field inside the laser cavity. Additionally mode-locking (ML) without the presence of a saturable absorber has been observed in single-section Fabry-Pérot lasers based on quantum well [2], quantum dash [3] and quantum dot [4] gain media. Although this regime is commonly attributed to the effect of nonlinear optical processes in the gain medium, such as four wave mixing (FWM), which cause phase correlation and equal frequency spacing between the longitudinal cavity modes, the phenomenon is still not fully understood. Recently Rosales *et al.* [3] showed that, even in the presence of phase correlation between the longitudinal modes, optical pulses can be observed at the output of single-section lasers only if the normal intracavity dispersion cumulated by the optical signal is compensated.

In this work we report to our knowledge on the first investigation of pulses generated by InAs/InP quantum dash based single-section MLLs using second-harmonic generation frequency-resolved optical gating (SHG-FROG) [5]. This technique, known to be very effective for the detailed investigation of semiconductor MLL pulses [6] is used here to study the dependence of the pulse profile of single-section MLLs as function of single-mode fiber (SMF) length and injection current.

II. LASER FABRICATION AND CHARACTERISTICS

The lasers were grown by Gas Source Molecular Beam Epitaxy (GSMBE) on an S-doped (001) InP substrate and consist of 9 layers of InAs quantum dashes separated by InGaAsP barriers (dashes-in-a-barrier design [7]). A modal gain of ~50 cm^{-1} and internal loss of ~18 cm^{-1} could be extracted from the measurement on broad area lasers processed from the grown structure. Single transverse mode buried ridge stripe lasers with ridge width of 1.5 μm and as cleaved facets were then fabricated. Devices were mounted on copper bases for thermal management purposes. Lasers with cavity lengths of 890 μm and 1820 μm were characterized using SHG-FROG. The lasers exhibit large emission spectra in excess of 14 nm full width at half maximum (FWHM) and narrow radio frequency linewidths on the order of few tens to hundreds of kHz over the injection current range from 100 to 400 mA, which implies low phase noise and good mode-locking characteristics.

III. FREQUENCY RESOLVED OPTICAL GATING MEASUREMENTS

SHG-FROG is a pulse characterization technique which is based on spectrally resolving the output of a SHG autocorrelator (SHG-AC), such that a frequency resolved convoluted intensity of the pulse as a function of time delay, known as spectrogram, is obtained (see inset of fig. 1). Extracting the intensity and phase profile from the spectrogram represents a two-dimensional phase retrieval problem, which can be solved by an iterative algorithm [5].

Fig. 1 shows the intensity and chirp of the pulse generated by a single-section QDash laser with cavity length L=890 μm for an injection current of 300 mA, after propagation through 66 m of SMF for compensating the intracavity dispersion. For such a fiber length and operating conditions the laser

TuD2-4 (Oral)
12:00 - 12:15

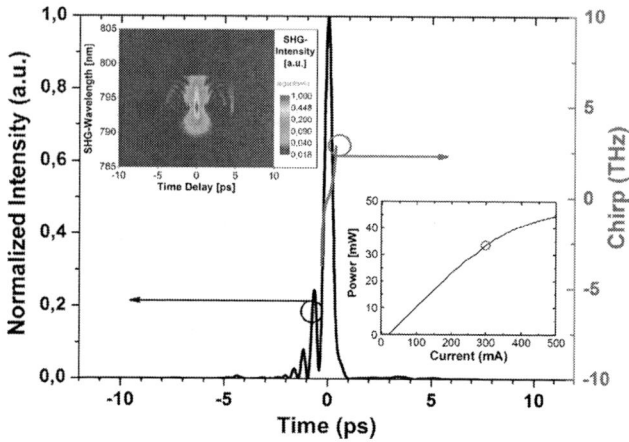

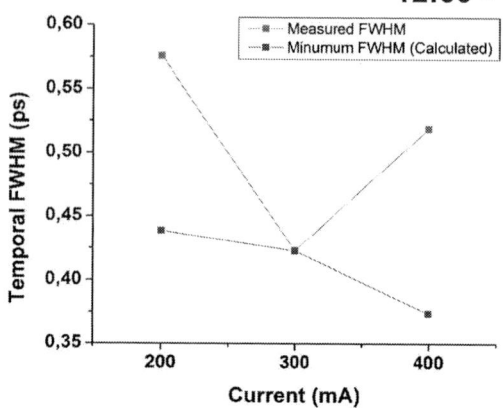

Figure 1. Intensity and chirp profile of pulse generated by a 890-μm-long single-section QDash MLL for an injection current of 300 mA, after dispersion compensation using 66 m of SMF. In the insets: (left) measured FROG spectrogram, (right) light-current characteristic of the laser.

Figure 2. Measured (red) and minimum FWHM duration (calculated by inducing dispersion to the retrieved pulses from the FROG measurements using Split-Step Fourier Method) (blue) of the pulses generated by a 890-μm-long single-section QDash MLL as a function of injection current for a SMF length of 66 m.

intracavity dispersion is fully compensated and a pulse duration (FWHM) of 423 fs and pulse peak power of about 100 mW (including setup losses) are measured. However, a few tens of meters deviation from the given fiber length induces severe pulse broadening and distortion of the pulse shape, due to the wide emission spectrum of the laser and the nonlinear chirp, not compensated by the SMF. Such broadening may affect the FROG measurement if the full pulse duration is larger than half of the pulse repetition interval, causing the spectrogram not to be zero at the edges, unlike what required by the FROG algorithm, and thus leading to large retrieval errors [6].

An analogous effect can be observed by keeping the length of SMF constant and varying the injection current. In fact, since the emission spectrum is red shifted as the current is increased due to thermal heating of the laser junction, because of the wavelength dependent dispersion coefficient of the SMF, the amount of group delay dispersion introduced by a fixed length of fiber will vary as a function of injection current (see fig. 2). It is interesting to note, however, that the minimum pulse FWHM achievable by fully compensating the intracavity dispersion is expected to decrease as a function of current. This fact is attributed to the increase with current of the optical spectrum FWHM, due to band filling of QDash ensembles, and to the higher efficiency of the nonlinear optical effects leading to ML in single-section semiconductor lasers.

Such behavior at high injection currents is observed also for the laser with cavity length L=1820 μm and represents an advantage over conventional two-section lasers, which are limited by quantum confined Stark effect (QCSE) at high currents and large reverse biases [6].

IV. CONCLUSIONS

Frequency-resolved optical gating is a valuable tool to analyze the performances of monolithic semiconductor mode-locked lasers. Single-section InAs/InP QDash MLLs, in particular, show very interesting properties in terms of pulse width (FWHM ~430 fs) and pulse peak power (~100 mW) once the laser intracavity dispersion is fully compensated using a suitable length of SMF.

ACKNOWLEDGMENT

This work has been supported by the ITN PROPHET, SFB 787 of DFG and ANR TELDOT.

REFERENCES

[1] E. A. Avrutin, J. H. Marsh, and E. L. Portnoi, "Monolithic and multi-gigahertz mode-locked semiconductor lasers: constructions, experiments, models and applications," IEE Proc.-Optoelectron. 147(4), pp. 251–278, 2000.

[2] Y. Nomura, et al., "Mode locking in Fabry-Perot semiconductor lasers," Phys. Rev. A 65, 043807, 2002.

[3] R. Rosales, et al., "High performance mode locking characteristics of single section quantum dash lasers," Opt. Express 20(8), pp. 8649–8657, 2012.

[4] Z. G. Lu, et al., "312 fs pulse generation from a passive C-band InAs/InP quantum dot mode-locked laser," Opt. Express 16(14), pp. 10835–10840, 2008.

[5] R. Trebino, Frequency-Resolved Optical Gating: The Measurement of Ultrashort Laser Pulses, Kluver Academic, 2002.

[6] H. Schmeckebier, G. Fiol, C. Meuer, D. Arsenijević, and D. Bimberg, "Complete pulse characterization of quantum dot mode-locked lasers suitable for optical communication up to 160 Gbit/s," Opt. Express 18(4), pp. 3415–3425, 2010.

[7] F. Lelarge, et al., "Recent advances on InAs/InP quantum dash based semiconductor lasers and optical amplifiers operating at 1.55 μm," IEEE J. Sel. Top. Quantum. Electron. 13(1), pp. 111–124, 2007.

1480nm InGaAsP LOC Broad-Area-Lasers with >18W Pulsed Output Power at 20°C

D. Fendler, M. Möhrle, M. Spiegelberg, W. Rehbein, W. Passenberg and N. Grote

Fraunhofer Institute for Telecommunications, Heinrich-Hertz-Institute

Berlin, Germany

david.fendler@hhi.fraunhofer.de

Abstract—1480nm InGaAsP large optical cavity broad-area laser diodes were developed and optimized for pulsed operation showing optical output power of >18W at 20°C. Furthermore the μs pulse duration regime was investigated with respect to power saturation and self-heating at elevated operating currents.

Keywords— *high power, NIR, InGaAsP, semiconductor laser*

I. INTRODUCTION

High-power broad-area (BA) laser diodes in the NIR around 1500 nm (eye safe) become more and more interesting for medical and security applications, pumping of crystals to realise solid state lasers, and environmental applications like trace gas detection and LIDAR. Pulsed output power >10W and a narrow optical far-field for efficient coupling to fibers or efficient lens focusing is a crucial requirement in all these applications. With BA laser diodes using a conventional MQW-active structure it is not possible to fulfill these requirements. To achieve both high optical pulse output power and narrow optical far-field use of large-optical cavity (LOC) laser structure has been proposed [1]. In this paper 1480 nm InGaAsP LOC BA-lasers have been fabricated and optimised with respect to maximum optical pulsed output power. More than 18 W of pulsed optical power has been achieved for a single emitter at room temperature. . The influence of cavity length, pulse width, repetition rate and temperature on the output power will be investigated.

II. INGAASP LOC BA DEVICE STRUCTURE

The laser diode consists of an InGaAsP-MQW-structure embedded in a LOC InGaAsP waveguide (Fig. 1a). The LOC design serves to decrease the confinement of the optical mode in the highly absorptive p-InP cladding and p-contact layer and to simultaneously increase the optical mode size in order to obtain a narrow optical far-field. Careful optimisation of the LOC-structure conducting extensive mode field simulations regarding the trade-off between optical confinement in the QW layers, the vertical far-field and the optical losses in the p-region have been carried out. Furthermore doping levels of the optical confinement layers where adjusted to reduce electron leakage into the p-region to increase internal quantum efficiency without a substantial increase of absorption losses. The p-doped InP-ridge structure is etched down to the p-side waveguide providing lateral index guiding of the optical mode. The p- and n-doping levels and the thickness of the p-InP

cladding layers have been optimised so as to accomplish low optical losses and low series resistance. The p-contact is Au-electroplated for efficient heat dissipation in upside-down configuration. The n-contact metallisation is deposited on the wafer backside after wafer thinning to about 150 μm. Significantly improved characteristics are obtained with this optimised LOC BA-structure (Fig. 1a). For high pulse power characterisation facet coated LOC BA laser devices were mounted p-side-down on a copper heat sink to guarantee efficient heat dissipation (Fig. 1b).

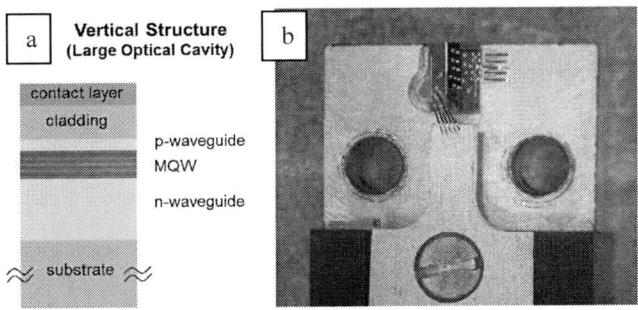

Figure 1. Vertical epitaxial LOC layer structure (a), laser diode mounted upside down on AlN-Submount and copper heat sink (b).

In Tab. 1 the determined basic laser parameters of the developed LOC BA-laser structure are compiled in comparison to a BA laser structure using an identical QW-structure but a thin n-waveguide layer, as used e.g. in common ridge-waveguide single mode transmitter lasers.

Table 1. Comparison of optimized 1480nm LOC BA laser width conventional laser structure (width: 50μm, length 1mm, as-cleaved facets).

	LOC BA-laser	Ref. conv. BA-laser
optical losses α / cm^{-1}	7.4	8.1
internal quantum effciency η_{in} / %	0.99	0.75
transparency current density J_{th} / kA/cm^2	0.5	0.8
threshold current I_{th} (L=1mm) / A	0.37	0.51
slope efficiency SLE (L=1mm) / W/A	0.27	0.19

To maximise the pulsed optical output power but maintain flexibility for coupling purposes 100 μm wide stripes where used. Optimisation of cavity-length and facet coating leads to

978-1-4673-6130-9/13 $31.00 © 2013 IEEE

further increase in output power. Best results were achieved for 2 mm long devices with a back-facet reflectivity of 90% and a front-facet reflectivity of 2 %. Measurements were made using a AV-Tech laser diode driver, a Tektronix oscilloscope triggered by the driver and a germanium photodetector with a calibrated integration sphere. The measurement setup and the laser diode driver where limited to pulse durations between 300 ns and 2 µs and repetition rates between 1 Hz and 1 kHz. The achieved maximum optical output power of >18 W at 20°C (fig.2) are among the best values published so far [2].

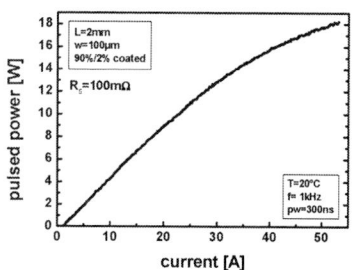

Figure 2. LI-curve of an upside-down mounted laser diode in pulsed operation

For high operating current over 15 A saturation of the output power can be observed even at pulse durations of only 300 ns. To analyse this behaviour and its impact on pulsed operation, LI-measurements up to 25 A have been performed.

III. INFLUENCE OF PULSE WIDTH ON OUTPUT POWER

Fig. 3 shows LI-curves for different pulse durations with a repetition rate of 1 kHz at 20°C to analyse the behaviour of the maximum pulsed output power of these devices in more detail. Identical threshold current and slope efficiency values obtained for all investigated pulse widths indicate that no noticeable thermal effects occur up to currents of some 15 A. The decrease of maximum optical output power with increasing pulse width is mainly attributed to self-heating because of joule heating [3] within the pulse duration. Self–heating tends to lead to considerably increased losses mainly because of increased auger recombination, optical gain reduction and leakage current at elevated temperatures [3],[4] which in turn lead to a decrease of the output power at elevated operating currents for this devices.

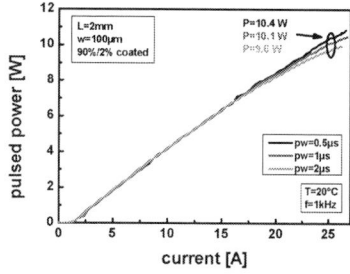

Figure 3. LI-curve of an upside-down mounted laser diode with a repetition rate of 1 kHz and pulse durations from 0.5 to 2 µs

---

This work was funded by the Land Berlin/Investitionsbank BerlinBB under the ProFIT program co-sponsored by the European Fund for Regional Development (EFRE).

The pulse width has also a significant influence on the temperature performance and spectrum of the lasers, characterized by a saturation of the output power and a redshift.

IV. INFLUENCE OF REPETITION RATE ON OUTPUT POWER

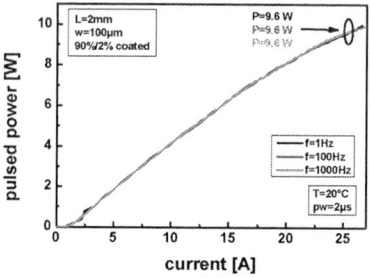

Figure 4. LI-curve of an upside-down mounted laser diode with repetition rates between 1 Hz and 1 kHz and a pulse duration of 2 µs

The curves depicted in Fig. 4 show that there is virtually no influence of the repetition rate on the maximum optical output power between 1 Hz and 1 kHz despite the drastically changing duty cycle. This demonstrates that the main self-heating effects occur during the short pulses. Because of the limitation of the driver we could only validate this for repetition rates of up to 1 kHz. For higher repetition rates the optical power decreases eventually leading to relatively moderate CW-output power levels of around 3 to 5 W because the duration between the pulses becomes too short for full heat dissipation.

V. CONCLUSION

We have demonstrated pulsed optical output power of >18 W at pulse widths of 300 ns and a repetition rate of 1 kHz at 20 °C obtained from a 100 µm wide InGaAsP LOC Broad-Area-Laser single emitter. The impact of self-heating effects on maximum output power for short pulses was studied in more detail. Optimisation of doping levels, stripe structure and series resistance will be undertaken to increase the pulsed output power and temperature stability even further. The results can be directly transferred to laser diodes designed for CW-applications.

REFERENCES

[1] H.F. Lockwood, H. Kressel, H.S. Sommers, F.Z. Hawrylo, "Efficient Large Optical Cavity Injection Lasers," Applied Physics Letters, vol. 17, pp. 499-& (1970)

[2] J.F. Boucher, J.J. Callahan, "Ultra-high-intensity 1550nm single junction pulsed laser diodes," Conference on the Laser Technology for Defense and Security VII, Orlando, FL, Apr 25-27 (2011)

[3] Piprek, J.; White, J.K.; SpringThorpe, A.J.; , "What limits the maximum output power of long-wavelength AlGaInAs/InP laser diodes?," Quantum Electronics, IEEE Journal of , vol.38, no.9, pp. 1253- 1259, Sep 2002

[4] Piprek, J.; Abraham, P.; Bowers, J.E.; , "Self-consistent analysis of high-temperature effects on strained-layer multiquantum-well InGaAsP-InP lasers," Quantum Electronics, IEEE Journal of , vol.36, no.3, pp.366-374, March 2000

TuD3-1 (Invited)
14:00 - 14:30

Monolithically Integrated Optical Link Using Photonic Crystal Laser and Photodetector

[1,2]Shinji Matsuo

[1]NTT Photonics Labs., [2]Nanophotonics Center
NTT Corporation
3-1 Morinosato-Wakamiya, Atsugi, Kanagawa 243-0198, Japan
e-mail address: matsuo.shinji@lab.ntt.co.jp

Abstract—We have demonstrated monolithic integration of photonic crystal (PhC) laser and photodetector. By employing an embedded active region structure, PhC laser exhibits ultra-low threshold current. Integrated optical link exhibits ultra-low operating energy of 28.5 fJ/bit.

Keywords— photonic crystal laser; nanocavity laser; optical interconnects; direct modulation

I. INTRODUCTION

Photonic crystal (PhC) nanocavity lasers have attracted much attention to construct the off-chip and on-chip interconnections for microprocessors. In these applications, extremely small energy cost is required for transmitter. Thus, directly modulated lasers with ultra-compact active region are essential to achieve both ultra-low operating energy and high-speed modulation. In this context, we have developed electrically-driven PhC nanocavity laser, in which active region is embedded with InP-based PhC line-defect waveguide that we call a lambda-scale embedded active-region PhC laser or LEAP laser [1,2]. This structure provides us to achieve the effective confinement both carrier and photon and the reduction of thermal resistance. Thus, we achieved ultra-low threshold current of 14 μA and continuous-wave lasing up to 95°C [3].

In this report, we describe monolithic integration of LEAP laser and photodetector (PD) for on-chip interconnect. For this photonic integrated circuit (PIC), the power consumption is a critical issue as several thousand lasers, photodetectors, and switches must be integrated on a single chip. Therefore, it is important to reduce the coupling loss between the laser and waveguide and the waveguide loss. In this context, we have proposed photonic network chip consisting of monolithically integrated InP-based devices, which is hybrid integrated with silicon CMOS [4]. This is because buried InP layer is transparent for lasing wavelength and, therefore, we can expect to construct a low-loss photonic network on CMOS chip.

II. DEVICE STRUCTURE AND FABRICATION

Figure 1(a) shows a schematic diagram of LEAP laser. The active region is embedded within the InP-based PhC line-defect waveguide. This structure allows us to greatly

Part of this work was supported by the New Energy and Industrial Technology Development Organization (NEDO).

reduce the thermal resistance and realize the strong confinement of both photons and carriers in the active region, because the thermal conductivity and bandgap of the InP layer are larger than those of the active region [1]. The output waveguide is placed in an offset position with respect to the line-defect waveguide including the active region to obtain effective coupling between the cavity and output waveguide [4].

Figure 1(b) shows a schematic diagram of PD. We used the laser active region as an absorption layer of PD. Thus, the PD also employs a PhC line-defect waveguide with an air-bridge structure to increase the optical confinement of the absorption layer.

To fabricate a lateral p-i-n current injection structure, we employ Zn diffusion and Si ion implantation, respectively, for the p- and n-type doping of an undoped InP layer. A fabrication process that employs impurity implantation and diffusion will result in greater performance repeatability and increased yield because it is the same fabrication technique that is used for silicon CMOS.

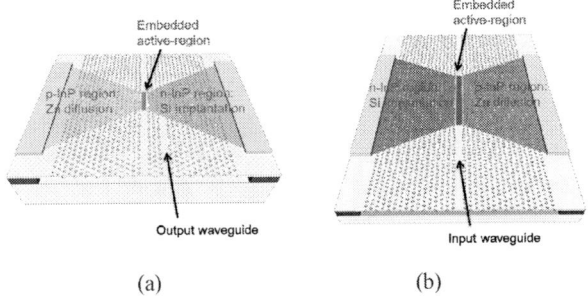

(a)   (b)

Figure 1. Schematic diagrams of (a) LEAP laser and (b) PD.

Figure 2(a) is an image of a fabricated monolithically integrated optical link consisting of LEAP laser, waveguide, and photodetector (PD). The waveguide between the laser and the PD is 3 μm wide and 500 μm long. A tapered waveguide is used to reduce the coupling loss between the line-defect waveguide and the 3-μm wide waveguide. The PD absorption region is 17 μm long.

A cross-sectional SEM image of fabricated LEAP laser is shown in Fig. 2(b). The InGaAlAs-based 3QW active

978-1-4673-6130-9/13 $31.00 © 2013 IEEE   115

TuD3-1 (Invited)
14:00 - 14:30

region is embedded within the InP-based PhC line-defect waveguide. The active volume is $3.4 \times 0.3 \times 0.15 \ \mu m^3$ (0.15 μm^3). As shown in this, we achieved flat-top surface and smooth side-wall of air holes. In addition, we used an InAlAs sacrificial layer with a larger band gap than InP to reduce the leakage current through the sacrificial layer and substrate.

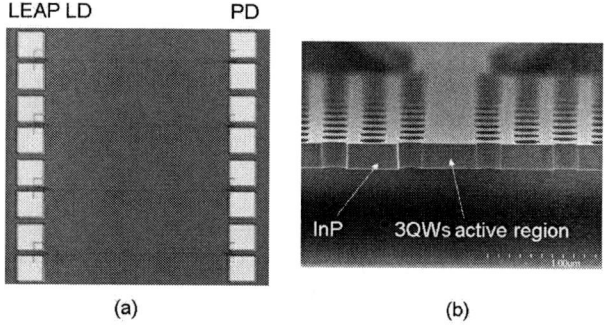

Figure 2. (a) Image of fabricated device. (b) Cross-sectional SEM image of LEAP laser.

III. DEVICE CHARACTERISTICS [5]

Figure 3 shows the photocurrent of the PD and conversion efficiency as a function of the injection current into the LEAP laser. We defined the conversion efficiency as the ratio of the photocurrent to the injection current, I_{PD}/I_{LD}. The device was operated in a RT-CW condition and the reverse bias voltage of the PD was -1 V. The lasing wavelength was 1540 nm. As shown in this figure, the laser threshold current was 22 μA and the maximum conversion efficiency was 8.5% when the injection current of the LEAP laser was 75 μA.

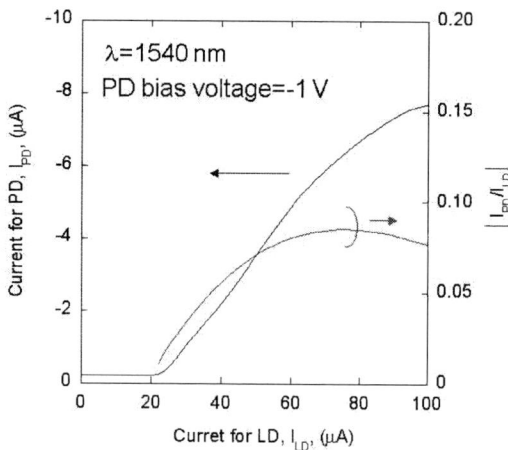

Figure 3. Static characteristic of fabricated optical link.

Figure 4 shows the signal responses of the PD when the laser was directly modulated at 4 Gbit/s. The LEAP laser was directly modulated with a bias voltage of 1.7 V and a bias current of 67 μA. The modulation voltage was 0.7 V_{pp}. The bias voltage of the PD was -2 V. Air coplanar probes were used to drive the laser and PD without a 50-Ω termination. As shown in these figures, on-chip optical data transmission was realized, and the energy cost was 28.5 fJ/bit when the laser was operated at 4 Gbit/s.

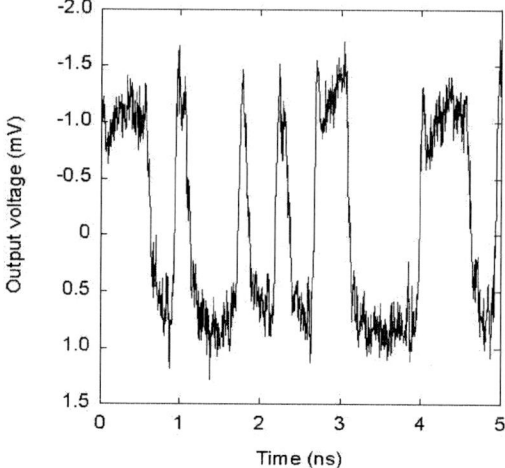

Figure 4. Dynamic response of fabricated optical link.

IV. CONCLUSION

We have demonstrated a monolithic integration of LEAP laser and PD. The LEAP laser exhibited a threshold current of 22 μA and an energy cost of 28.5 fJ/bit when the device directly modulated at 4 Gbit/s. The maximum conversion efficiency from laser injection current to PD photocurrent was 8.5%. These results indicate that a LEAP laser is a promising device for constructing a large scale InP-based PIC.

REFERENCES

[1] S. Matsuo et al., "High-speed ultracompact buried heterostructure photonic-crystal laser with 13 fJ of energy consumed per bit transmitted," *Nature Photon.*, 4, pp. 648-654, 2010.

[2] S. Matsuo et al., "Room-temperature continuous-wave operation of lateral current injection wavelength-scale embedded active-region photonic-crystal laser," *Opt. Express*, vol. 20, pp. 3773-3780, 2012.

[3] T. Sato et al., "95°C CW operation of InGaAlAs multiple-quantum-well photonic-crystal nanocavity laser with ultra-low threshold current," *IEEE Photonics Conference (IPC)*, WF2, 2012.

[4] S. Matsuo et al., "20-Gbit/s directly modulated photonic crystal nanocavity laser with ultra-low power consumption," *Opt. Express*, vol. 19, pp. 2242-2250, 2011.

[5] S. Matsuo et al., "28.5-fJ/bit on-chip optical interconnect using monolithically integrated photonic crystal laser and photodetector," *ECOC2012*. Th.3.B.2, 2012.

978-1-4673-6130-9/13 $31.00 © 2013 IEEE

TuD3-2 (Oral)
14:30 - 14:45

Low Crosstalk and High Modulation Bandwidth 100GbE Optical Transmitter Using Flip-Chip Interconnects

Shigeru Kanazawa, Takeshi Fujisawa, Kiyoto Takahata, Akira Ohki, Ryuzo Iga, and Hiroyuki Ishii

NTT Photonics Laboratories.
NTT Corporation
3-1 Morinosato-Wakamiya, Atsugi, Kanagawa 243-0198, Japan
Kanazawa.shigeru@lab.ntt.co.jp

Abstract— We developed the first compact 100GbE optical transmitter to use flip-chip interconnects for the first time. The flip-chip interconnects provide low crosstalk and a high modulation bandwidth. Under four-channel simultaneous operation, the 100GBASE-LR4 mask margin of the flip-chip interconnection module was improved by 10% to 27% compared with that of a wire interconnection module.

Index Terms— DFB laser, EAM, Photonic integrated circuits, 100 Gigabit Ethernet, EADFB laser array, flip-chip.

I. INTRODUCTION

100-Gbit/s Ethernet (100GbE) [1] has been standardized to handle rapidly increasing volume of data traffic. And a centum form-factor pluggable (CFP) transceiver [2] has been defined as a 100GbE transceiver for use over distances greater than 100 m (100GBASE-LR4 (10-km reach) / -ER4 (40-km reach)). The current 100GbE CFP transceivers comprise many discrete components. For example, their transmitter section consists of four 25.8-Gbit/s DFB lasers integrated with electro-absorption modulator (EADFB lasers) modules [3-7] and an optical multiplexer (MUX). Therefore, the current 100GbE transceiver is rather large and needs downsizing. Next-generation 100GbE transceivers have been under discussion in relation to the CFP Multi-Source Agreement (CFP MSA) as compact transceivers.

The optical section of next-generation 100GbE transceivers must be miniaturized. To meet this demand, a monolithically integrated transmitter chip has recently been developed on which four 25.8-Gbit/s EADFB lasers and a 4x1 multi-mode interference (MMI) optical MUX are integrated [8], and this chip has been installed in a compact optical module [9]. However we cannot obtain sufficiently clear eye opening. There are two main reasons for eye degradation. One is bandwidth degradation caused by long bonding wires. The bonding wires have parasitic inductances, and these inductances degrade the bandwidth. The other reason is adjacent channel electrical crosstalk caused by the bonding wires. The electrical crosstalk is increased by the increase in inductive coupling between the bonding wires.

In this work, we developed the first compact 100GbE

optical transmitter to use flip-chip interconnects. The flip-chip technique provides low crosstalk and a high modulation bandwidth because wire interconnects are unnecessary. The fabricated optical transmitter achieved 25.8-Gbit/s operation with sufficiently clear eye opening for a back-to-back (BtoB) transmission.

II. DEVICE AND MODULE DESIGN

Figure 1 (a) is a photograph of the EADFB laser array chip. The chip consists of four 1.3-μm InGaAlAs multiquantum-well EADFB lasers with monitor photodiodes (MPDs) and an MMI coupler as their optical MUX. The fabricated chip is only 2.4 mm x 3.3 mm in size.

Figure 1 (b) shows a schematic of a bridge-type RF circuit board and an EADFB laser array chip using conventional wire interconnects. In this figure and Fig. 1 (c), the termination circuit is omitted. To shorten the bonding wires, the RF circuit board is placed over the laser array, and supported by spacers [9]. However, the bonding wires degrade the modulation bandwidth and increase the electrical crosstalk. To solve these problems, we propose an RF circuit that employs flip-chip interconnects as shown in Fig. 1 (c). The RF signals are transmitted from upper to lower signal lines through the RF vias. And the EA modulator (EAM) pads are connected to the

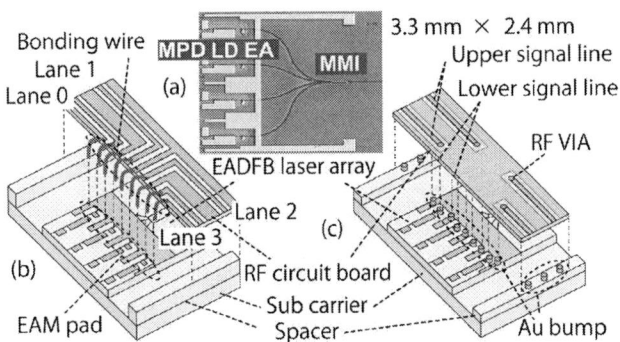

Fig. 1 (a) Photograph of EADFB laser array chip. Schematic structure of RF circuit using (b) wire interconnects and (c) flip-chip interconnects.

978-1-4673-6130-9/13 $31.00 © 2013 IEEE

lower signal lines through 30μm high Au bumps. As a result, we require no bonding wires, so we are able to reduce the electrical crosstalk and increase the modulation bandwidth.

III. MODULE PERFORMANCE

The performance of the 100GbE optical transmitters was measured at 25 °C. For all the measurements, the LD bias currents of all four lanes were 50 mA, and the EA bias voltages of lanes 0, 1, 2, and 3 were -1.2, -1.4, -1.6, and -1.8 V, respectively. Figure 2 shows the small-signal frequency responses of the transmitters using (a) the wire interconnects and (b) the flip-chip interconnects. The EADFB lasers were discretely operated. For all four lanes, the 3-dB frequency bandwidths of the flip-chip interconnection module exceeded 30 GHz, which constitutes an improvement of up to 5 GHz compared with a wire interconnection module. And the small-signal frequency response of the flip-chip interconnection module is flatter than that of a wire interconnection module. Figure 3 shows the crosstalk characteristics of Lane 1 in (a) the wire interconnection module and (b) the flip-chip interconnection module. By using the flip-chip interconnects, the electrical crosstalk was suppressed to less than -20 dB at 20 GHz. And this value is about 5 dB less than that of a wire interconnection module at around 15 GHz.

Figure 4 shows eye diagrams of the Lane 1 optical signals obtained through a fourth-order Bessel filter for a BtoB transmission. The mask margins (MMs) were measured using a 100GBASE-LR4 eye mask. The four EAMs were driven by 2.0 Vpp, 25.78125 Gbit/s non-return to zero (NRZ), PRBS 2^{31}-1 signals. We obtained clearer eye opening with the flip-chip interconnection module than with a wire interconnection module under single channel and four-channel simultaneous operation. The dynamic extinction ratios for all the modules

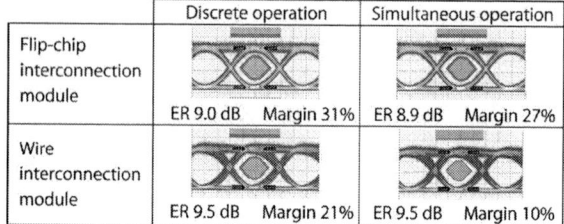

Fig. 4. Eye diagrams for BtoB transmission

exceeded 8.5 dB thus meeting the 100GBASE-LR4 and -ER4 specifications. Under simultaneous operation, the MM of the wire interconnection module was degraded by more than 10% compared with that under single channel operation. Under simultaneous operation, the MM of the flip-chip interconnection module was degraded by only 4% compared with that for single channel operation. These results show that the electrical crosstalk of the flip-chip interconnection module was sufficiently suppressed and that there were no significant problems in the optical transmitter when generating 100GbE optical signals.

IV. CONCLUSION

The first compact 100GbE optical transmitter to use the flip-chip interconnects was developed. The flip-chip interconnects enabled us to achieve a 3-dB frequency bandwidth of 30 GHz, which is an improvement of up to 5 GHz compared with a wire interconnection module. Under simultaneous operation, the MM degradation of the flip-chip interconnection module was suppressed by only 4% compared with that under single channel operation. The proposed flip-chip interconnection optical transmitter is a promising candidate for use as a compact and high-performance 4 x 25.8-Gbit/s optical transmitter in next generation small-form 100GbE transceivers.

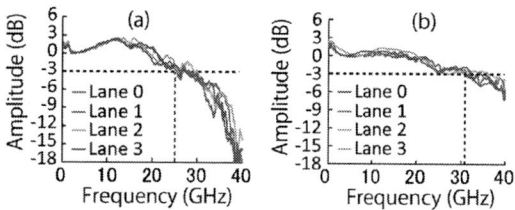

Fig. 2. Small-signal frequency response of (a) wire interconnection module and (b) flip-chip interconnection module.

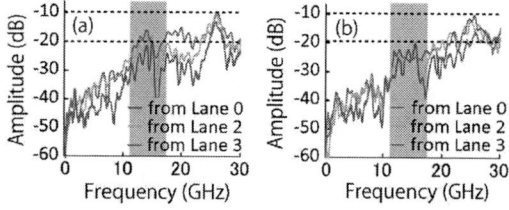

Fig. 3. Crosstalk characteristics of (a) wire interconnection module and (b) flip-chip interconnection module.

REFERENCES

[1] C. Cole, "Beyond 100G client optics," IEEE Communications Magazine, pp. 558-566, Feb. 2012.
[2] http://www.cfp-msa.org/
[3] T. Fujisawa et al., " 25 Gbit/s 1.3 μm InGaAlAs-based electroabsorption modulator integrated with DFB laser for metro-area (40 km) 100 Gbit/s Ethernet system," Electron. Lett., vol. 45, no. 17, pp. 900-902, 2009.
[4] S. Makino et al., "Uncooled CWDM 25-Gbps EA/DFB lasers for cost-effective 100-Gbps Ethernet transceiver over 10-km SMF transmission," in Proc. OFC, 2008, PDP21.
[5] H. Oomori et al., "An extremely compact electro-absorption modulator integrated DFB laser module for 100Gbps Ethernet over 75km SMF reach," in Proc. ECOC, 2008, P.2.07.
[6] H. Takahashi et al., "High-power 25-Gb/s electroabsorption modulator integrated with a laser diode," IEEE Photon. Technol. Lett., vol. 21, no. 10, pp. 633-635, MAY. 2009.
[7] M. Moehrle et al., "Low-cost 25Gb/s 1300nm electroabsorption -modulated InGaAlAs RW-DFB-laser," in Proc. IPRM, 2011, Mo-1.1.2.
[8] T. Fujisawa et al., "1.3-μm 4 x 25-Gb/s monolithically integrated light source for metro area 100-Gb/s Ethernet," IEEE Photon. Technol. Lett., vol. 23, no. 6, pp. 356-358, Mar. 2011.
[9] S. Kanazawa et al., "A compact EADFB laser array module for a future 100-Gbit/s Ethernet transceiver," IEEE J. Sel. Topics Quantum Electron., vol. 17, no. 5, pp. 1191-1197, Sep. 2011.

Non-blocking 4x4 InAlGaAs/InAlAs Mach-Zehnder-Type Optical Switch Fabric

N. KOYAMA [1], H. KOUKETSU [1], S. KAWASAKI [1], A. TAKEI [2], T. TANIGUCHI [2], Y. MATSUSHIMA [3],

and K. UTAKA [1]

[1] Faculty of Science and Engineering, Waseda University, Tokyo, Japan
[2] Central Research Laboratory, Hitachi Ltd., Tokyo, Japan
[3] Green Computing System Research Organization, Waseda University, Tokyo, Japan
E-mail: kh.monsanmisyeru@suou.waseda.jp

Abstract— **We demonstrated the full-interport connection of a non-blocking 4x4 InAlGaAs/InAlAs Mach-Zehnder-type optical switch (MZ-OS) fabric. This switch fabric is consisted of six 2x2 MZ-OS elements in a cascading configuration, and it successfully operated with high-extinction ratios of about 20dB and polarization independence.**

Keywords—InAlGaAs/InAlAs compound semiconductors; Mach-Zehnder-type optical switch; rearrangeable non-blocking;

I. INTRODUCTION

Recently high-performance optical switches have been spotlighted for the applications to optical packet switching (OPS) in next generation optical networks and to optical interconnections in data centers in order to meet the explosive increase of data traffics and consequently the power consumption issue. To solve these problems, required characteristics to optical switches (OSs) are a fast response time of a few nanoseconds (ns), a low switching power consumption and polarization-independent operation. In addition, these optical switches have to be in a multi-port configuration for the expansion of practical network scale. [1],[2]

So far, we reported a 2x2 InAlGaAs/InAlAs Mach-Zehnder -type optical switch (MZ-OS) with low power consumption and low polarization-dependent operation [3], and fundamental switching operation of a 4x4 OS fabric by using a three-stage cascading approach with five 2x2 MZ-OS elements [4]. In this previous study, low current operation of about 5mA and a high-speed response time of about 3ns were attained, but this configuration was not non-blocking and limited numbers of optical path were confirmed.

This paper reports the realization of a non-blocking 4x4 MZ-OS fabric, and full-interport connection operation is exhibited by improving the fabrication process.

II. DEVICE ARCHITECTURE & FABRICATION

A layout diagram of the non-blocking 4x4 OS fabric developed is shown in Fig.1. This fabric is composed of four cascading stages with six 2x2 OSs, and exactly speaking, it works as a rearrangeable non-blocking switch to reduce the total number of optical switches [5]. As for each 2x2 OS element, the 2x2 MZ-OS which we have developed was used, and it is consisted of two multi-mode interference (MMI) 3dB couplers and a phase-shifter by carrier injection, as schematically shown in Fig.2 [3]. Our conventional 2x2 MZ-OS has two different etching-depth mesa structures, so-called the hybrid-waveguide structure, with a deep-etching one in 3dB couplers called "high-mesa waveguide" for low crosstalk operation and a middle-etching one in a phase shifter called "middle-mesa waveguide" for polarization independent operation.

The MZ-OS fabric was fabricated by using an i-line stepper as photolithography for waveguide patterning and an inductively-coupled-plasma reactive-ion etching (ICP-RIE) for waveguide formation, which may be ones of the most important processes for low loss and well-designed characteristics. As a result, a scanning electron microscope (SEM) image of an MMI part of the central switch element fabricated on an InP substrate by employing the same process as to epitaxial wafers for the MZ-OSs is shown in Fig.3. And, a top-view photograph of the fabricated 4x4 MZ-OS fabric is shown in Fig.4. This device has 6mm length and 1mm width including IO ports, curved access waveguides and six MZ-OS elements.

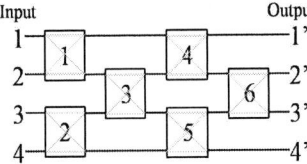

Figure 1. Layout diagram of the rearrangeable non-blocking 4x4 OS fabric.

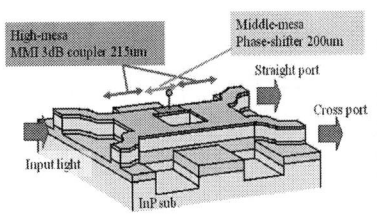

Figure 2. Schematic structure of the hybrid-waveguide-type 2x2 MZ-OS.

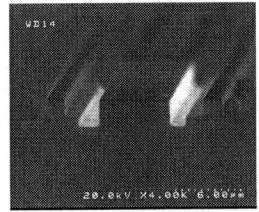

Figure 3. SEM image of an MMI part fabricated on an InP substrate.

This work was partly supported by NEDO and the SORTE project.

TuD3-3 (Oral)
14:45 - 15:00

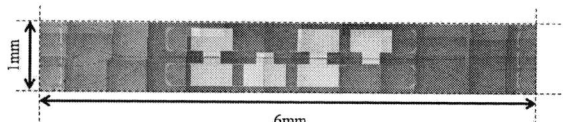

Figure 4. Top-view photograph of the fabricated 4x4 MZ-OS fabric.

III. CHARACTERISTICS MEASUREMENT

A structure and characteristics of one of the 2x2 MZ-OS elements formed on the same wafer as the 4x4 OS fabric is shown in Fig.5: (a) a diagram, (b) a top-view photograph and (c) switching characteristics from each output port for an input light with a wavelength of 1520nm as a function of injection current. This device performed in polarization independence, but this switch was found to need an offset bias current of 6.5mA to be optimized for low crosstalk at off state, which may be due to the mismatch of phases between two waveguides of the phase shifter part.

As for the 4x4 MZ-OS fabric, near-field patterns from the output facet were observed for each input, from port 1 to port 4, realizing full-interport connection operation, as shown in Fig.6. Fig.7 shows the switching characteristics for two cases of modulation of the switches, (a) OS-3 and (b) OS-4, keeping the other OS statuses constant, such as OS-(1,2,3,4,5,6) = (a) (-, on, mod, on, on, off) and (b) (-, on off, mod, -, off) with an input of port 3 from all the output ports and for both polarizations. Here, "mod" means to vary an injection current. These measurements were carried out under the adjusted bias currents to other routing OS elements. Consequently, Fig.7(a) and (b) exhibit the switching from output port 3' to port 2' and to port 1', respectively. As a whole, relatively large extinction ratios (ERs) of about 20dB were observed with almost the same crosstalks.

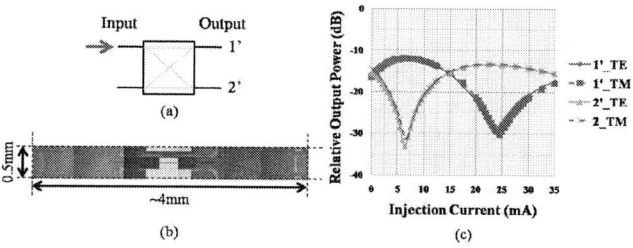

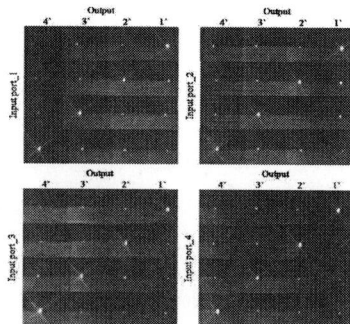

Figure 5. On the 2x2 MZ-OS element: (a) a diagram, (b) a top-view photograph and (c) measured switching characteristics for TE and TM-mode input lights as a function of injection current.

Figure 6. Near-field output patterns of the fabricated 4x4 MZ-OS fabric for full-interport connection operations.

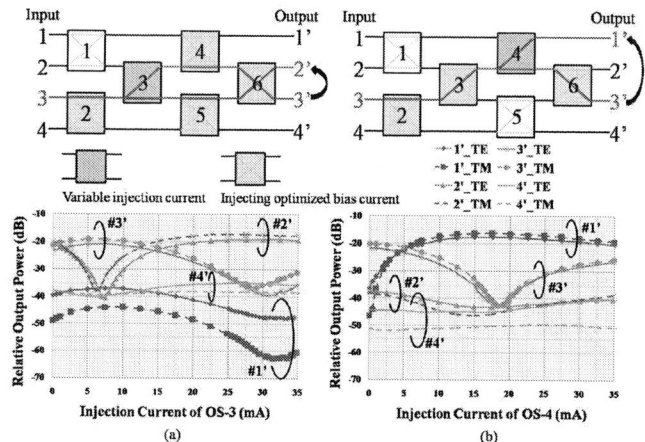

Figure 7. Measured switching characteristics of the fabricated 4x4 MZ-OS fabric for the input light of port 3 under modulation of the 2x2 OSs of (a) No. 3 (OS-3) and (b) No.4 (OS-4).

IV. CONCLUSION

The non-blocking 4x4 InAlGaAs/InAlAs MZ-OS fabric was fabricated by the improved process, and full-interport connection operation was successfully realized. The ERs were up to about 20 dB under the adjusted bias currents for each routing switching elements. The performances can be improved by further precisely optimizing the bias currents for all OS elements and/or by fabricating the device with higher precision.

From this work , it is expected that the hybrid-waveguide-type InAlGaAs/InAlAs MZ-OS fabric is a promising device to realize high-speed, low-power and large-scale optical switching for future highly-efficient photonic network.

REFERENCES

[1] S. C. Lee, R. Varrazza, and S. Yu, "Advanced optical packet switching functions using active vertical-couplers-based optical switch matrix," J. of Sel. Topics in Quantum Electron., vol. 12, no. 4, pp. 817-827, Jul./Aug., 2006

[2] M. Yang, W. M. J. Green, S. Assefa, J. V. Campenhout, B. G. Lee, C. V. Jahnes, F. E. Doany, C. L. Schow, J. A. Kash, and Y. A. Vlasov, "Non-blocking 4x4 electro-optic silicon switch for on-chip photonic networks," Opt. Exp., vol. 19, no. 1, pp. 47-54, Jan., 2011

[3] Y. Ueda, S. Nakamura, S. Fujimoto, H. Yamada, K. Utaka, T. Shiota, and T. Kitatani, "Polarization-independent low-crosstalk operation of InAlGaAs/InAlAs Mach-Zehnder interferometer-type photonic switch with hybrid-waveguide structure," IEEE Photon. Technol. Lett., vol. 21, no. 16, pp. 1118-1120, Aug., 2009

[4] Y. Ueda, N. Koyama, K. Kambayashi, S. Fujimoto, K. Utaka, T. Shiota, and T. Kitatani, "4x4 InAlGaAs/InAlAs optical-switch fabric by cassading Mach-Zehnder interferometer-type optical switches with low-power and low-polarization-dependent operation," IEEE Photon. Technol. Lett., vol. 24, no. 9, pp. 757-759, May, 2012

[5] R. A. Spanke and V. E. Benes, "N-stage planar optical permutation network," Appl. Opt., vol. 26, no. 7, pp. 1226-1229, Apr., 1987

Design of Multi-Functional GaInAsP/Si Hybrid Semiconductor Optical Amplifier Array with AlInAs-Oxide Current Confinement Layer

Yusuke HAYASHI[1], Keita FUKUDA[1], Ryo OSABE[1], Jun-ichi SUZUKI[1], Yuki ATSUMI[1],
JoonHyun KANG[1], Nobuhiko NISHIYAMA[1], and Shigehisa ARAI[1,2]

[1]Department of Electrical and Electronic Engineering, Tokyo Institute of Technology
[2]Quantum Nanoelectronics Research Center, Tokyo Institute of Technology
2-12-1-S9-5 O-okayama, Meguro-ku, Tokyo 152-8552, Japan
E-mail:hayashi.y.ao@m.titech.ac.jp

Abstract— **Novel configuration of multi-functional SOA array by III-V/Si hybrid technique using one-time III-V/Si wafer bonding is proposed. A III-V/Si hybrid structure can realize multi-functional SOA array with different gain characteristics by tuning Si waveguide width.**

Keywords— *Semiconductor optical amplifier; III-V/Si hybrid integration; Si photonics; AlInAs oxidation;*

I. INTRODUCTION

High performance and large scale optical routers are required to increase network transmission and transaction speed [1]. Although extensive researches and developments have been carried out in order to realize such high performance routers, the footprints of these systems tend to become large since large amount of optical devices are needed, such as semiconductor optical amplifiers (SOAs), arrayed waveguide gratings (AWGs), matrix switches and tunable laser diodes (TLDs) and so on. To realize large-scale routers with small footprint and low cost, the integration of these devices is crucial.

Using Si optical circuits on the silicon-on-insulator (SOI) is promising for the large-scale integration because of well-developed fabrication technology through CMOS process and the possibility of electrical-optical integration in one chip [2]. The problem is such routers require many kinds of active functions, and it is difficult to obtain highly efficient optical gain by Si itself due to the indirect band-gap properties. To overcome the problem, III-V/Si wafer bonding technology is an attractive way [3], [4]. We suggest Si-based devices should be used for passive components, such as waveguides, filters and AWGs. For the active devices such as optical sources, amplifiers, and wavelength converters (which use nonlinear optical gain), the III-V/Si hybrid devices should be used. In this case, the important point is that all active devices may require different optical gain characteristics in the gain section. For examples, lasers need high optical gain with small current, amplifiers need linear optical gain at high power level, and wavelength converters need nonlinear optical gain at small power level. These gain properties can be controlled by changing III-V structure, but it is not good idea to bond different III-V wafers for each function.

In this paper, we propose multi-functional GaInAsP/Si SOA array with different saturation output power levels by controlling Si waveguide width, which can be realized by "single" wafer bonding process.

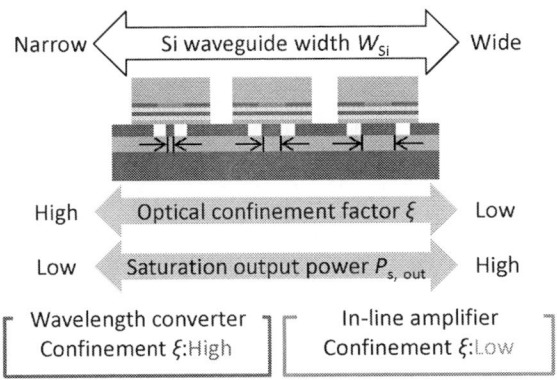

Fig. 1 The image of controlling saturation output power by tuning Si waveguide width. Please note that the device length is also one parameter.

II. DESIGN AND SIMULATION

For the in-line amplifier, distortion-free amplification in the linear amplification regime is required. This means the SOA requires high saturation output power. On the other hand, for the wavelength converter, gain saturation with low power is required. If we set our target of the saturation output power $P_{s,\,out}$ is more than 10 mW for the linear amplifier and less than 3 mW for wavelength converter, the optical confinement factor in the GaInAsP 5 quantum wells (QWs) ξ should be controlled in the range of $\xi < 2\%$ ($P_{s,\,out} \approx 10$ mW) and $\xi > 5\%$ ($P_{s,\,out} \approx 3$ mW), respectively.

To control the optical confinement factor without changing III-V wafer structure, Si waveguide width dependence was investigated, as shown in Fig. 1, with the finite difference method (FDM). As shown in Fig. 2, we assumed a structure where III-V layers with GaInAsP 5QWs were bonded on SOI substrates. Instead of conventional current confinement structure by ion implantation reported in other groups[3], AlInAs-oxide current confinement was used for its advantages of high controllability of current confinement area and strong optical confinement [5], [6]. AlInAs unoxidized width and the refractive-index of AlInAs-oxide were set to be 2.0 µm and 1.8. As shown in Fig. 2(b), we investigated ξ dependence on the height of Si waveguide H_{Si} and GaInAsP/InP superlattice (SL) H_{SL}, where the SL layer (refractive-index of 3.3) was introduced in order to mitigate threading dislocations at bonding interfaces [7] as well as control optical mode profile. To check the fabrication

TuD3-4 (Oral)
15:00 - 15:15

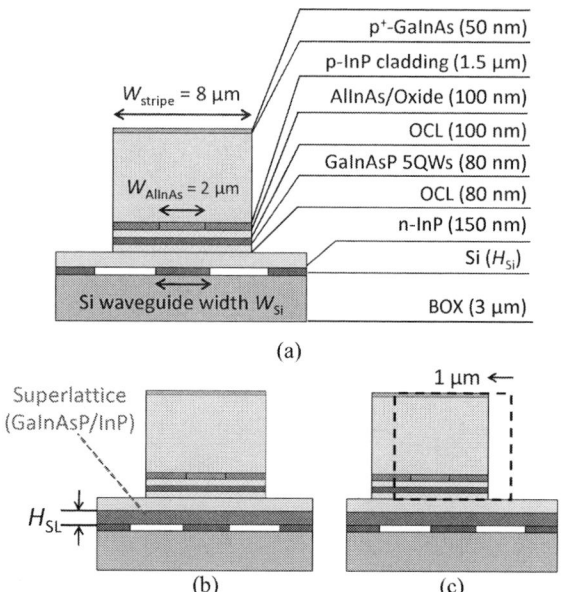

(a)

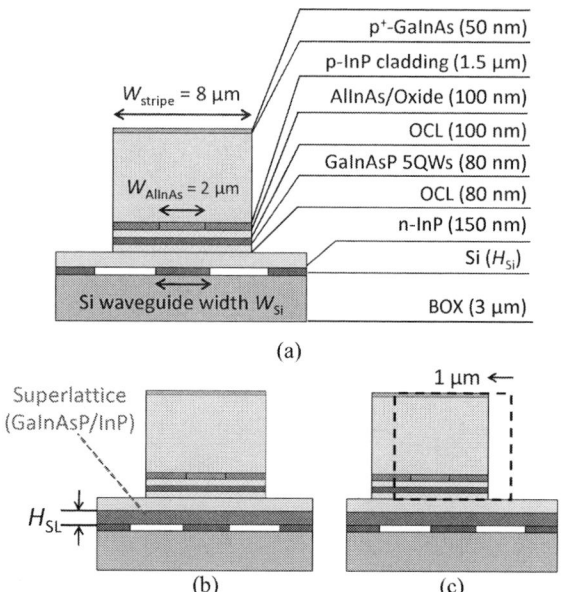

(b)   (c)

Fig. 2 Calculation model of GaInAsP/Si hybrid SOA with AlInAs-oxide layer (a) initial structure, (b) introduction of superlattice, and (c) 1-μm misalignment case.

tolerance, the structure with 1-μm misalignment between center of III-V stripe and Si waveguide was also investigated as shown in Fig. 2(c). The calculated wavelength was 1.55 μm. ζ as a function of Si waveguide is shown in Fig. 3. At H_{Si} = 220 nm, only 0.55% change of ζ was observed. However, introduction of 200-nm height SL layer brought over 1.5% change of ζ due to the high-reflective-index layer near the Si waveguide. Additionally, by increasing H_{Si} to 300 nm, ζ = 5.7% was achieved at W_{Si} < 0.5 μm, meanwhile ζ = 1.9% was achieved at W_{Si} > 5 μm. As shown in Fig. 4(a), mode distribution was strongly attracted to Si side by an introduction of 200-nm height SL layer, and 300-nm height and 4-μm width Si waveguide. Because of so high optical confinement in Si side, large difference did not appeared in the case of 1-μm misalignment as shown in Fig. 3 and Fig. 4(b). From these results, we indicated possibility of multi-functional integration of III-V/Si hybrid SOA by an introduction of AlInAs-oxide current confinement layer and SL layer.

III. CONCLUSION

For multi-functional integration by a GaInAsP/Si hybrid SOA array on a SOI substrate, controllability of optical confinement factor in active region by tuning Si waveguide width was investigated by FDM. In order to enlarge effect of Si waveguide width, AlInAs-oxide confinement layer, 200-nm height GaInAsP/InP superlattice layer, and 300-nm height Si waveguide were introduced. Then, we showed optical confinement factor could be controlled in the range of ζ < 2% and ζ > 5% by tuning Si waveguide width from 0.5 μm to 5 μm.

ACKNOWLEDGMENT

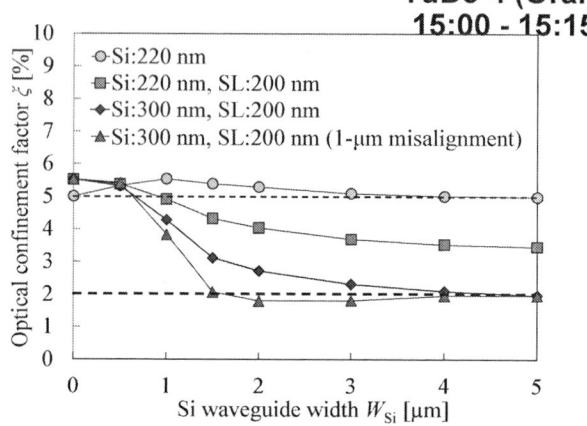

Fig. 3 The optical confinement factor in QWs as a function of Si waveguide width.

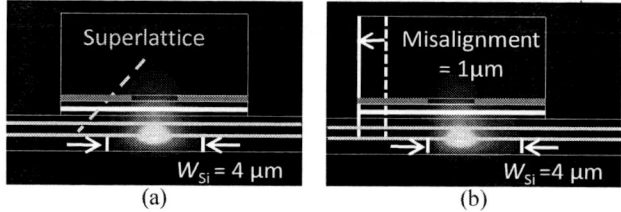

(a)   (b)

Fig. 4 Optical mode profiles in the case of (a) H_{SL} = 220 nm, H_{Si} = 300 nm, and W_{Si} = 4 μm (b) H_{SL} = 200 nm, H_{Si} = 300 nm, and W_{Si} = 4 μm (1-μm misalignment).

This research was supported by JSPS KAKENHI (grant numbers 22360138, 21226010, 24246061, and 11J08863) the Council for Science and Technology Policy (CSTP) under the FIRST program, the Ministry of Internal Affairs and Communications under the SCOPE program and Ministry of Economy, Trade and Industry (METI) under the Pioneering Project.

REFERENCES

[1] S. Yao, S. J. Ben Yoo, and B. Mukherjee, "All-Optical Packet Switching for Metropolitan Area Networks: Opportunities and Challenges," *IEEE Commun. Magn.*, vol. 39, no. 3, pp. 142–148, Mar. 2001.

[2] R. Soref, "The Past, Present, and Future of Silicon Photonics," *IEEE J. Sel. Top. Quantum Electron.*, vol. 12, no. 6, pp. 1678–1687, Nov. 2006.

[3] H. Park, A. W. Fang, O. Cohen, R. Jones, M. J. Paniccia, and J. E. Bowers, "A Hybrid AlGaInAs – Silicon Evanescent Amplifier," *IEEE Photon. Technol. Lett.*, vol. 19, no. 4, pp. 230–232, Feb. 2007.

[4] T. Okumura, T. Maruyama, H. Yonezawa, N. Nishiyama, and S. Arai, "Injection-Type GaInAsP–InP–Si Distributed-Feedback Laser Directly Bonded on Silicon-on-Insulator Substrate," *IEEE Photon. Technol. Lett.*, vol. 21, no. 5, pp. 283–285, Mar. 2009.

[5] N. Iwai, N. Yamanaka, S. Arakawa, H. Shimizu, and A. Kasukawa, "GaInAsP SL-QW Al-oxide Confined Inner Stripe Lasers on p-InP Substrate with AlInAs-oxide Confinement Layer," *Electron. Lett.*, vol. 34, no. 14, pp. 1427–1428, June1998.

[6] Y. Hayashi, R. Osabe, K. Fukuda, N. Nishiyama, and S. Arai, "AlInAs Selective Oxidation for GaInAsP/Si Hybrid Semiconductor Laser using Surface Activated Bonding," *Low Temperature Bonding for 3D Integration (LTB-3D 2012)*, p. 93, May 2012.

[7] S. Kondo, T. Okumura, R. Osabe, N. Nishiyama, and S. Arai, "Investigation of Bonding Strength and Photoluminescence Properties of InP/Si Surface Activated Bonding," *Indium Phosphide & Related Materials (IPRM 2010)*,Takamatsu, Japan, WeP28, pp. 273–276, May/Jun. 2010.

978-1-4673-6130-9/13 $31.00 © 2013 IEEE

TuD3-5 (Oral)
15:15 - 15:30

New Fabrication Method of Trapezoidal Polarization Converters for InP-Based Photonic Integrated Circuits

Dzmitry O. Dzibrou, Jos J. G. M. van der Tol, Meint K. Smit

Group of Photonic Integration
Eindhoven University of Technology
Eindhoven, the Netherlands
d.o.dzibrou@tue.nl

Abstract—**This paper presents a new way to make trapezoidal polarization converters. The fabrication process has four steps, two steps less than the standard fabrication. The fabricated converters have a 97.9% polarization conversion and loss below 0.5 dB at a wavelength of 1.55 μm.**

Keywords—polarization converter; photonic integrated circuits; InP.

I. INTRODUCTION

Polarization converters are needed to control polarization in photonic integrated circuits [1]. There are several ways to make polarization converters for the photonic circuits. In [2], a polarization converter had eleven sections of a trapezoidal cross section. At a 1.55-μm wavelength, the efficiency of polarization conversion was 93 %, the loss was 1.1 dB, and the length was about 1 mm. In [3], the authors made a converter with curved waveguides. At a 1.55-μm wavelength, its polarization conversion was 85 %, the loss was 2.7 dB, and the size was 975 × 83 μm. In [4], the converter was a one-dimensional grating. At a 1.55-μm wavelength, its polarization conversion was 95 %, the loss 3 dB, and the length 1.6 μm. Right-angled trapezoidal polarization converters suit the InGaAsP-InP photonic circuits the best because of a polarization conversion larger than 95 %, loss below 1.5 dB, and length below 800 μm [5-7]. The standard fabrication of the converters, however, introduces an underetch (up to 650 nm [5]) which may destroy the performance of the device. We report a new way to make right-angled trapezoidal polarization converters. This fabrication process has two steps less than the standard fabrication and introduces almost no underetch.

II. CONVERTER AND STANDARD FABRICATION

Figure 1 shows the cross section and top view of the polarization converter. The layer stack of the converter is a 1.03-μm-thick InGaAsP (bandgap wavelength 1.06 μm) layer on an undoped InP substrate. To have a 95 % polarization conversion, the top width of the converter should be 1.24 ± 0.05 μm, the length 275 ± 40 μm, the slope angle 55 ° defined by wet etching, and the width of the input-output waveguides 1.69 μm. The standard fabrication of the converter has six main steps (Fig. 2a) [6]. Step 1 is masking the top width of the converter. Step 2 is protection of one side of the converter with a photoresist. Step 3 is etching the straight wall of the converter. Step 4 is deposition of a masking material to protect the

The research leading to these results has received funding from the European Community's Seventh Framework Program FP7/2007-2013 under grant agreement ICT 257210 PARADIGM. We thank Photon Design for giving us the FIMMWAVE/FIMMPROP simulation tool.

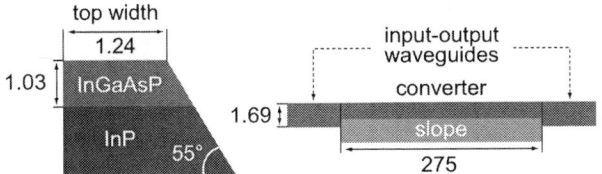

Figure 1. Cross section and top views of polarization converter.

straight wall during the following wet etching. Step 5 is dry etching this mask material on flat surfaces. Step 6 is wet etching the slope of the converter. The problem is in step 5: the masking material on the straight wall creates strain in the top width mask causing the latter to lift. This lift allows the etchant to go under the mask, so the underetch appears [6]. This underetch may be up to 650 nm [5]. The tolerance of the converter to variation in the top width is ± 50 nm, so ideally, we want to avoid the underetch. This fabrication uses symmetric mask coverage at all fabrication steps, avoids the lifting, and therefore, introduces almost no underetch.

III. NEW FABRICATION

Figure 2b shows the new fabrication method. It has four main steps, two steps less than the standard fabrication. Step 1: masking the converter's top. We deposit 600 nm of SiN_x and evaporate 50 nm of Cr to be the mask for the SiN_x etching. We then spin the ZEP520A resist and bake it. An electron beam lithography machine writes an array of lines. Several lines have straight waveguides only; the other lines have converters between the tapers that are connected to straight waveguides. We define 1.24-μm-wide and 275-μm-long converters, 1.24-μm-wide converters of different lengths, and 275-μm-long converters of different top widths. We develop ZEP520A, etch the Cr layer by Cl_2-O_2 inductively-coupled plasma, and etch the SiN_x layer by reactive ion etching. Step 2 is wet etching of the slopes on both sides of the waveguides and converters by Br_2-CH_3OH (0.17 vol. %). The slope height is 2 μm. Step 3 is protection of the slope of the converter with a photoresist. We spin the AZ4533 photoresist, bake it, expose it by an optical mask aligner so that only the areas of the converter's slopes are covered, develop the resist, and bake it at 200 °C for 20 min to make it solid. Step 4 is CH_4-H_2 inductively-coupled plasma etching of the straight walls for the waveguides, tapers and converters. The etching depth is 2 μm. Figure 2b also shows a scanning electron microscope image of the cross section

978-1-4673-6130-9/13 $31.00 © 2013 IEEE

TuD3-5 (Oral)
15:15 - 15:30

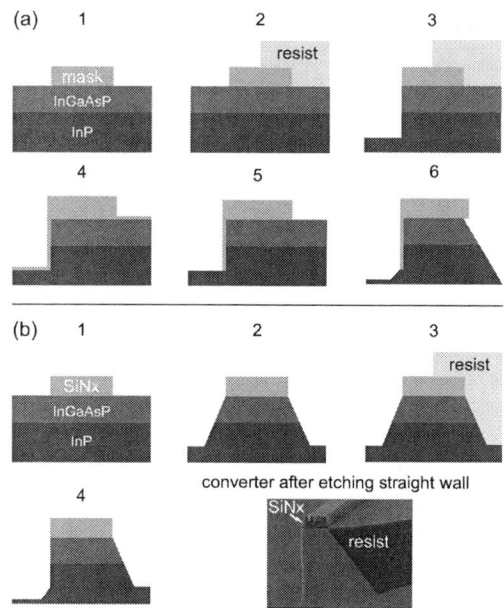

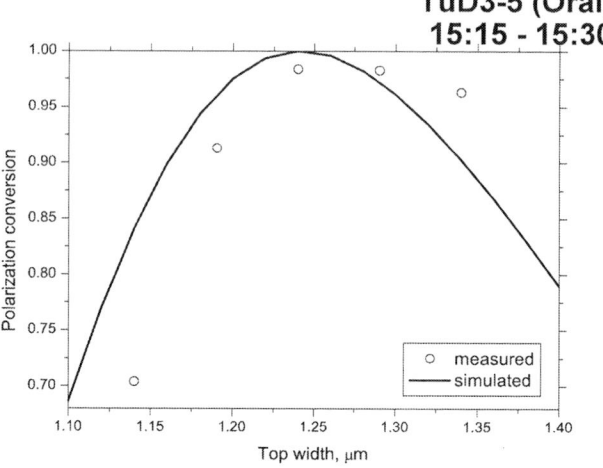

Figure 4. Polarization conversion as function of top width of 275-μm-long converters.

growth, and fabrication imperfections such as curvature of the straight wall of the converter. At this stage, we could not determine to what extent these factors influence the difference.

V. CONCLUSIONS

We showed a new way to make trapezoidal polarization converters for InGaAsP-InP photonic integrated circuits. The fabrication method introduces an underetch of less than 20 nm which gives sufficient control over the top width for a conversion > 95 %. Polarization conversion of the fabricated converters is 97.9 % and the loss is below 0.5 dB at a 1.55-μm wavelength.

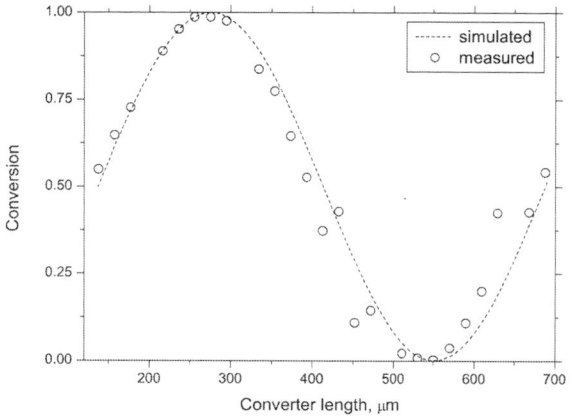

Figure 2. Fabrication process of right-angled trapezoidal polarization converters: (a) standard, (b) new.

through the converter made. A zoom-in on the area of the mask shows an underetch of less than 20 nm.

IV. MEASUREMENTS

We use a Fabry-Pérot setup to measure polarization conversion and loss of the converters at a wavelength of 1.55 μm and at the room temperature. The measurement setup and the measurements are described in [8]. The 1.24-μm-wide and 275-μm-long converters have an averaged polarization conversion of 97.9 %. Figure 3 shows polarization conversion as a function of length of 1.24-μm-wide polarization converters. The simulation result agrees with the measurements well. We also simulated the polarization conversion as a function of the top width of the 275-μm-long converters. Figure 4 shows the result. The measured and simulated results differ. The difference may be because of the deviation from the designed top width during the fabrication, different thickness of the InGaAsP layer due to some deviation during the layer stack

REFERENCES

[1] M. Smit, X. Leijtens, E. Bente, J. van der Tol, H. Ambrosius, D. Robbins, M. Wale, N. Grote, and M. Schell, "Generic foundry model for InP-based photonics," IET Optoelectron., vol. 5, pp. 187--194, 2011.

[2] J. J. G. M. van der Tol, J. W. Pedersen, E. G. Metaal, F. Hakimzadeh, Y. S. Oei, F. H. Groen, and I. Moerman, "Realization of short integrated optic passive polarization converter," IEEE Photon. Techonol. Lett., vol. 7, pp. 893—895, 1995.

[3] L. H. S. C. van Dam, F. P. G. M. van Ham, F. H. Groen, J. J. G. M. van der Tol, I. Moerman, W. W. Pascher, M. Hamacher, H. Heidrich, C. M. Weinert, and M. K. Smit, "Novel compact polarization converters based on ultra short bends," IEEE Photon. Technol. Lett., vol. 8, pp. 1346–1348, 1996.

[4] M. V. Kotlyar, L. Bolla, M. Midrio, L. O'Faolain, and T. F. Krauss, "Compact polarization converter in InP-based material," Opt. Lett., vol. 13, pp. 5040–5045, 2005.

[5] H. El-Refaei, D. Yevick, and T. Jones, "Slanted-rib waveguide InGaAsP-InP polarization converters," J. Lightwave Technol., vol. 22, pp. 1352–1357, 2004.

[6] L. M. Augustin, J. J. G. M. van der Tol, E. J. Geluk, and M. K. Smit, "Short polarization converter optimized for active-passive integration in InGaAsP-InP," IEEE Photon. Technol. Lett., vol. 20, pp. 1673–1675 2007.

[7] J. Z. Huang, R. Scarmozzino, G. Nagy, M. J. Steel, and J. R. M. Osgood, "Realization of a compact and single-mode optical passive polarization converter," IEEE Photon. Technol. Lett., vol. 12, pp. 317–319, 2000.

[8] D. O. Dzibrou, J. J. G. M. van der Tol, M. K. Smit, "Improved fabrication process of low-loss and efficient polarization converters in generic InP-based integration technology," submitted to Optics Letters.

Figure 3. Polarization conversion as a function of length of 1.24-μm-wide converters.

TuD3-6 (Oral)
15:30 - 15:45

IPRM2013, May 19 - 23, 2013, Kobe, Japan
The 25th International Conference on Indium Phosphide and Related Materials

Butt-Joint Built-in (BJB) Structure for Membrane Photonic Integration

Daisuke Inoue[1], Jieun Lee[1], Takahiko Shindo[2], Mitsuaki Futami[1], Kyohei Doi[1],
Tomohiro Amemiya[2], Nobuhiko Nishiyama[1], and Shigehisa Arai[1,2]

[1]Department of Electrical and Electronic Engineering, Tokyo Institute of Technology
[2]Quantum Nanoelectronics Research Center, Tokyo Institute of Technology
2-12-1-S9-5 O-okayama, Meguro-ku, Tokyo, 152-8552, Japan
E-mail: inoue.d.ac@m.titech.ac.jp

Abstract—On-chip optical interconnections have potential for replace global copper wires on LSI chips. In this work, as an integration method, an OMVPE butt-joint regrowth of 175-nm thick GaInAsP/InP was conducted toward an integration of active and passive components. In the numerical calculation, a coupling efficiency and residual reflection of designed butt-joint coupling were estimated to be 98% and -40dB, respectively. In the experimental method, we investigated the dependence of butt-joint interface morphology and regrown surface flatness on the side etch depth and the mesa angle. As a result, a flat regrown surface without degradation in crystalline quality was obtained.

Keywords—butt-joint regrowth; OMVPE; photonic integrated circuit; membrane structure

I. INTRODUCTION

An introduction of optical interconnections to replace copper global wires is considered as a promising solution of performance limitation of Si-LSI [1]. To realize the optical interconnects, we have proposed GaInAsP/InP membrane photonic integrated circuits on Si substrate using benzocyclobutene (BCB) bonding [2]. The Membrane structure consists of a thin semiconductor core layer sandwiched by low refractive index claddings such as SiO_2, BCB, and the air. A large refractive-index difference between the thin semiconductor core layer and dielectric cladding layers leads to the strong optical confinement to the active region hence an extremely low-threshold laser with high-speed direct modulation capability is expected. In our previous researches, we have reported lateral current injection membrane lasers [3], GaInAsP wire waveguides [4] and lateral junction membrane photodetectors [5].

In order to integrate each membrane component, high coupling efficiency and low diffraction losses as well as low optical absorption at the passive region are necessary, which can be achieved with sufficient thickness controllability and surface smoothness. There has been several approaches to photonic integration, such as butt-joint built-in (BJB) structure [6], selective area growth (SAG) [7], offset quantum-well [8], and quantum-well intermixing (QWI) [9] to name just a few. However, it is difficult to fabricate with membrane structure or hard to achieve rapid band gap change at the active-passive interface. Therefore, in this work, we employed the butt-joint built-in (BJB) structure using organo-metallic vapor phase epitaxy (OMVPE) as an integration method, and demonstrated it satisfies the requirements above for membrane integration. Moreover in the calculation, the coupling efficiency and the

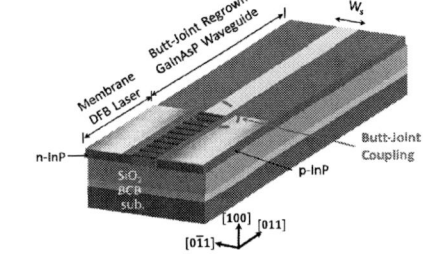

Fig.1 Schematic structure of the membrane DFB laser integrated with GaInAsP waveguide.

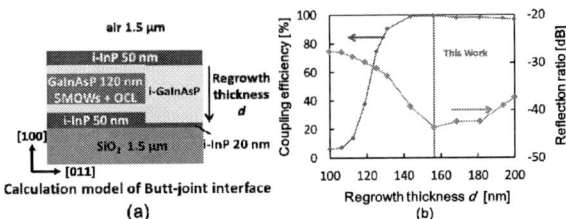

Fig. 2 Regrown GaInAsP thickness dependence of butt-joint coupling efficiency and reflection ratio.

residual reflection of the butt-joint interface was estimated with the finite difference method (FDM) and the eigen-mode expansion method (EME).

II. NUMERICAL ANALYSIS

Figure 1 shows the schematic structure of the membrane DFB laser integrated with a GaInAsP waveguide. The BJB coupling characteristics, which are determined by the equivalent refractive-index and the mode profile difference between the active and passive regions, depend on the regrown GaInAsP/InP thickness. In order to calculate the coupling efficiency and the residual reflection of the BJB coupling, a simple simulation model was used and the simulation was carried out by the FDM and EME methods. Figure 2 (a) shows the calculation model of the BJB interface. The parameters were set as the following; the regrowth thickness of i-GaInAsP (with constant i-InP cap thickness of 20 nm) varies from 100 nm to 200 nm. The coupling efficiency and the residual reflection at each thickness were plotted on graph in Fig. 2 (b). The maximum coupling efficiency of around 98% and the minimum residual reflection of -42dB can be achieved simultaneously at the thickness of 155 nm. At this thickness, the equivalent index at the laser and passive sections are 2.66 and 2.63, respectively. The coupling efficiency decreases rapidly when regrowth thickness becomes less than 130 nm due to a decrease in the optical confinement factor in the

978-1-4673-6130-9/13 $31.00 © 2013 IEEE   125

regrown waveguide. Therefore, i-GaInAsP regrown thicker than the optimal value has little effect, whereas i-GaInAsP regrown thinner than the optimal value causes significant coupling loss.

III. EXPERIMANTAL RESULTS AND DISCUSSION

The initial epitaxial layers consist of GaInAsP 5QWs (90 nm) sandwiched by GaInAsP optical confinement layers (15 nm) and i-InP layers (50 nm), which were grown by OMVPE on an (100) n-InP substrate. The unregrown island region was defined by photolithography, using a 50 nm SiO_2 mask. The island masks are 13 μm wide and 10 to 300 μm long, oriented in the [011] direction. Each island mask is spaced apart by 250 μm. The 50 nm i-InP surface was etched by CH_4/H_2 reactive-ion-etching (RIE), and GaInAsP core layers were etched by selective wet etching ($H_2SO_4:H_2O_2:H_2O = 1:1:40$ at 20°C for 11 min.). Then an InP buffer layer, undoped GaInAsP ($\lambda_g = 1.22$ μm) and an InP cap layer were grown by OMVPE. The growth conditions are summarized in Table 1. In previous studies, it was found that the BJB interface in the [011] direction has a tendency to cause an air gap between the as-grown region and regrown region. For high efficiency coupling, it is necessary to get rid of the air gaps on the BJB interface. Therefore we investigated the change in regrowth morphology at the BJB interface, by using the island mesa angle and under etch of the SiO_2 mask with wet chemical etching.

TABLE I. Growth Conditions

	Material	Thickness (nm)	Time	Growth Rate(nm/min.)	Growth Temperature (°C)
Step 1	i-InP	3	10 sec.	20	600
Step 2	i-GaInAsP	155	8 min.	19	650
Step 3	i-InP	20	1 min.	20	650

Figure 3 shows an SEM image of the regrown BJB interface observed from the <01$\bar{1}$> direction. A smooth surface and a gapless BJB coupling were achieved. The island mesa angle of 50° and the under etch depth underneath the SiO_2 mask of 160 nm were observed from this SEM picture. The deviation in surface height near the BJB interface is estimated to be less than 5 nm.

Then we evaluated the regrown GaInAsP luminescence properties by photoluminescence (PL) measurement. PL mapping was generated by stepping the 3 μm × 3 μm field of view; with the regrowth mask region in a 200-μm-wide and 200 μm-long square in the center. Figure 4 (a) and (b) show the PL peak intensity and PL peak wavelength mapping result respectively. The peak wavelengths were 1.22 μm and 1.55 μm in the regrown GaInAsP region and as-grown active region respectively. The full width at half maximum of regrown region was 49.4 meV in the measured regrown area. Compared with the bulk growth in our previous result, the value of it was 57.7 meV. From the results of the PL measurement, the quality of regrown layers near the mask can as good as that far from the mask.

IV. CONCLUSION

BJB regrowth using OMVPE was demonstrated for membrane photonic integration. According to a numerical

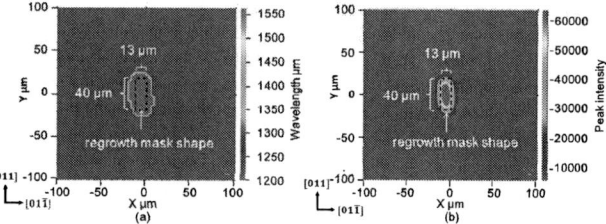

Fig. 3 SEM picture of the butt-joint interface observed from [01$\bar{1}$] direction.

Fig. 4 Photoluminescence mapping of (a)peak intensity, (b)peak wavelength around regrowth mask (13 μm wide and 40 μm long).

calculation, a coupling efficiency of 98% and a residual reflection of -42dB can be achieved at an optimum thickness of the membrane structure. In the experiments, the gapless and flat regrown surface were obtained with fairy good photoluminescence properties.

ACKNOWLEDGMENT

This work was supported by JSPS KAKENHI (Grant numbers 24246061, 24656046, 22360138, 21226010, 23760305, and 10J08973), the Council for Science and Technology Policy (CSTP) under the FIRST program, and the Ministry of Internal Affairs and Communications under the SCOPE program.

REFERENCES

[1] D. A. B. Miller, "Device requirements of optical interconnects to silicon chips," Proc. IEEE, Vol. 97, No. 7, pp. 1166-1185, July 2009.

[2] S. Arai, N. Nishiyama, T. Maruyama, T. Okumura, "GaInAsP/InP Membrane Lasers for Optical interconnects," IEEE J. Sel. Top. Quantum Electron, Vol. 17, No. 5, pp. 1381-1389, Oct. 2011.

[3] M. Futami, et al., "Low-Threshold Operation of LCI-Membrane-DFB Lasers with Be-doped GaInAs Contact Layer," Proc. Int. Conf. Indium Phosphide and Related Materials, Th-2C, Aug. 2012.

[4] J. Lee, et al., "Low-Loss GaInAsP Wire Waveguide on Si Substrate with Benzocyclobutene Adhesive Wafer Bonding for Membrane Photonic Circuits," J. App. Phys., vol. 51, No.4, pp. 042201-1-042201-5, Apr. 2012.

[5] Y. Yamahara, et al., "Characterization of GaInAsP Lateral Junction Waveguide Type Membrane Photodiode," IEICE, C-4-15, Sep. 2012.

[6] Y. Abe, K. Kishino, Y. Suematsu, S. Arai, "GaInAsP/InP integrated laser with butt-jointed built-in distributed-Bragg-reflection waveguide," IEEE Electron. Lett., Vol. 17, No.25-26, pp. 945-947, Dec. 1981.

[7] M. Aoki, et al., "Novel structure MQW electroabsorption modulator/DFB-laser integrated device fabricated by selective area MOCVD growth," Electron. Lett., Vol. 27, No. 23, pp. 2138-2140, Nov. 1991.

[8] B. Mason, G. A. Fish, S. P. DenBaars, L. A. Colden, "Ridge waveguide sampled grating DBR lasers with 22-nm quasi-continuous tuning range," IEEE Photon. Technol. Lett., vol. 10, no. 9, pp. 1211–1213, Sep. 1998.

[9] D. Hofstetter, B. Maisenholder, H. P Zappe, "Quantum-well intermixing for fabrication of lasers and photonic integrated circuits," IEEE J. Sel. Top. Vol. 4 , No. 4, pp. 794-802, Jul. 1998.

TuD3-7LN (Oral - Late News)
15:45 - 16:00

Tunable InP Photonic Integrated Circuit for Millimeter Wave Generation

Marco Lamponi[1], Mourad Chtioui[2], François Lelarge[1], Gaël Kervella[1], Efthymios Rouvalis[3], Cyril Renaud[3], Martyn Fice[3], Guillermo Carpintero[4], Frederic van Dijk[1]

[1] III-V Lab, a joint Laboratory of "Alcatel Lucent Bell Labs", "Thales Research & Technology" and "CEA-LETI", Palaiseau, France (e-mail : marco.lamponi@3-5lab.fr)

[2] Thales Air Systems, 91470 Limours, France

[3] Department of Electronic and Electrical Engineering, UCL, Torrington Place, WC1E 7JE, United Kingdom

[4] Universidad Carlos III de Madrid, Av de la Universidad, 30 Leganes 28911 Madrid, Spain

Abstract—We demonstrate a fully integrated tunable continuous-wave (CW) millimeter wave heterodyne synthesizer. DFB lasers, SOA amplifiers, passive waveguides, beam combiners, high speed photodetectors have been integrated on the same InP-based platform. Millimeter wave generation up to 110 GHz has been demonstrated.

Keywords—*millimeter wave generation; semiconductor lasers; photonic integrated circuits*

I. INTRODUCTION

Systems suitable for the generation and transmission of millimeter wave frequencies (30 GHz - 300 GHz) are of great interest in the development of broadband wireless communication systems [1]. Millimeter-wave photonic devices capable of generate carrier frequencies from 30 GHz to nearly 200 Ghz have already been demonstrated [2-4]. Many of the solutions presented up to now involved the use of separate components in a free space optics configuration [5]. More recently hybrid integration has been employed in order to reduce the packaging dimension and the optical delays inside the chip [6]. In the long term, monolithic integration promises to allow cost effective production, even tighter packaging, and fewer alignment issues [7-8]. However the challenge of monolithically integration is to combine the needed high performances of two tunable lasers with a high bandwidth photodiode on the same substrate.

In previous works very high bandwidth Uni-Traveling Carrier Photodiodes (UTC-PDs) compatible with active/passive integration were demonstrated [2]. We have also demonstrated monolithic integration of two narrow linewidth DFB lasers with passive optical waveguides and couplers [4]. In this paper we demonstrate active/passive integration that allows Distributed Feedback (DFB) Lasers, Semiconductor Optical Amplifiers (SOAs), Multimode Interference (MMI) couplers and high speed UTC-PDs on the same chip.

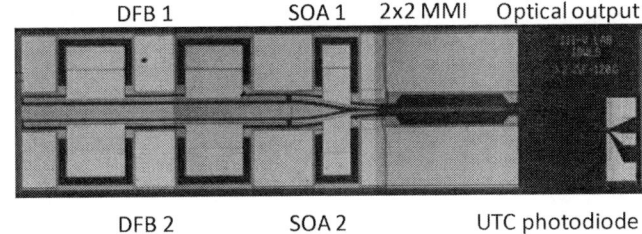

Figure 1: Optical image of the fabricated chip

II. DEVICE DESCRIPTION

The fabricated device is composed of two 1mm-long DFB lasers with a phase shift in the middle of the Bragg grating in order to guarantee single mode operation (see). The outputs from the two lasers are combined, after passing through bent SOAs, using a multimode interference (MMI) coupler. One of the two outputs of the coupler is evanescently coupled to an UTC photodiode. The other output is sent to the edge of the chip to optically collect the two optical wavelengths.

In this device each DFB laser generates an optical tone from which the output power can be adjusted using the independently electrically connected SOA. The two optical tones are sent to the high speed UTC photodiode in which the optical signal is converted into a high speed electrical signal at a frequency equal to the difference between the two optical frequencies.

III. DEVICE FABRICATION

In order to reduce the parasitic capacitance and have a very large detection bandwidth of the photodiodes, epitaxial growth was performed on a semi-insulating InP substrate. Active/passive integration was achieved using a butt-joint process. The active layers consisted in 6 InGaAsP quantum wells. The Bragg grating was formed in an InGaAsP layer placed above the quantum wells and defined by e-beam lithography. The UTC layers, that are similar to the layers used in [2], are grown above the passive waveguide. The fabrication needed 3 epitaxial growth steps. After wafer thinning and back metal deposition a fist set of measurements

---

The authors acknowledge cooperation within and financial support by the EC project iPHOS

were performed directly on the wafer. After these first measurements chips were cleaved from the wafer and mounted on AlN submounts.

IV. RESULTS AND DISCUSSION

We have made some preliminary electrical measurements on the photodiodes showing a dark current lower than 1µW. Measurements on a standing alone test laser showed a lasing threshold current of 80 mA and an generated optical power of 6 mW at 200 mA bias current.

We have measures the optical spectrum of one of the fabricated devices. For these measurements the SOAs were not biased. DFB2 was biased at 180 mA and the current of DFB1 was adjusted in order to check the tuning range. The resulting optical spectra are presented in . A tuning range of more than 90 GHz was obtained.

Before cleaving the wafer into chips we have measured the electrical signal obtained from the photodiode biased at -3V while the two SOA sections were biased with the same current supply at 30mA. DFB1 was biased at 182 mA and the bias current of DFB2 was changed in order to check the tuning range of the device. The photodiode was probed using an ACP 65 probe from Cascade Microtech, biased through an Anritsu 65GHz bandwidth bias tee. The electrical spectrum was measured using a 67 GHz Rohde&Schwarz electrical spectrum analyzer. For measurements above 70 GHz, a Rohde&Schwartz FS-Z90 external mixer was added.

As can be seen in Figure 3, we were able to observe a beat note covering a frequency range from around 10 GHz to more than 110 GHz. This is, to our knowledge, the highest frequency range ever obtained from a monolithic heterodyne source integrating the high speed photodetector.

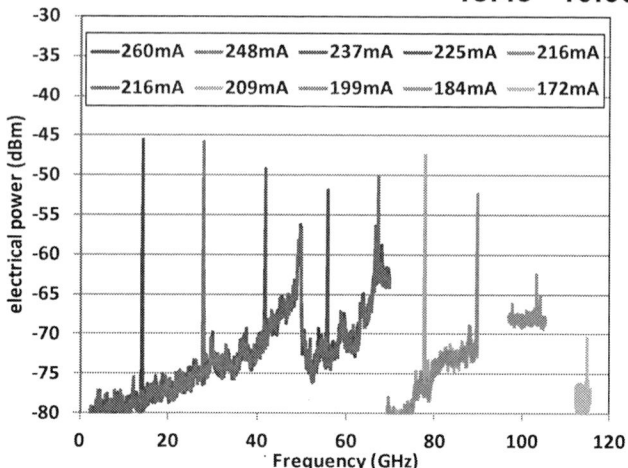

Figure 3: electrical spectra obtained for different laser biasing conditions

V. CONCLUSIONS

We have demonstrated the integration of a widely tunable heterodyne system able to generate millimeter wave signals up to 113 Ghz. The integration of DFB lasers, SOAs, passive waveguides and couplers, UTC-type high bandwidth photodetectors have been achieved. Preliminary characterizations show low photodiodes dark current, 90 Ghz DFB lasers tuning range, reasonable threshold and output power. Electrical spectra measurements show more than 100 Ghz tunability range. To the best of our knowledge this is the first demonstration of a high frequency fully integrated tunable monolithic heterodyne source, monolithically integrating two DFBs, SOAs, passive waveguides and high speed UTC-PDs, reaching frequencies above 100 GHz.

REFERENCES

[1] A. Stöhr, "Photonic Millimeter-Wave Generation and its Applications in High Data Rate Wireless Access", IEEE International Topical Meeting on Microwave Photonics 2010.

[2] E. Rouvalis, M.Cthioui, F. van Dijk, et al., "170 GHz Photodiodes for InP-based photonic integrated circuits", IEEE Photonics Conference (IPC), 2012.

[3] S. Ristic, A. Bhardwaj, M.J. Rodwell, L.A. Coldren, L.A. Johansson, "An Optical Phase-Locked Loop Photonic Integrated Circuit", Journal of Lightwave Technology, Volume: 28 , Issue: 4, Page(s): 526 – 538, 2010.

[4] F. Van Dijk, A. Accard, A. Enard, O. Drisse, D. Make, F. Lelarge, "Monolithic dual wavelength DFB lasers for narrow linewidth heterodyne beat-note generation", International Topical Meeting on Microwave Photonics 2011, MWP/APMP 2011 Page(s): 73 – 76.

[5] L. N. Langley, et al., "Packaged semiconductor laser optical phase-locked loop (OPLL) for photonic generation, processing and transmission of microwave signals," IEEE trans. on Microw. Theory and Techn. 47(7), 1257–1264, 1999.

[6] R. J. Steed, et al., "Hybrid integrated optical phase-lock loops for photonic terahertz sources," J. Sel. Top. Quantum Electron. 17(1), 210–217, 2011.

[7] M. Hamacher, et al., "Fabrication of a heterodyne receiver OEIC with optimized integration process using three MOVPE growth steps only", IEEE Photonics Technology Letters, Volume: 8 , Issue: 1, 1996.

[8] M. Lu, et al., "Monolithic Integration of a High-speed Widely-tunable Optical Coherent Receiver", IEEE Photonics Technology Letters, "in press", 2013.

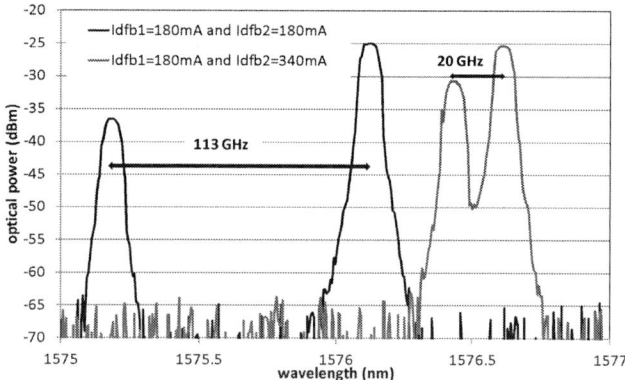

Figure 2: optical output spectra for different bias currents

TuD4-1 (Invited)
16:30 - 17:00

AlGaInAs Selective Area Growth
for high-speed EAM-based PIC Sources

Jean Decobert[1], Pierre-Yves Lagrée[2], Hugues Guerault[3], Christophe Kazmierski[1],

[1] III-V lab, Route de Nozay, 91460 Marcoussis, France
[2] CNRS, UPMC Univ Paris 06, IJLRA, F-75005 Paris, France
[3] BrukerAXS GmbH, O. Rheinbrueckenstr. 49, 76187 Karlsruhe, Germany
jean.decobert@3-5lab.fr

Abstract—We present a generic integration platform based on the SAG of AlGaInAs/InP MQW material. For efficient bandgap engineering of the different areas of the PICs, active and passive function heterostructures are precisely modeled and characterized in the SAG zones. This approach has allowed to design novel high-speed InP-based PIC.

Keywords—*MOVPE, Selective Area Growth (SAG), Photonic Integrated Circuit (PIC), AlGaInAs multiple quantum wells (MQW), Electro-absorption Modulators.*

I. INTRODUCTION

Selective area growth (SAG) by metalorganic vapor phase epitaxy (MOVPE) has been studied for about 30 years. This technique continues to attract considerable interest, especially at the nano-scale, for many opto- and micro-electronic applications based on various semiconductor substrates. Since the earliest works, considerable efforts have been done to analyze the experimental data and to model the growth mechanisms on dielectric patterned substrates and finally to serve more and more complex integration schemes [1-6]. For InP-based photonic integrated circuits (PICs), driven by the telecommunication needs, SAG remains one of the most powerful tools to concentre on the same wafer an increasing number of active and passive optoelectronic functions. In SAG, the active precursors do not nucleate on dielectric amorphous surfaces. They diffuse in the vapor phase and induce a growth rate enhancement in the vicinity of the masked zone. By a precise design of the mask geometry, nearby the area of interest, different structures, in terms of thicknesses and compositions, can be grown on a same wafer. The main effect is a local band-gap variation, especially for MQW structures where the electronic transitions mainly depend on the well's thickness. An appropriate design of the mask patterns then allows the integration of different band-gap areas on a same wafer.

SAG has been widely used for the "longitudinal" integration, along the optical waveguide, of active and passive components such as electro-absorption modulators (EAM), lasers, semi-conductor optical amplifiers (SOA) and spot size converters [7-9]. SAG has also demonstrated its high potential for "transversal" integration, with large wavelength shift of numerous juxtaposed devices such as lasers arrays [10-12].

We have developed a technology platform for efficient photonic integration. It is based on AlGaInAs MQW material for its high electronic confinement and high temperature operation ability [13]. The SAG of this material is completed by a semi-insulating buried hetero-structure (SIBH) in order to achieve high-speed modulators, low loss waveguides and low thermal resistance [14]. We have recently proposed novel devices for complex modulation formats based on electro-absorption modulator (EAM) switching of prefixed optical phases [14]. This leads to reduction in size of circuits 20 to 30 times versus previously proposed solutions. Also, EAMs allow lower power consumption and scalability in speed including by the growing complexity of circuits. In this paper, focusing on the AlGaInAs/InP-based material system, we will describe how the two pillars of SAG, saying the modeling and the characterization, can make a powerful "tool box" for band-gap engineering and device designing. As a demonstration, we will present a 4-arms QPSK interferometer based on AlGaInAs MQW structures.

II. MODELING AND CHARACTERIZATION

A modeling and predictive tool of SAG effects is essential. We first define each function hetero-structure in the PIC, then we calculate the corresponding mask stripe geometry and finally we extract the "reference" hetero-structure that should be effectively grown. That is why a full three dimensional model, where all material parameters - such as composition, strain, thickness - are calculated for the complete stack of layers everywhere on the substrate, is mandatory to design the engineered bandgaps.

A simple approach is to consider only vapor phase diffusion (VPD) as the source of material supply. This is particularly efficient in the case of wide-stripe SAG, where the devices are far away from the masked areas and surface migration effects. This simplifies drastically the modeling and enhances its reliability when tuning the growth conditions [4-5]. The only adjustable parameters of the model are the diffusion lengths of the group III elements of the AlGaInAs quaternary. These parameters must be precisely quantified in exactly the same growth conditions as the final device.

To refine and calibrate the SAG model, we have conducted a parametrical study on dedicated samples, using synchrotron based x-ray microbeam techniques [15-16]. Using the ID01 beamline at the European Synchrotron Radiation Facility (ESRF), we have measured, between the oxide mask stripes, the diffraction curves from AlGaInAs-based selectively grown MQW structures. It has allowed detailed mapping of the material strain and precise extraction of Al, Ga and In diffusion

TuD4-1 (Invited)
16:30 - 17:00

lengths [12]. Nevertheless, an accurate and non-destructive structural characterization of all the daily grown samples is highly desirable to feed-back growth parameters and to qualify the delivered devices at the SAG scale. Recently, we have reported the latest improvements performed in the laboratory with a new high-resolution diffractometer on SAG samples [17]. The setup (D8 Discover from Bruker) is equipped with a microfocus x-ray source ($I\mu s$), achieving a sub-millimeter ($50x100\ \mu m^2$) spotsize on the surface. Accurate lateral positioning of the sample is achieved with a laser-video microscope (Fig.1-inset). Using this technique, it is possible to access the structural information of the central SAG area. As shown in Fig. 1, the computed profile extracted from lab-source is very close to the μ-XRD measurement [17].

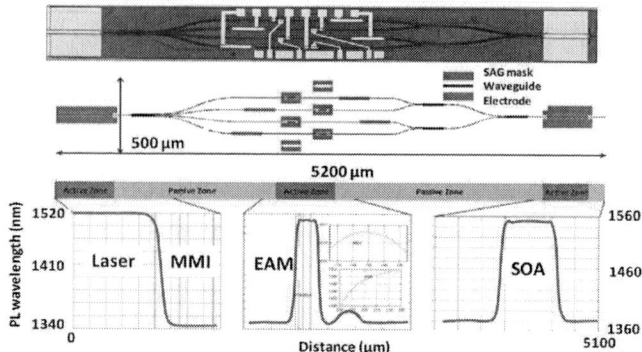

Figure 2.   QPSK TX chip photography (upper), layout (middle) and SAG engineering of the different material gaps (lower)

IV.   CONCLUSION

We presented a technology platform based on the SAG of AlGaInAs MQW core material. Good modeling predictability and non-destructive daily access to the structural information in the SAG zone are key for the efficient bandgap engineering of our circuits. With this flexible PIC technology, we integrated monolithically a novel QPSK transmitter source based on prefixed phase switching by fast EAMs. We believe that this work demonstrates the viability of this SAG technology platform for advanced modulation format PICs.

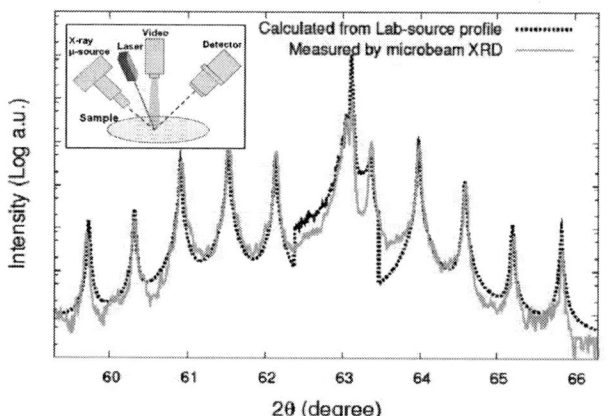

Figure 1.   Measured profile (green) using synchrotron radiation compared to extracted profile (black) using lab-diffractometer with a sub-millimeter focused x-ray source and a laser-video microscope (inset).

III.   COMPONENT DESIGN AND REALIZATION

Using the AlGaInAs MQWs SAG-SIBH platform, we have fabricated several PICs [14,18]. In this paper, we present the first full-monolithic transmitter chip integrating on InP, a DFB laser emitting at 1,550 μm and a 4-arms QPSK (2xBPSK) interferometer based on the EAM phase switching concept. The PIC layout in Fig. 2 shows its architecture integrating numerous actives functions: a DFB laser, which emission is power-splitted by 1:4 MMI, four phase shifters and four EAMs in an interferometric arrangement and a boost SOA at the single output, after the four waveguides have combined in two-stage 2:1 MMI. All chip components (5 passives and 14 actives) were operational. The chip footprint is 5200 x 500 μm^2 making it probably one of the smallest reported PSK transmitter.

For each active and passive section of the device, predictive SAG simulation of the photoluminescence wavelength has been calculated as reported on Fig. 2. Unlike the Butt-joint technique, self-aligned and genuinely continuous SAG guide avoids reflection between each area whatever was the number of engineered bandgap. The obtained component properties attested the correct operation of the circuit and allowed us to demonstrate 28GB/s BPSK experiments using PDM emulator providing 56Gb/s total data rate [19].

REFERENCES

[1]   R. Bhat, J. Crystal Growth. 120 (1992) 362.

[2]   M. Gibbon et al. , semicon. Science and Tech., 8 (1993) 998-1010.

[3]   H. Oh, M. Sugiyama, Y. Nakano, Y. Shimogaki, Jpn J. of Appl. Physics, 42, 10, pp. 6284 (2003).

[4]   J. Décobert, N. Dupuis, P.Y. Lagrée, N. Lagay, A. Ramdane, A. Ougazzaden, C. Kazmierski, J. Crystal Growth. 298 (2007) 28.

[5]   N. Dupuis, J. Décobert, P-Y. Lagrée, et al. J. of Applied Physics 103, 113113 (2008)

[6]   M. Sugiyama, Proc IPRM 2010, 31 may- 4 june 2010, Kagawa, Japan.

[7]   D. Delprat, A. Ramdane, L. Silvestre, A. Ougazzaden, F. Delorme, and S. Slempkes, IEEE Photon. Technol. Lett. 9, 898 (1997).

[8]   B. Mason, A. Ougazzaden et al., IEEE Photon. Technol. Lett., vol. 14, 1 (2002).

[9]   N. Dupuis, J. Decobert, et al., IEEE Photon. Technol. Lett., 20, 21(2008) 1808-1810.

[10]   J. Darja, M.J. Chan, S.R. Wang, M. Sugiyama, Y. Nakano, IEICE Trans. On Elec., vol. E90-C, pp.1111-1117, 2007.

[11]   H. Hatakeyama, K. Kudo, Y. Yokoyama, K. Naniwae, T. Sasaki,. IEEE J. Sel. Top. Quantum Electron., 8(6) : pp. 1341–1348, Nov.–Dec. 2002.

[12]   R. Guillamet, N. Lagay, C. Mocuta, P.Y. Lagrée, G. Carbone, J. Décobert, J. of Crystal Growth. 26 November 2012, in press.

[13]   M. Le Pallec, et al., Proc IPRM 2004, 3pp 577-580, 1 may- 4 june 2004.

[14]   C. Kazmierski, J. Opt. Commun. Netw. 4, (2012),9 , pp. A8-A16.

[15]   M. A. Alam et al., Appl. Phys. Lett. 74, 2617 (1999).

[16]   A. A. Sirenko et al., J. Appl. Phys., 97, 063512 (2005).

[17]   J. Décobert, R. Guillamet, C. Mocuta, G. Carbone, H. Guerault, J. of Crystal Growth, 15 June 2012, in press.

[18]   D. Carrara, C. Kazmierski, J-G. Provost, R. Guillamet, R.Berruée and A. Ramdane, ECIO 2012, Barcelona, 18-20 April 2012

[19]   C. Kazmierski, N. Chimot, F. Blache, J. Decobert, Proc. IPRM 2013, Kobe, 19-23 may 2013.

TuD4-2 (Oral)
17:00 - 17:15

56Gb/s PDM-BPSK Experiment with a Novel InP-Monolithic Source Based on Prefixed Optical Phase Switching

C. Kazmierski[1], N. Chimot[1], F. Blache[1], J. Decobert[1], F. Alexandre[1], J. Honecker[2], C. Leonhardt[2], A. Steffan[2], O. Bertran-Pardo[3], H. Mardoyan[3], J. Renaudier[3] and G. Charlet[3]

[1]III-V Lab-Common laboratory of Alcatel-Lucent Bell Labs France', 'Thales Research and Technology' and 'CEA Leti'
Route de Nozay, 91460 Marcoussis, France
[2]U2T Photonics AG, Reuchlinstrasse 10-11, 10553 Berlin, Germany
[3]Alcatel-Lucent, Bell Labs, Route de Villejust, 91620, Nozay, France
christophe.kazmierski@3-5lab.fr

Abstract—**A novel monolithic QPSK-ready transmitter source based on prefixed phase switching by fast EAMs has been realized on InP using a flexible photonic integrating circuit technology. It has been used up to 56Gb/s in DPSK coherent transmission experiments.**

Keywords— Waveguide modulators, Semiconductor lasers, Photonic Integrated Circuits, AlGaInAs quantum wells, BPSK

I. INTRODUCTION

Photonic Integrated Circuits (PICs) are currently transforming the optoelectronic industry. Monolithic integration on InP is a way to lower footprint, cost, consumption, and to facilitate the design of complex photonic circuits with multiple functions. InP monolithic circuits for spectrally efficient modulation format were previously inspired by LiNbO3 Mach-Zehnder architectures. However, they showed issues with a large footprint, important optical losses, high power consumption and integration compatibility with a laser [1]. A novel, very simple way, to generate the advanced formats has been proposed by Inuk Kang who used short Electro-Absorption Modulators (EAM) in an interferometric arrangement [2]. Low power consumption EAMs acting as a prefixed optical phase switch were shown to simplify and reduce the circuit size. Several subsequent demonstrations of such modulators working at 80 Gb/s DQPSK and a 43 Gb/s 16-QAM signals [3,4] have been realized by C.R. Doerr. However, those PIC applications were limited by their high optical losses and lacking of full source integration with a single frequency laser.

In this work we present the first full-monolithic QPSK-ready transmitter chip integrating on InP a DFB laser emitting at 1.550 µm and a 4-arms QPSK (2 x BPSK) interferometer based on the prefixed phase switching concept. To the date, the monolithic transmitter PIC has been used in BPSK coherent transmission experiments up to 2 x 28 Gb/s

II. COMPONENT DESIGN, REALIZATION AND PACKAGING

The QPSK PIC operation principle is represented on upper part of Fig1. A DFB laser emission is splitted by 1:4 MMI.

Optical phases 0°-90°-180°-270° are prefixed by the MMI and DC-current controlled phase shifters. EAM on each arm switches on/off appropriate phase reflecting a modulation data. The four waveguides are then combined in two-stage 2:1 MMI into a single output. The total chip size is 5200µm length and 500µm width.

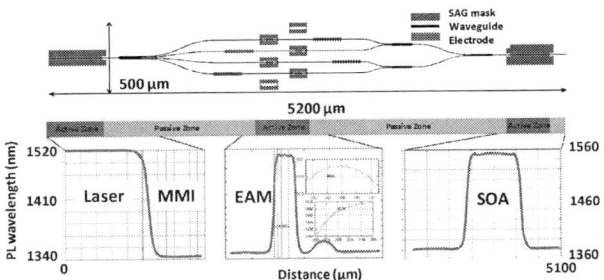

Figure 1. QPSK TX design principle (upper) and SAG engineering of different material gaps (lower)

In order to simplify the fabrication and target its high yield, we choose a Selective Area (SAG) Growth for bandgap engineering of our circuit integrating 5 different bandgap regions made of AlGaInAs MQWs [5]. Unlike the butt-joint technique, self-aligned and genuinely continuous SAG guide avoids reflections between each area whatever was the number of engineered bandgaps. Fig 1 shows also our predictive SAG simulation of the photoluminescence wavelength for each PIC component. We used Semi Insulating Buried Heterostructure (SI-BH) providing only one single p-type regrowth for all PIC components and low RC product for Electro-Absorption Modulators (EAM) [6].

In the circuit design and realization a care has been taken to reduce passive waveguide optical losses. The loss split is as follows: 6.5dB for linear waveguide, 3x0.5dB for waveguide bends, 3x0.5dB for MMI excess loss and 0.5dB for on-state EAMs. Therefore, for all EAMs in on-state about 0dBm output is expected for 10dBm laser output without SOA amplification.

An actual photograph of the realized circuit is shown in Fig 2. The PIC has been soldered on a submount with DC/RF

978-1-4673-6130-9/13 $31.00 © 2013 IEEE

ceramics and assembled into OIF standard module intended for LiNbO3 QPSK-MZM shown in Fig 2. Our InP-monolithic QPSK-source is "lost inside" emphasizing well 50 times footprint gain.

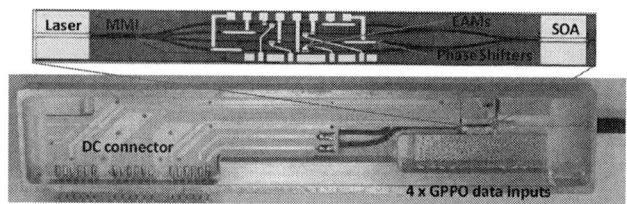

Figure 2. Actual TX PIC chip photograph (upper). The PIC is assembled into OIF standard QPSK package designed for LiNbO3 component. Our InP QPSK-source is "lost inside" imaging well 50 times fooprint gain.

All chip components (5 passives and 14 actives) were operational with excellent fabrication yield. Phase shifters provided up to 2π phase change in 0-30mA current range, EAMs have 15dB Extinction Ratio and above 20GHz bandwidth. Good power balance in the arms provided up to 25dB power extinction in BPSK destructive interference state.

III. TRANSMISSION EXPERIMENTS

The PIC TX module is fed by $(2^{15}-1)$-bit-long pseudo-random bit at 28 Gb/s, including 12% overhead for forward error correction (FEC) and protocol, to produce BPSK data at 28 Gb/s (Fig 3). Polarization division multiplexing (PDM) is performed by dividing, decorrelating and recombining the BPSK data through a polarization beam combiner (PBC), yielding data stream at 56 Gb/s. The emission from the test channel is sent to a flexible filter with tunable optical profile acting as a 33-GHz interleaver. Afterwards, noise loading is performed through a variable attenuator and a double stage amplifier including a 4-nm bandwidth optical filter (to reduce out of band noise). Then the light is divided through a 3dB power splitter to be simultaneously sent to the coherent receiver and an optical spectrum analyzer for optical signal-to-noise ratio (OSNR) measurement in 0.1 nm.

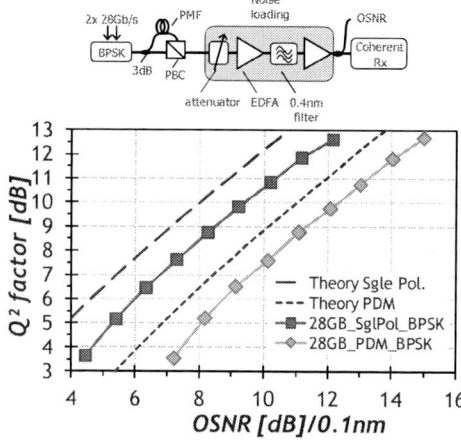

Figure 3. Experimental setup for noise sensitivity measurement (upper) and optical noise sensitivity of 28-Gb/s BPSK and 56 Gb/s PDM-BPSK (lower)

At the receiver side, we use a coherent receiver described in detail in [7]. The 8 outputs of the coherent mixer are detected by four balanced photodiodes and sampled at 80 GS/s using a digital oscilloscope with a 30-GHz electrical bandwidth. The recovered data are then decoded, the bit-error ratio (BER) performed and converted into Q^2-factor. In Fig. 3, we provide the measured back-to-back Q^2 factor versus optical signal-to-noise ratio (OSNR, in 0.1 nm) of single-polarization 25-Gb/s BPSK (actual 28 Gb/s) and 50-Gb/s PDM-BPSK (actual 56 Gb/s) signals respectively. The experimental results are then compared to the theoretical OSNR curve for BPSK formats showing only about 1 dB penalty.

IV. CONCLUSION

We integrated monolithically a novel QPSK-ready transmitter source based on prefixed phase switching by fast EAMs using a flexible photonic integrating circuit technology. To the date we realized 28GB/s BPSK experiments using PDM emulator providing 56Gb/s total data rate. Slightly above 1dB OSNR penalty versus theory was found. Simulations suggest the PIC ability to digital complex signal modulations (QAM) as well as to Side Band Suppressed or Carrier Suppressed digital/analog formats. We believe that this work demonstrates the smallest full monolithic QPSK/BPSK transmitter and confirms the viability of the novel advanced format PICs. Speed scalability, small footprint and low power drive of the EAM based integrated sources are attractive for advanced format migration towards low-cost applications.

ACKNOWLEDGMENT

This work has received support from IST-257980 MIRTHE CEE project. We acknowledge the work on chip design and technology from David Carrara, Romain Berruée and Ronan Guillamet and a processing assistance from Francis Poingt, Nadine Lagay, Florence Martin and Michel Goix.

REFERENCES

[1] N. Kikuchi, Y. Shibata, K. Tsuzuki, H. Sanjoh, T. Sato, E. Yamada, T. Ishibashi and H. Yasaka, "80-Gb/s Low-Driving-Voltage InP DQPSK Modulator With an n-p-i-n Structure", in IEEE Photonics Technology Letters, IEEE, vol. 21, no. 12, pp. 787–789, Jun. 2009

[2] I. Kang, "Phase-shift-keying and on-off-keying with improved performances using electroabsorption modulators with interferometric effects", in Opt. Express, vol. 15, no. 4, pp. 1467–1473, 2007

[3] C.R. Doerr, L. Zhang, A. L. Adamiecki, N.J. Sauer, J.H. Sinsky, and P.J. Winzer, "Compact EAM-Based InP DQPSK Modulator and Demonstration at 80 Gb/s," OFC 2007, PDP33 (2007).

[4] C.R. Doerr, L. Zhang, P.J. Winzer, and A. H. Gnauck, "28-Gbaud InP Square or Hexagonal 16-QAM Modulator", OFC/NFOEC2011, OMU2

[5] C. Kazmierski, "Electro-absorption-based fast photonic integrated circuit sources for next network capacity scaling", J. Opt. Commun. Netw. vol. 4, no. 9, pp. A8-A16, 2012, invited paper.

[6] M. Le Pallec, J. Decobert, C. Kazmierski, A Ramdane, N. El Dahdah, F. Blache, J.-G. Provost, J. Landreau, F. Barthe, N. Lagay, "42 GHz bandwidth InGaAlAs/InP electro absorption modulator with sub-volt modulation drive capability in a 50 nm spectral range," Proc. IPRM 2004, pp. 577 – 580, 31 May – 4 June 2004

[7] O.Bertran-Pardo, J.Renaudier, G.Charlet,H.Mardoyan, P.Tran, M.Salsi, and S.Bigo," Overlaying 10 Gb/s Legacy Optical Networks With 40 and 100 Gb/s Coherent Terminals", J.Lightwave Technology, vol 30, No14, July 2012

TuD4-3 (Oral)
17:15 - 17:30

IPRM2013, May 19 - 23, 2013, Kobe, Japan
The 25th International Conference on Indium Phosphide and Related Materials

InP-based Compact Reflection-Type Transversal Filter

Y. Ueda, T. Fujisawa, K. Takahata, M. Kohtoku, H. Takahashi, and H. Ishii

NTT Photonics Laboratories
NTT Corporation
3-1, Morinosato-Wakamiya, Atsugi, Kanagawa, 243-0198, Japan.
ueda.yuta@lab.ntt.co.jp

Abstract— **We developed a reflection-type 4x1 transversal filter (TF) on an InP substrate as a compact and low-loss wavelength multiplexer (MUX) for monolithically integrated light source arrays. The filter has a satisfactory MUX function and is very compact of about 900 μm x 50 μm which is one twentieth of a size of a conventional 4x1 TF. It is suitable for a MUX of a monolithically integrated light source array thanks to its compactness and input/output-port arrangement.**

Keywords— *transeversal filter; multi-mode interference coupler; reflection mirror; monolithically integrated light source array*

I. INTRODUCTION

100Gbit Ethernet (100GbE) is being strongly promoted as a way of dealing with the recent rapid increase in the data rate of data communication systems. The 100GbE specifications stipulate the use of 4-channel 25-Gbit/s signals over four wavelengths in the 1.3-μm band, and we have developed a monolithically integrated light source array (MILSA) that consists of four DFB lasers, four electroabsorption modulators (EAMs) operating at 25 Gbit/s and one 4x1 multi-mode interference coupler (MMI) as a wavelength multiplexer (MUX) [1]-[3]. The wavelength interval of each DFB laser is 800 GHz (approximately 4.5 nm in the 1.3-μm band) in accordance with the 100GbE specifications. Our MILSA for DFB lasers integrated with EAMs (EADFB laser array) shows good 40-km-transmission characteristics thanks to the clear EAM waveforms. However, an issue with the EADFB laser array chip is that it has an on-chip loss of about 7 dB due to the MUX, including the principal 6-dB loss of the 4x1 MMI. Since such MUX-induced loss is common in MILSAs, MUXs are key devices in various kinds of MILSAs, including both our 100GbE light source and other wavelength division multiplexing light sources such as tunable lasers. A transversal filter (TF) based on two MMIs is a possible candidate as a MUX device. An MMI-based TF has already been demonstrated as a demultiplexer (DEMUX) for an optical orthogonal frequency division multiplexing receiver on a silica planar lightwave circuit [4]. It has the advantage of a compact size compared with arrayed waveguide grating filters (AWGs) when the number of channels is limited to several. But, we should make it smaller not to increase the chip size of our conventional MILSA using a 4x1 MMI as a MUX.

Therefore, to reduce TF size we newly propose a reflection-type TF (RTF), which is formed with only one MMI and shorter delay lines than those of conventional TFs. In this paper, we describe the design of the RTF and the experimental

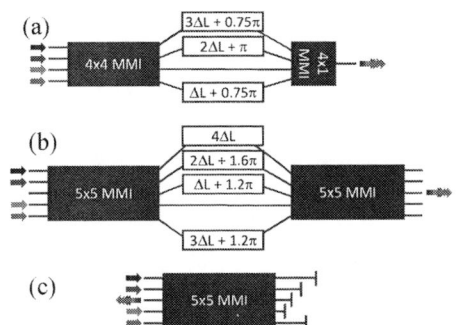

Fig. 1 Schematic structure of TFs as MUX devices.
(a) Conventional 4x1 TF using a 4x4 and a 4x1 MMI.
(b) Conventional 5x1 TF using two 5x5 MMIs.
(c) Newly designed 4x1 RTF using one 5x5 MMI.

characteristics of a fabricated RTF which exhibited satisfactory filter characteristics as a MUX. The RTF size is about 900 μm x 50 μm for a 4x1-port configuration and it is very compact compared to conventional MUX filters. It needs no additional footprint for a MUX when we adopt it to MILSAs because of its compactness and input/output-port arrangement.

II. DEVICE DESIGN AND FABRICATION

The design of our newly created 4x1 RTF is similar to that of a conventional 4x1 TF (fig.1 (a)) as described in [4]. When we consider the 4x1 MUX TF to be a "1x4" DEMUX TF for ease of understanding, the 1x4 TF consists of a 1x4 MMI as a splitter, four delay lines and a 4x4 MMI as a discrete-Fourier-transform (DFT) circuit. The length of each delay line is chosen in terms of both the free spectral range (FSR) of the TF and the phase characteristics of 1x4 and 4x4 MMIs. Namely, the length of each delay line changes approximately in steps of $\Delta L = c / n_g \nu_{FSR}$. Here, c, n_g and ν_{FSR} are the speed of light in a vacuum, the group refractive index in the delay lines and the FSR of the TF, respectively. In addition, the fine length adjustment of delay lines for phase compensation, as shown in fig. 1, is needed so that the phases of the output beams from the 1x4 MMI are the same and the transfer matrix of a 4x4 MMI can correspond to that of the 4x4-DFT circuit. To convert a 4x1 TF to a 4x1 RTF, we design a "symmetric 5x1" TF using the same approach as that employed for the 4x1 TF. That is, a 5x5 MMI is used for the output-side coupler instead of a 5x1 MMI as shown in fig.1 (b). Since both the first and second couplers are 5x5 MMIs, the device structure is symmetric. Hence, by folding the 5x1 TF at its center position we can obtain the 5x1 RTF (fig. 1 (c)). When we input light whose wavelength

978-1-4673-6130-9/13 $31.00 © 2013 IEEE

133

TuD4-3 (Oral)
17:15 - 17:30

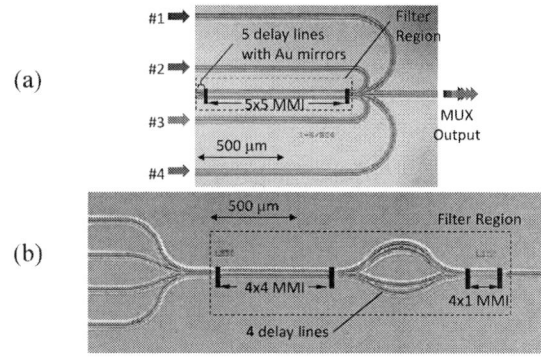

Fig. 2 Photographs of fabricated TFs
(a) 4x1 RTF (b) conventional 4x1 TF.

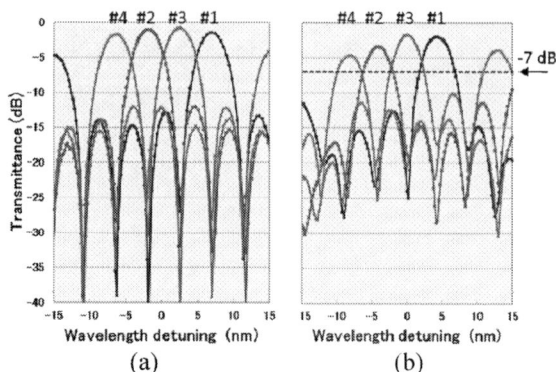

Fig. 3 Transmittance of a 4x1 RTF
(a) designed (b) experiment.

matches the transmittance peak of the 5x1 RTF into only four of the five input ports we can use the 5x1 MUX RTF as a 4x1 MUX RTF.

We fabricated an RTF with an InP/InGaAsP (1.15Q, 300 nm)/InP wafer by conventional photolithography and dry etching in the form of inductively coupled plasma reactive ion etching (ICP-RIE). The waveguide has a high-mesa structure which is advantageous for the compact device size. At the etched facet of the delay lines we employ Au mirrors providing high reflectance [5]. Figure 2 shows photographs of (a) a fabricated 4x1 RTF and (b) a conventional 4x1 TF to allow a size comparison. The filter region of the fabricated RTF is about 900 μm x 50 μm and it is about one twentieth compared with the conventional TF of about 1700 μm x 400 μm. For the conventional TF, the length of the shortest reference delay line is determined by the distance between two MMIs. In fig. 2 (b), the distance is more than 500 μm. On the other hand, for the RTF, no reference delay line is needed. As regards the widths of the two types of TFs, the RTF needs a width of less than 50 μm, which is determined only by the 5x5-MMI width, and the conventional TF needs a width of ~400 μm because of its delay line. The width difference is also derived from the delay lines of each type. Namely, the RTF does not need bent waveguides for the delay lines.

III. CHARACTERISTICS MEASUREMENT

Figure 3 shows (a) the designed and (b) the experimental transmission characteristics of a 4x1 RTF. In fig. 3 (b) the reference wavelength is 1285 nm. Labels #1 to #4 correspond to the numbers in fig. 2 (a). For the designed RTF, peak intervals among four channels are set to 4.5 nm as the wavelength grid of 100GbE. In the ~9-nm interval between channels #1 and #4, at the detuning of plus-minus ~11 nm, "reflectance" characteristics should appear at the MUX output port, which can be understood by comparing fig. 1(b) and (c). From fig.3 (a), it is found that the designed RTF has ~1dB loss even at the peak wavelength of each channel. It is caused by optical decay at the Au mirror and coupling loss between the MMI and input/output ports. Although the former is intrinsic loss of the Au mirror, the later can be improved by devising the junction between the MMI and its input/output ports.

From fig. 3 (b), it is found that the peak intervals correspond to designed ones, which indicates that each of the

five delay lines provides the propagating beams with an accurate phase change. The Au mirrors at the etched facets of the delay lines are also correctly formed since the reflectance loss at the boundary between the InP-based waveguide and air is larger than 5 dB. The difference of each transmittance peak between designed and fabricated RTF means that the fabricated RTF has larger wavelength dependent loss of the MMI than that of the designed one. It is caused by the fabrication error of the MMI, resulting in unintended shift of the central wavelength of the MMI transmittance to long-wavelength side from the designed one.

The dashed line in fig. 3 (b) indicates a transmittance of -7 dB, which is the experimental excess loss we obtained for a 4x1 MMI as a MUX. Therefore, we can improve MUX loss more than 2 dB when we adopt the RTF as a MUX to a 4-channel MILSA instead of a 4x1 MMI. And, the advantage of the RTF is its compactness as indicated in fig. 2 (a). It needs no additional footprint when we use the RTF as a MUX for MILSAs because we can place the 5x5 MMI of the RTF between the light sources. This technique for saving length relies on both the compactness and the input/output-port arrangement of the RTF.

IV. CONCLUSION

We proposed a novel type TF using reflection as a low-loss and compact MUX for MILSAs. We designed the RTF as a MUX for our developing 100GbE light source and fabricated it with InP-based high-mesa waveguides. The RTF exhibited a basic filtering function. And, it needs no additional chip footprint because of its compact size of about 900 μm x 50 μm and input/output-port arrangement. We believe that our RTF can help to enhance the output power of various kinds of MILSA devices.

REFERENCES

[1] T. Fujisawa, et al., *IEEE Photon. Technol. Lett.*, **23**, p. 356, 2011.

[2] S. Kanazawa, et al., *IEEE J. Sel. Topics. Quantum Electron.*, **17**, p. 1191, 2011.

[3] T. Fujisawa, et al., *Opt. Express*, **20**, p. 614, 2012.

[4] K. Takiguchi, et al., in Proc. of *OFC/NFOEC 2012*, OM3J.6, 2012.

[5] T. Segawa, et al., *IEICE Trans. Electron.*, **E94-C**, p. 1439, 2011.

TuD4-4 (Oral)
17:30 - 17:45

Transmitter PIC for THz Applications Based on Generic Integration Technology

F.M. Soares, J. Kreissl, M. Theurer, E. Bitincka*,T. Goebel, M. Moehrle, N. Grote

Photonics Components Dept.
Fraunhofer Institute for Telecommunications, HHI
10587 Berlin, Germany
norbert.grote@hhi.fraunhofer.de

Abstract—**A generic InP based monolithic photonic integration platform is introduced that is capable of simultaneously incorporating transmitter, receiver and passive-optical functionalities. On this basis, an integrated transmitter component for THz applications has been implemented.**

Keywords—generic PIC technology; building blocks; butt-joint technology; component; integrated DFB laser; THz transmitter

I. INTRODUCTION

In the past, the dominant working method in photonics basically relied on developing optimized fabrication processes for each individual application, starting from the specifications of the product. In contrast to this approach, development of generic photonic integration technology [1] is being undertaken in the frame work of the European projects *EuroPIC* [2] and *PARADIGM* [3] to be capable of realising complex InP based photonic integrated circuits (PIC) from a small set of basic building blocks. In the *EuroPIC* project Fraunhofer HHI has successfully developed a semi-insulating substrate based monolithic integration platform that was restricted to mere receiver (Rx) type PICs [4,5]. This work is being extended in *PARADIGM* to incorporate transmit (Tx) type active photonic functionalities as well. In this paper the integration scheme applied will be outlined, and early results will be presented on integrated laser devices and, in particular, on a transmitter PIC designed for THz applications.

II. GENERIC INTEGRATION SCHEME

The Tx/Rx integration scheme was contrived as extension of the previously developed generic Rx-platform without impacting its established building blocks (BB) and fabrication technology. In brief, these comprise an evanescent-coupled pin photodiode of 40 GHz capability placed on top of a weekly guiding Fe-doped ridge waveguide; two more semi-insulating waveguide variants differing in their guiding strength; associated transition elements; waveguide crossings; circular arcs; and a waveguide spot size converter ("10 µm" mode size). Further available waveguide-based passive composite BB include various MMI couplers, Y-junctions, and thermo-optic phase shifters.

To combine those elements with active Tx structures (basically DFB/DBR laser diodes, semiconductor optical amplifiers (SOA) and electro-absorption devices) as seamlessly as possible butt-joint integration technology has been adopted. To this end, the Tx layer stack is grown first, and after patterning the resulting mesa structures are regrown with the Rx layer stack using adapted selective area MOVPE. In total, three growth steps are required. Because the Rx platform was built on InP:Fe substrate (to enable very high electrical bandwidth) the Tx devices had to be redesigned accordingly in that not only the p- but also the n-contact is formed on top. As to the bottom n-InP access layer, a thickness of 1 µm was experimentally ascertained to be sufficient for not affecting the series resistance significantly. All Tx-BBs are designed as RW structures. A schematic cross-section of the Tx/Rx generic integration platform is depicted in Fig. 1 showing the different classes of functionalities incorporated.

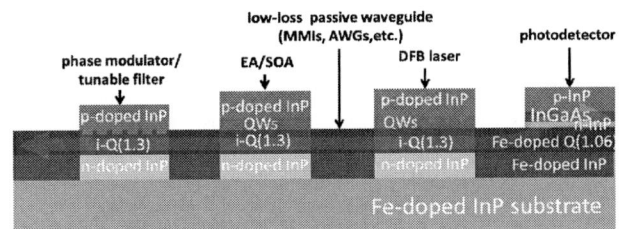

Figure 1. Schematic of generic Tx/Rx PIC platform

III. PERFORMANCE OF INTEGRATED LASER DIODES

To verify the feasibility of the chosen integration approach, the first structure that was characterized was a DFB laser integrated with the key Rx building blocks, i.e. the semi-insulating (weakly guiding) optical waveguide and the photodiode. The waveguide facets are angled and AR coated in order to minimize back reflections into the laser. The DFB laser itself, featuring a complex-coupled grating, is composed of an active region of strained InGaAsP/InGaAsP MQW layers sandwiched between an upper grating layer and a lower undoped 1.3 µm-quaternary waveguide layer. This stack is aligned to the Fe-doped interconnecting waveguide such as to provide optimum coupling efficiency. Theoretically, as derived from optical simulations, coupling losses of as low as 1.6 dB are achievable with non-tapered waveguides, and even < 0.1 dB when using appropriate tapering. The laser width was

---

The research leading to these results has received funding from the European Commission's 7th Framework Programme FP7/2007-2013 under grant agreement ICT 257210 PARADIGM
*on leave from: COBRA Res. Inst., Technische Universiteit Eindhoven, 5600 MB Eindhoven, The Netherlands

varied from 2.0μm to 2.4μm, whereas the length was varied from 200 μm to 400 μm.

The basic performance of such integrated DFB lasers is exemplarily represented in Fig. 2. PI curves measured at the non-PD output side of a passive waveguide section (1.2 mm long) indicate threshold currents in the range from about 8 mA to about 17 mA as the length increases from 200 μm to 400 μm. The shortest DFB lasers of 200 μm showed the highest output power, as to be expected from the DFB coupling strength which was optimized for the 200 μm long devices. Light from the rear side was directly coupled into the PD and measured via the photocurrent. Given a responsivity of 90%, a coupling loss at the (non-tapered) laser/waveguide interface of about 3 dB can be estimated which is still a factor of 2 higher than the calculated value. Small-signal rf-measurements carried out on-chip show that the 3 dB-bandwidth exceeds 12 GHz at 22 mA bias current which renders these lasers highly suited for 10 Gb/s modulation and even above.

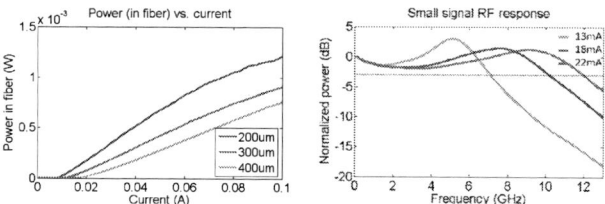

Figure 2. PI characteristics vs length of integrated complex-coupled DFB lasers with light coupled from passive waveguide output into fiber (left), and on-chip small-signal high-frequency curves for 200μm long laser (right)

IV. THz TRANSMITTER PIC

Based on the technology approach outline afore, a transmitter PIC was realized to form an integrated optical beat source for photomixing-based cw THz systems [6]. A photograph of the fabricated PIC is shown in Fig. 3.

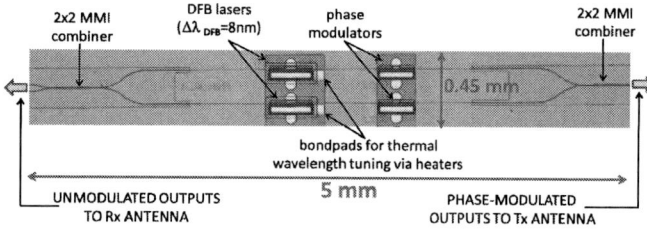

Figure 3. Photograph of the fabricated transmitter chip for THz applications.

This PIC contains the following BBs: two DFB lasers with on-chip electrical heaters for thermal wavelength tuning, two current-injection based phase modulators, and two 2x2 MMI combiners. The operating principle is as follows: The optical signals generated by each DFB laser are combined in a 2x2 MMI to create a heterodyne output signal with an envelope frequency equal to the optical frequency difference between both lasers. The integrated thermo-optic heaters on each DFB laser are used to adjust the wavelengths and sweep the frequency of the heterodyne signal. Furthermore, the PIC contains phase modulators on the right-hand side of the DFB lasers. This allows for access of the THz phase via standard

electronics. The left-hand side output is used on the receiver side of the THz antenna for coherent detection.

Fig. 4 shows measured optical spectra of the output from the THz generator PIC for various heater currents showing well-behaved single-mode characteristics over the full heater current range (90 mA). The output power from these first PICs measured at the chip output was still moderate in the 10-100 μW range. The phase modulator of 250 μm length allows for a π-shift at 5 mA of injected current.

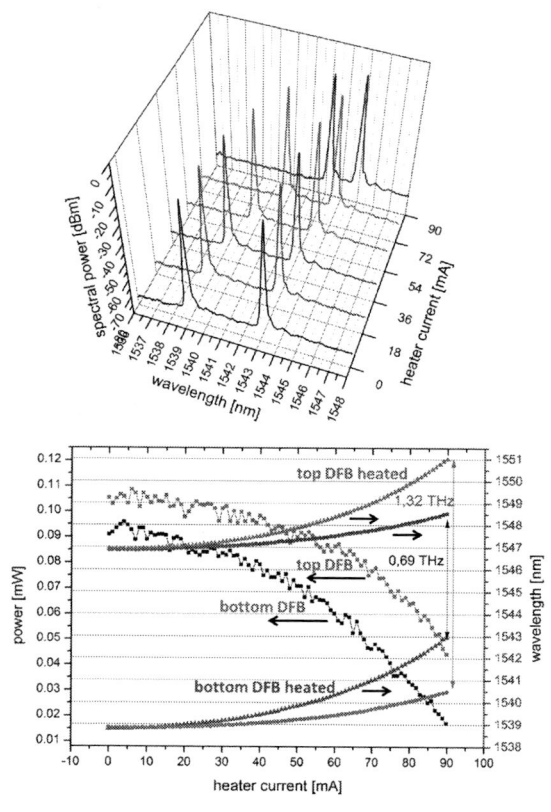

Figure 4. Measured output spectra of the THz generator PIC for various DFB heater currents (top), and a plot of the output power and the wavelengths of both DFB lasers versus the heater current (bottom).

REFERENCES

[1] M. Smit, X. Leijtens, E. Bente, J. v. d. Tol, H. Ambrosius, D. Robbins, M. Wale, N. Grote, and M. Schell, "Generic Foundry model for InP-based photonics", IET OPtoelectron., 2011, vol. 5, p. 187-194

[2] www.europic.jeppix.eu

[3] www.paradigm.jeppix.eu

[4] F. M. Soares, K. Janiak, N.Grote, D.Szymanski, and M.J.Wale, "Generic InP-Based Monolithic Photonic Integration Platforms", Proc. 16th Europ. Conf. on Integrated Optics (ECIO 2012), April 2012, Sitges (Spain), paper #177

[5] F. M. Soares, K. Janiak, A. Seeger, R. G. Broeke, and N. Grote. "An InP-Based Generic Integration Technology Platform", IEEE Photon. Conf. 2012, Burlingame, California, U. S. A., Sept. 2012, paper ThZ1.

[6] D. Stanze et al.: "Compact CW terahertz spectrometer pumped at 1.5 μm wavelength," Journal of Infrared, Millimeter and Terahertz Waves, vol. 32, no. (2), p. 225-232 (2011).

TuD4-5 (Oral)
17:45 - 18:00

Intermixing of Highly-Stacked InAs/InGaAlAs Quantum Dots Grown on InP (311)B Substrate by SiO₂ Sputtering and Annealing Technique

A. MATSUSHITA[1], A. MATSUMOTO[1], K. AKAHANE[2], Y. MATSUSHIMA[3], and K. UTAKA[1]

[1] Faculty of Science and Engineering, Waseda University, Tokyo, Japan
[2] National Institute of Information and Communications Technology, Tokyo, Japan
[3] Green Computing System Research Organization, Waseda University, Tokyo, Japan

e-mail: asuka-matsushita@ruri.waseda.jp

Abstract—**We have studied the intermixing of highly-stacked InAs/InGaAlAs quantum dots grown on InP (311)B substrates by SiO₂ sputtering and annealing technique with a low temperature of 650 □ to find a large PL spectral blue-shift by about 60 nm. This result suggests that the low temperature intermixing technique is promising for easy formation of monolithic integrated circuits with the highly-stacked QD structures.**

Keywords— InAs/InGaAlAs; quantum dots; intermixing; PL wavelength

I. INTRODUCTION

Quantum dots (QDs) are very promising materials for high performance photonic devices such as laser diodes, semiconductor optical amplifiers, nonlinear functional devices and solar cells from viewpoints of high gain, low threshold, low noise, ultra-fast response and so on, which are originated from inherent narrow density of states. Especially highly-stacked InAs/InGaAlAs QD structures grown on InP (311)B substrates by molecular beam epitaxy (MBE) with strain compensation technique [1] seem interesting since the operating wavelength is in a range of 1550 nm suitable for telecom application, and highly-stacked QD layers are capable of further enhancement of these performances mentioned above.

As for the application of functional photonic devices using the highly-stacked QD structure, we have analytically revealed the potential performance of an ultra-fast optical logic-gate device integrated with a QD semiconductor optical amplifier (QD-SOA) and ring-resonators operated at 160Gb/s [2]. Here, these semiconductor integrated devices are required to be consisted of the same kind of materials in order to avoid possible drawbacks such as inferior optical coupling at the different material interfaces due to refraction index difference, and also easy fabrication without crystal regrowth should be pursued at the same time. From this point, an intermixing technique seems promising, and this technique has been applied to quantum well structures, known as quantum well intermixing (QWI) [3],[4], since the QWI technique gives wider bandgap materials suitable for a passive waveguide. So far, QD intermixing (QDI) technique was also reported for

This work was partly supported by the SORTE project.

InGaAs/GaAs/AlGaAs QDs on GaAs substrates with high temperature annealing of 900 □ exhibiting a large bandgap blue-shift of 280nm [5].

In this paper we studied the QDI technique for highly-stacked InAs/InGaAlAs QDs on InP substrates with low temperature annealing of 650 □ in order to realize integrated photonic devices with QD-SOAs and passive waveguides. In the field of QWI, there are many ways of intermixing such as impurity induced disordering (IID), impurity free vacancy disordering (IFVD), laser induced disordering (LID), ion implantation induced interdiffusion (IIII), and universal damage induced technology (UDIT). We focused on UDIT which has the characteristics of high crystal quality and low optical propagation loss because it does not intentionally use impurities. Therefore, we introduced damages into the QD wafer by repetitive SiO₂ sputtering, and the QD wafer was annealed in the end. The effect of QDI was evaluated by measuring PL emission spectra to successfully find its blue-shift.

II. EXPERIMENTAL METHODS

The highly-stacked InAs/InGaAlAs QD structure used in the experiment was basically the one for a laser diode application, and it was grown on an InP (311)B substrate by MBE using the strain compensation technique [1]. The wafer consisted of a 100-nm-thick lattice-matched Si-doped n-type InAlAs buffer layer, an active layer with 30 stacked layers of 3ML InAs QDs and 15-nm-thick InGaAlAs spacer layers, a 2000-nm-thick lattice-matched Be-doped p-type InAlAs cladding layer and a p-type InGaAs contact layer. The QDs were of uniform size and in ordered arrays without coalescence, as shown in Fig.1. The average lateral size and height of QDs were 66 and 7.5 nm, respectively, and the total QD density was ~2.5x10¹² cm⁻².

The QDI procedures were as follows. To realize selective QDI, a SiO₂ film was deposited by CVD to protect the surface from damage. In order to introduce damages into the wafer another SiO₂ film was sputtered on the cap layer for a couple of times. The conditions of sputtering were an Ar flow rate of 20 sccm, an O₂ flow rate of 3 sccm, a RF power of 400 W, and a pressure of 0.5 Pa. A deposited thickness of the SiO₂ film was 220 nm. Each time the deposited SiO₂ film was eliminated by buffered HF (BHF) solution before depositing it again so that the damage would not be alleviated by the SiO$_2$ deposition previously. Finally the sample was annealed with and without the last deposited SiO$_2$ film. The annealing was done in the condition of N$_2$ gas atmosphere at 650 ☐ for 1 minute. Afterwards, these samples were evaluated by measuring photoluminescence (PL) spectra.

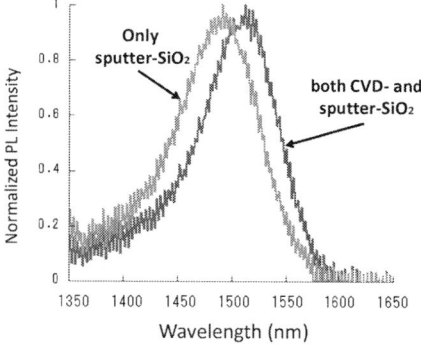

Fig.1 STEM image of highly-stacked InAs/InGaAlAs quantum dots

III. RESULTS AND DISCUSSION

Fig.2 shows the PL spectra for the samples with both CVD- and sputter-deposited SiO$_2$ films and only a sputter-deposited SiO$_2$ film. The annealing was treated after removing all the films for both cases. It was confirmed that the PL spectrum for the sample with only sputter-deposited SiO$_2$ film was blue-shifted by about 20 nm compared with the one for the sample with both SiO$_2$ films, in which the CVD-deposited SiO$_2$ film is said to have the function to protect the substrate surface from the introduction of damages at the time of the sputter deposition of SiO$_2$.

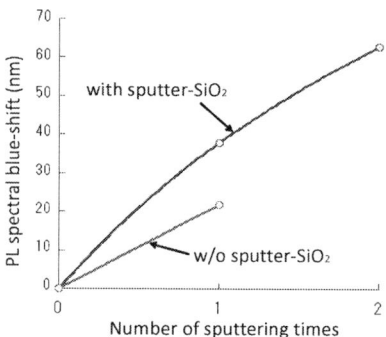

Fig.2 PL spectra for the samples with both CVD- and sputter-deposited SiO$_2$ films and only sputter-deposited one.

Fig.3 shows the effect of the sputtering time on the amount of PL spectral blue-shift. Due to the existence of a cap layer PL signal levels were rather low, but the effect of the QDI seems clear showing the blue-shifts of the spectra depending on the sputtering time of SiO$_2$. The relationship between the sputtering time and the amount of PL spectral blue-shift is plotted in Fig.4. From Fig. 4 it can be said that the more times SiO$_2$ is sputtered, the more the PL peak wavelength is blue-shifted. This may be because the repeating of sputtering increased the damages introduced into the substrate even though the distance of the sample surface and QD layers was

about 2000 nm. It could also be said that the PL spectral blue-shift was larger for the case that the sputter-deposited SiO$_2$ film was left when annealing. The sputter-deposited SiO$_2$ film is thought to have a role in stimulating the intermixing through the diffusion of defects to the substrate direction. A large bandgap blue-shift of about 60 nm was observed even for low temperature annealing of 650 ☐ compared to ref. [5].

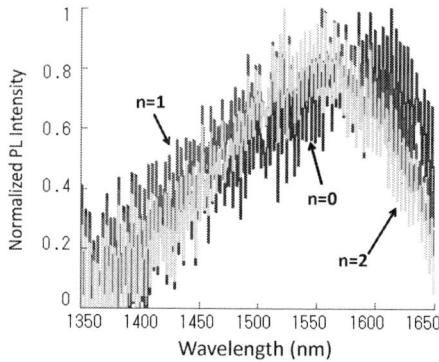

Fig.3 PL spectra for the samples for various SiO$_2$ sputtering times of n = 0, 1 and 2.

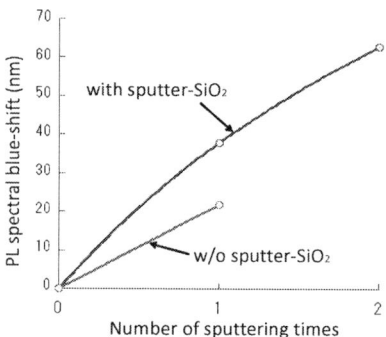

Fig.4 Relationships between the number of SiO$_2$ sputtering time and the amount of PL spectra blue-shift for the annealing with and without the last sputter-deposited SiO$_2$ film.

IV. CONCLUSION

A PL spectral blue-shift as large as 60 nm was observed for a highly-stacked InAs/InGaAlAs QDs by QDI with a twice SiO$_2$ sputtering and low-temperature annealing at 650 ☐. A much larger shift would be expected by improving the QDI technique such as under a thinner cladding layer and more sputtering times for low loss waveguides and high performance photonic integrated circuits using QDs.

REFERENCES

[1] K.Akahane, et al., Photon. Technol. Lett., vol.22, No.2, pp.103-105, 2010.

[2] A.Matsumoto, et al.,23rd IPRM 2011, We-7.2.5, 2011.

[3] D.A.Yanson, et al., IPRM 2005, pp.504-507, 2005.

[4] O.P. Kowalski, et al., Appl. Phys. Lett., vol.72, No.5, pp.581-583, 1998.

[5] A.C.Bryce, et al, Proc. 5th Conf. on Nanotech., pp.1122-1124, 2005.

WeD1-1 (Oral)
8:30 - 8:45

High Transconductance Surface Channel In$_{0.53}$Ga$_{0.47}$As MOSFETs Using MBE Source-Drain Regrowth and Surface Digital Etching

Sanghoon Lee[1], Cheng-Ying Huang[1], Andrew D. Carter[1], Jeremy J. M. Law[1], Doron C. Elias[1], Varistha Chobpattana[2], Brian J. Thibeault[1], William Mitchell[1], Susanne Stemmer[2], Arthur C. Gossard[2], and Mark J. W. Rodwell[1]

[1]Department of Electrical and Computer Engineering, [2]Materials Department, University of California, Santa Barbara, CA 93106 USA,

Email: sanghoon_lee@ece.ucsb.edu

Abstract — We demonstrate In$_{0.53}$Ga$_{0.47}$As surface channel MOSFETs using a gate-last process and MBE source/drain (S/D) regrowth. The structure uses a sacrificial N+ InGaAs channel cap layer between the regrown S/D contact layer and the channel, which is removed in the channel region by a "digital" etch process incorporating UV ozone oxidation and surface stripping in dilute HCl. A device with 65 nm-L_g and 1.2 nm EOT shows 1.6 mS/μm peak transconductance at V_{ds} = 0.5 V and 120 mV/dec SS at V_{ds} = 0.05 V, while 535 nm-L_g devices show 95 mV/dec SS at at V_{ds} = 0.1 V

Keywords—InGaAs MOSFETs; source-drain regrowth; digital etching style; substitutional-gate; surface channel

I. INTRODUCTION

Because of the potential for increased on-state current, In$_{1-x}$Ga$_x$As MOSFETs have been widely studied for potential future application in VLSI [1]-[6]. High performance MOSFETs require ultra-thin EOT gate dielectrics with low interface trap density (D_{it}). In addition, heavily-doped source/drain (S/D) regions are required for adequate carrier supply and for low S/D access resistance given roadmap-compliant (~20-50 nm) S/D contact pitch. MOSFETs with heavily-doped S/D regions can be formed by first growing an N+ S/D contact layer above the channel and subsequently recess etching through this layer in the region where the gate will be placed [1]-[3]. To be viable in VLSI at the 8-22 nm nodes, this recess etch would need to provide ~0.5-2 nm precision depth control within gate openings of 8-22 nm length. N+ S/D regions can also be formed by epitaxial regrowth [4]-[6], in either a gate-first or a substitutional-gate process. In such devices, performance can be limited by defects at the junction between regrown S/D and the channel, or by damage to the channel surface during regrowth.

Here, we report InGaAs MOSFETs with an N+ source-drain contact layer regrown by molecular beam epitaxy (MBE). These MOSFETs incorporates a 5 nm InGaAs N+ epitaxial cap layer grown above the InGaAs channel. The S/D contact layers are regrown on this cap layer, and the cap layer is then removed in the channel region using digital etch process [3], [7] incorporating UV ozone oxidation and surface stripping in dilute HCl. A device with 65 nm-L_g shows an excellent peak transconductance of ~1.60 mS/μm

at V_{ds} = 0.5 V and 120 mV/dec sub-threshold swing (SS) at V_{ds} = 0.05 V.

II. DEVICE FABRICATION

The epitaxial layers, grown by MBE, consist of a InP (100) semi-insulating (S. I.) substrate, a 400 nm unintentionally doped (UID) In$_{0.52}$Al$_{0.48}$As buffer/barrier layer, a 3 nm Si-doped (3.9·10^{12} /cm^2) In$_{0.52}$Al$_{0.48}$As pulse doping layer, a 10 nm In$_{0.53}$Ga$_{0.47}$As channel, and a 5 nm Si-doped (4-5·10^{19} /cm^3) In$_{0.53}$Ga$_{0.47}$As cap layer. Dummy gates were defined by e-beam lithography using hydrogen silsesquioxane (HSQ) resist and an atomic layer deposition (ALD) Al$_2$O$_3$ adhesion layer. Prior to transferring into the MBE chamber, the semiconductor surface was oxidized by UV ozone exposure and then etched in 10:1 deionized water (DI):HCl. Approximately 50 nm Si-doped (5·10^{19} /cm^3) InAs was non-selectively grown on the N+ cap layer. Amorphous InAs growth on top of the dummy gates was removed by a planarization/etch process [8]. Device mesas were defined by wet-etching, and the dummy gates removed in buffered oxide etch (BOE) with Tergitol surfactant. The exposed InGaAs N+ cap layer was etched by 4 cycles of oxidation by UV ozone and surface removal by dilute HCl dip. AFM measurements indicate that the exposed channel surface has 0.15-0.16 nm RMS roughness after the etch. This is comparable to an unprocessed InGaAs surface. Immediately after removing the InGaAs native oxide in BOE, the sample was loaded into the ALD loadlock. 1 nm/4 nm Al$_2$O$_3$/HfO$_2$ (~1.2 nm EOT) gate dielectric was deposited by ALD. The sample was then annealed for 15 minutes at 400 °C in forming gas. 20 nm/100 nm Ni/Au was thermally deposited as the gate electrode. Subsequently, 20 nm/100 nm Ni/Au was lifted off for source/drain metallization by thermal evaporation. The schematic cross-section of the device is shown in Fig. 1.

III. RESULTS AND DISCUSSION

Fig. 2 shows transfer characteristics (I_d-V_{gs} and g_m-V_{gs}) for a device with L_g = 65 nm. The peak transconductance is 1.6 mS/μm at V_{ds} = 0.5 V, while its minimum SS is 120 mV/dec at V_{ds} = 0.05 V. Fig. 3 shows transfer and output characteristics a L_g = 75 nm device. This device exhibits ~1.4 mS/μm peak transconductance at V_{ds} = 0.5 V and ~0.65 mA/μm on-current at V_{gs}-V_{th} = 0.5 V and V_{ds} = 0.5 V, where the threshold voltage is determined to be ~0.15 V from linear extrapolation. From the I_d-V_{ds} plot in Fig. 4, its S/D on-resistance is approximately 300 Ω-μm at V_{gs} = 1.6 V. Fig. 4(a) and (b) show sub-threshold characteristics for both a short channel (L_g = 75 nm) and a long channel (L_g = 535 nm) device. The 75 nm-L_g device shows 124 mV/dec minimum SS at V_{ds} = 0.1 V and 375 mV/V DIBL, whereas a 535 nm-L_g device shows 95 mV/dec minimum SS and <10 mV/dec DIBL at the same V_{ds}.

In comparison with similar substitutional-gate MOSFETs [6] which do not use the recess etch, both the sub-threshold swing and on-state transconductance are considerably improved. The improved sub-threshold swing is strong evidence of decrease D_{it} in the recess-etch sample, suggesting the removal of surface damage associated with processing. Experiments are in progress to determine whether the increased transconductance is be a consequence of removal of this surface damage or of improved S/D access resistivity.

ACKNOWLEDGMENTS

This work was supported by the SRC Non-classical CMOS Research Center (Task 1437.006). A portion of this work was done in the UCSB Nanofabrication facility, part of the NSF funded NNIN network and MRL Central Facilities supported by the MRSEC Program of the NSF under award No. MR05-20415

REFERENCES

[1] M. Radosavljevic, et al., IEDM 2009
[2] D.-H. Kim, et al. IEDM 2012
[3] J. Jin, et al. IEDM 2012
[4] M. Egard, et al., IEDM 2011
[5] U. Singisetti, et al., IEEE Electron Device Lett. Vol. 30, 2009
[6] S. Lee, et al, IEEE Electron Device Lett. Vol. 33, 2012
[7] D. Koegh, et al. Journal of electronic material, Vol. 35, 2006
[8] G. J. Burek, et at., Journal of crystal Growth, Vol. 311, 2009

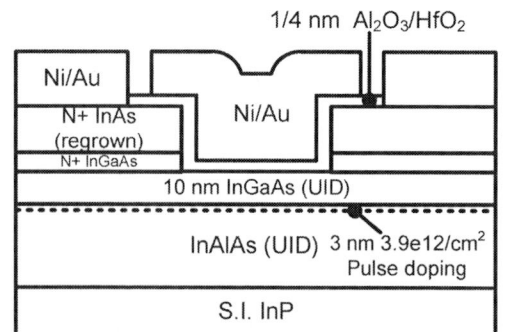

Fig. 1. Schematic cross-section of the substitutional-gate MOSFET.

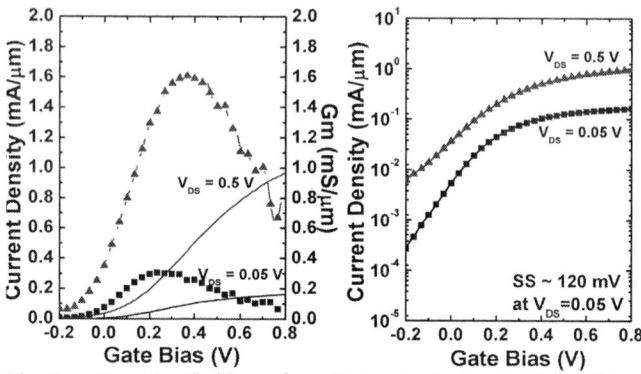

Fig. 2. Transfer (I_d-V_{gs} and g_m-V_{gs}) and sub-threshold ($\log(I_d)$-V_{gs}) characteristics of a 65 nm-L_g device. The device shows an excellent peak transconductance of ~1.60 mS/μm at Vds = 0.5 V and 120 mV/dec at Vgs = 0.05 V.

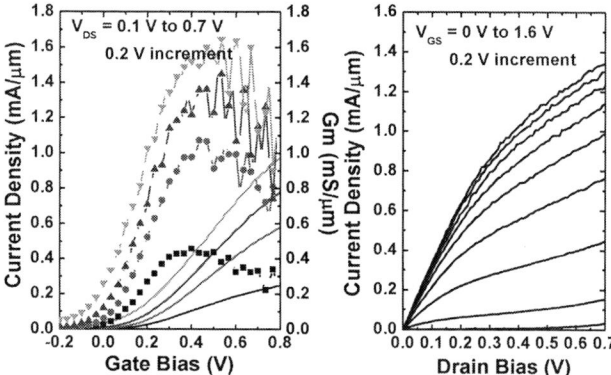

Fig. 3. Transfer (I_d-V_{gs} and g_m-V_{gs}) characteristics and output characteristics (I_d-V_{ds}) of a 75 nm-L_g device. Its on-resistance is extracted to be approximately 300 Ω-μm at V_{gs} = 1.6 V.

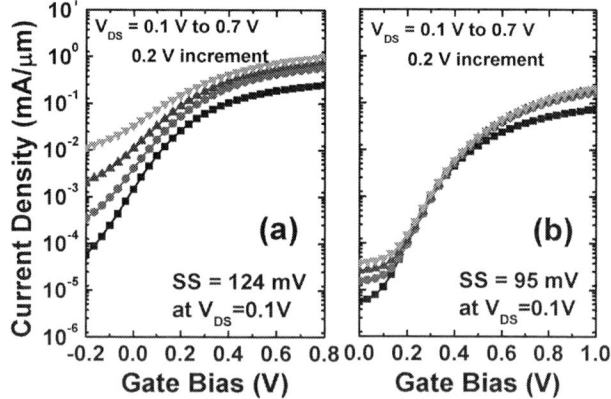

Fig. 4. Sub-threshold ($\log(I_d)$-V_{gs}) characteristics of a short channel device (75 nm-L_g) (a) and a long channel device (535 nm-L_g).

WeD1-2 (Oral)
8:45 - 9:00

Sub-50-nm InGaAs MOSFET with n-InP source on Si substrate

Atsushi KATO[1], Toru KANAZAWA[1], Eiji UEHARA[1], Yoshiharu YONAI, and Yasuyuki MIYAMOTO[1]

[1]*Department of Physical Electronics, Tokyo Institute of Technology,*
2-12-1-S9-2, O-okayama, Meguro-ku, Tokyo 152-8552, Japan

Abstract—We demonstrated a sub-50-nm InGaAs 5-nm/InP 5-nm MOSFET with an n-InP source on a Si substrate using a 5-nm Al$_2$O$_3$ dielectric. In the measurement of the fabricated device, the maximum drain current and the peak transconductance at V_D = 0.5 V were 0.9 mA/μm and 0.8 mS/m, respectively. The threshold voltage was 0.09 V, and the drain-induced barrier lowering was 378 mV/V. From the channel length dependence, clear suppression of the short channel effect by the 5-nm-thick Al$_2$O$_3$ gate dielectric and the extremely thin body III–V–OI structure was confirmed.

I. INTRODUCTION

According to ITRS 2011, the drain current of MOSFETs for high-performance logic circuits will reach approximately 2 A/mm with a low supply voltage (<0.7 V) after 2020 [1]. To realize such high drain current density, the simple introduction of high-mobility channel materials is insufficient. To realize a high drain current density of 2.4 A/mm at V_D = 0.5 V, we used a heavily doped n-InP source with a 50-nm-long InGaAs channel [2]. However, the threshold voltage roll-off and degradation of the drain-induced barrier lowering (DIBL) pose a problem, and extremely thin body (ETB) III–V–OI [3] or nanowire FET structures [4] are required to avoid the short channel effect (SCE). In this report, we fabricated ETB–OI InGaAs-MOSFETs with an n-InP source and a channel length of 45 nm on a Si substrate.

II. EXPERIMENT

Figure 1 shows the schematic image of the fabricated devices. Initially, an epitaxial structure with a composite channel of 5-nm i-InP and 5-nm i-InGaAs layers, 200-nm n-InP source, 10-nm n-InGaAs contact, 20-nm i-InP etch stop, and 200-nm i-InGaAs buffer was grown on an n-InP substrate by MOVPE. The carrier concentration of the n-InP source was 2×10^{19} cm^{-3}. Then, a 2-nm-thick Al$_2$O$_3$ dielectric was deposited by atomic layer deposition (ALD) to avoid the

degradation of the interface, followed by the deposition of a 200-nm-thick SiO$_2$ by plasma-enhanced chemical vapor deposition to ease the strain on the thin-film layer during the bonding process [5]. To transfer the epitaxial structure, the sample was bonded on the Si substrate coated with a 1.1-μm-thick benzocyclobutene (BCB) layer. Subsequently, the InP substrate, 200-nm i-InGaAs buffer, and 20-nm i-InP etch stop were removed by wet etching. After the transfer of the epitaxial layers on the Si substrate, a channel region was fabricated using Ti/Pd/Au source/drain electrodes as an etching mask [2]. The channel width was 2 μm. A 5-nm-thick Al$_2$O$_3$ gate dielectric was deposited by ALD after the (NH$_4$)$_2$S surface treatment. After the formation of the Ti/Au gate electrode, post-metallization annealing was performed. The 45-nm-gate length was verified by the cross-sectional image, as shown in Figure 2. Figure 3 shows the I_D–V_D characteristics. The drain current density (at V_G = 2 V and V_D = 0.5 V) was 0.93 A/mm, and the peak transconductance (at V_D = 0.5 V) was 0.8 S/mm. The threshold voltage, estimated by extrapolating the I_D–V_G characteristics at V_D = 50 mV, was 0.09 V. The DIBL was estimated as 378 mV/V from the difference of the I_D–V_G characteristics between V_D = 50 mV and V_D = 0.5 V.

III. DISCUSSION

Channel length dependence was observed, as shown in Figs. 4–7. The device was compared with an InGaAs-MOSFET with a 12-nm-thick InGaAs channel and a 10-nm-Al$_2$O$_3$ dielectric on a 300-nm buffer/p-InP substrate [2] and an InGaAs-MOSFET with a 5-nm-thick InGaAs channel and a 10-nm-Al$_2$O$_3$ dielectric on a 300-nm buffer/p-InP substrate. Figures 4–7 show the dependence of the drain current density at V_D = 0.5 V, maximum transconductance at V_D = 0.5 V,

Fig. 1. Schematic image.

Fig. 2. Cross-sectional view by SEM.

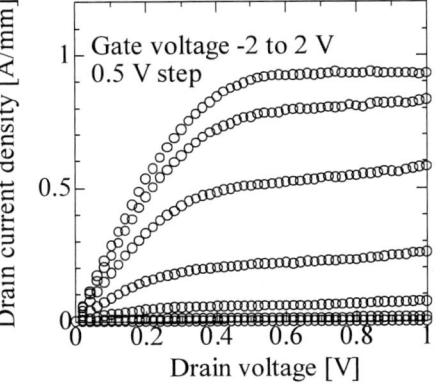

Fig. 3. I_D–V_D characteristics.

threshold voltage, and DIBL, respectively. The characteristics of the devices used in this study with the 5-nm-thick Al_2O_3 dielectric on the Si substrate are represented by the red squares, whereas the black circles and blue triangles show the characteristics of the device with the 5-nm- and 12-nm-thick channels, respectively. Fig. 4 shows that the reduction in the channel thickness from 12 nm to 5 nm reduces the drain current density attributed to the reduction in the energy difference between the Fermi level of the InP source and the quantized level of the channel. Thus, increasing the carrier concentration of the n-InP will improve the drain current density. A simple reduction in the channel decreases the transconductance because of low maximum drain current, whereas a thinner Al_2O_3 dielectric increases the transconductance, as shown in Fig. 5. Figs. 6 and 7 show the clear suppression of the threshold voltage roll-off and the degradation of the DIBL observed in the present device. The clear suppression can be explained by the thinner dielectric and the decrease in the relative permittivity in the region under the channel from $\varepsilon_{r:InAlAs} = 12.5$ to $\varepsilon_{r:SiO2} = 3.8$ by the III–V–OI structures.

IV. CONCLUSION

We have shown the fabrication of a sub-50-nm InGaAs 5-nm/InP 5-nm MOSFET with an n-InP source on a Si substrate using a 5-nm-thick Al_2O_3 dielectric. The device shows a drain current of 0.9 mA/µm at $V_D = 0.5$ V and a peak transconductance of 0.8 mS/m at $V_D = 0.5$ V. Clear suppression of the SCE is also confirmed from the channel length dependence of the threshold voltage and the DIBL. The suppression of the SCE can be explained by the thinner dielectric and the ETB III-V-OI structures.

ACKNOWLEDGMENT

This work was supported by the Grant-in Aid for Scientific Research by the MEXT/JSPS and SCOPE by MIC.

REFERENCES

[1] ITRS 2011 PIDS.

[2] Y. Yonai, T. Kanazawa, S. Ikeda, and Y. Miyamoto, "High drain current (>2 A/mm) InGaAs channel MOSFET at $V_D = 0.5V$ by shrinkage of channel length with InP anisotropic etching," IEDM 2011, pp. 307–310.

[3] S. H. Kim, M. Yokoyama, N. Taoka, R. Nakane, T. Yasuda, O. Ichikawa, N. Fukuhara, M. Hata, M. Takenaka and S. Takagi, "Sub-60 nm deeply-scaled channel length extremely-thin body $In_xGa_{1-x}As$-On-insulator MOSFETs on Si with Ni-InGaAs metal S/D and MOS interface buffer engineering," 2012 Symp.on VLSI Tech. pp. 177–178

[4] J. J. Gu, X. W. Wang, H. Wu, J. Shao, A. T. Neal, M. J. Manfra, R. G. Gordon, and P. D. Ye, "20–80-nm channel length InGaAs gate-all-around nanowire MOSFETs with EOT = 1.2 nm and lowest SS = 63 mV/dec," IEDM 2012, pp. 633–636.

[5] A. Kato, T. Kanazawa, S. Ikeda, Y. Yonai, and Y. Miyamoto, "Reduction of access resistance of InP/InGaAs composite-channel MOSFET with back-source electrode," IEICE Trans. on Electronics, vol. E95-C, no. 5 pp. 904–90, 2011.

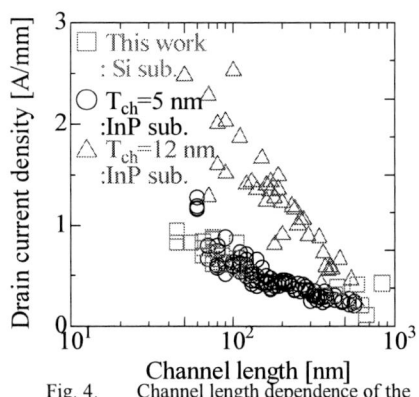

Fig. 4.   Channel length dependence of the drain current density.

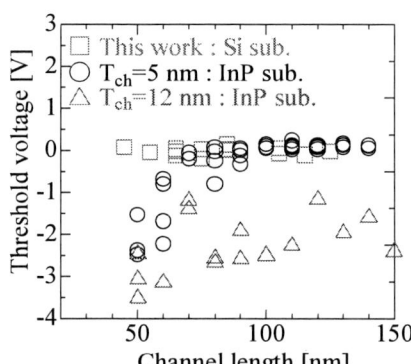

Fig. 6.   Channel length dependence of the threshold voltage.

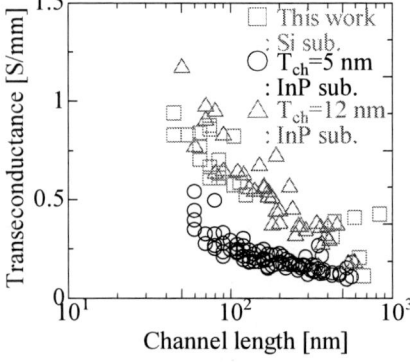

Fig. 5.   Channel length dependence of the transconductance.

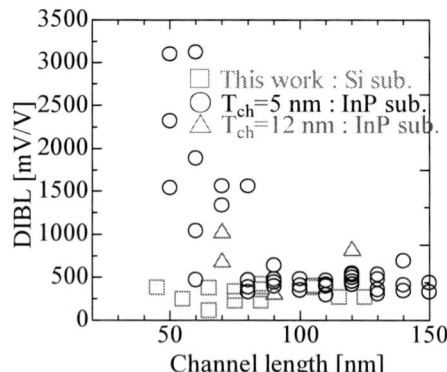

Fig. 7.   Channel length dependence of the DIBL.

WeD1-3 (Oral)
9:00 - 9:15

Analysis on channel thickness fluctuation scattering in InGaAs-OI MOSFETs

S. H. Kim[1,*], M. Yokoyama[1], R. Nakane[1], O. Ichikawa[2], T. Osada[2], M. Hata[2], M. Takenaka[1] and S. Takagi[1]

[1]The Univ. of Tokyo , [2]Sumitomo Chemical Co. Ltd.

*E-mail: dadembyora@mosfet.t.u-tokyo.ac.jp

I. INTRODUCTION

Recently, III-V compound semiconductors have attracted strong attention as a new channel material for post Si CMOS technologies. To fully utilize the channel properties at deeply scaled transistor node, short channel effects (SCEs) control is also very important. Therefore, we have developed extremely-thin-body (ETB) InGaAs-on-insulator (-OI) structures [1]-[4]. However, mobility degradation with a decrease of body thickness (T_{body}) is critical problem to realize high performance ETB InGaAs-OI MOSFETs. Therefore, physical understanding and mobility enhancement technology are quite important. Actually, we have introduced MOS interface buffer layer, which has higher bandgap than that of channel layer, between oxide and channel, and increased the Indium (In) content in channel layer, as shown in Fig. 1 [3]-[4]. As a result, we have achieved high mobility at around T_{body} of 10 nm and have clarified the mobility enhancement mechanism [4]. In this work, we have further analyzed the effect of channel-thickness-fluctuation scattering on mobility characteristics in ETB InGaAs-OI MOSFETs.

II. DEVICE FABRICATION

Schematic cross-section of MOSFETs have been studied in this work is shown in Fig. 1. MOSFETs were fabricated with Ni-InGaAs metal S/D (Fig. 2) on InGaAs-OI wafer fabricated by direct wafer bonding [3]-[4]. Here, we have changed existence of MOS interface buffer layer and In content in the channel layer (0.53, 0.7, and 1). We represent the thickness of the channel structure (a-nm-$In_{0.3}Ga_{0.7}As$/b-nm-$In_xGa_{1-x}As$ channel/c-nm- $In_{0.3}Ga_{0.7}As$) as a/b/c nm as shown in Fig. 1.

III. ELECTRICAL PROPERTIES

I_D-V_G and I_D-V_D characteristics of InAs-OI MOSFETs with buffer, gate length of 0.7 μm are shown in Fig. 3 (a) and (b), respectively. Good transfer and output characteristics were obtained. The effective electron mobility (μ_{eff}) characteristics of $In_xGa_{1-x}As$-OI MOSFETs with T_{body} of between 10-20 nm at room temperature (R.T.) are shown in Fig. 4. It is found that the $In_xGa_{1-x}As$ channels with higher In content shows higher μ_{eff} even though the higher In content channel has thinner T_{body}. Also, MOSFETs with the buffer layer shows higher μ_{eff}. The InAs-OI MOSFETs with $T_{channel}$ of 3 nm exhibits high peak mobility of 3180 cm^2/Vs. In order to analyze the scattering mechanism in our MOSFETs, temperature dependence was evaluated. To suppress phonon scattering, μ_{eff} characteristics of $In_xGa_{1-x}As$-OI MOSFETs were examined at low temperature of 150 K (Fig. 5). MOSFETs with MOS interface buffer layer shows higher mobility, indicating that mobility dominated by channel-thickness-fluctuation and/or surface roughness scattering was enhanced by inserting MOS interface buffer layer.

In order to quantitatively examine the effects of channel-thickness-fluctuation scattering on the mobility characteristics of the present devices, we have plotted the mobility limited by channel-thickness-fluctuation scattering, $\mu_{fluctuation}$ for the various devices as a function of $T_{channel}$ and T_{body}. Here, $\mu_{fluctuation}$ was determined by the mobility at 150 K and N_S of 1×10^{12} cm^{-2}. Also, T_{body} is defined as total thickness of the III-V semiconductors including the two MOS buffer layers, while

$T_{channel}$ is defined as the thickness of the central channels only, as shown in Fig. 1. Fig. 6 (a) and (b) show $\mu_{fluctuation}$-$T_{channel}$ and $\mu_{fluctuation}$-T_{body} characteristics, respectively. The clear T^6 dependence of $\mu_{fluctuation}$ is observed in the $\mu_{fluctuation}$-T_{body} plot than in the $\mu_{fluctuation}$-$T_{channel}$ plot [5]-[6]. These results indicate that contribution of channel-thickness- fluctuation scattering in the present devices is determined roughly by total T_{body} than $T_{channel}$ and that fluctuation of the total channel thickness is dominating the scattering probability more.

In order to more clearly show the effect of the MOS buffer layers, $\mu_{fluctuation}$ with different channel In contents with and without the MOS buffers is plotted under almost the same T_{body} as a function of the effective mass of the channel materials, m_{eff}. Generally, $\mu_{fluctuation}$ is proportional to m_{eff} in the single channel MOSFETs, as shown in Fig. 7 [6]. Fig. 8 shows $\mu_{fluctuation}$ of $In_xGa_{1-x}As$-OI MOSFETs at T_{body} of around 10 nm as a function of m_{eff}. It is found that MOSFETs with the MOS interface buffer layers exhibits higher $\mu_{fluctuation}$ than that in MOSFETs without the buffer layers and that $\mu_{fluctuation}$ still increases for $In_xGa_{1-x}As$-OI channels with higher In content and lower m_{eff} as opposite to the equation shown in Fig. 7. These findings can be explained by considering that the higher concentration of the electron wave function into the channel region due to the MOS interface buffers mitigates thickness fluctuation scattering and that this wave function modulation is enhanced by the increase in the conduction band offset against the buffer layers lead by higher In content in the channel. As a consequence, we can conclude that MOS interface buffer engineering enhances $\mu_{fluctuation}$ at a given T_{body} value even with thin effective $T_{channel}$. Moreover, even at a given T_{body} value, MOS interface buffer engineering allows us to provide better SCE control thanks to thinner $T_{channel}$ than in the single channel structure. Both effects can contribute to the superiority of III-V-OI MOSFETs with the MOS interface buffers to the single channel III-V-OI ones in terms of channel mobility and SCE control.

IV. CONCLUSION

Effects of channel-thickness-fluctuation scattering on electron mobility have been analyzed and it was found that channel-thickness-fluctuation scattering is important parameters decide mobility in ETB III-V-OI MOSFETs. Also, we have clarified that the introduction of MOS interface buffer layer is effective to enhance mobility through the increase of $\mu_{fluctuation}$.

ACKNOWLEDGMENT

This work was supported by Research and Development Program for Innovative Energy Efficiency Technology from NEDO. S. H. Kim would like to thank JSPS for research fellowship.

REFERENCES

[1] M. Yokoyama et al.,VLSI,p.242 (2009) [2] M. Yokoyama et al.,IEDM,p.46 (2010) [3] S. H. Kim et al., VLSI, p.58 (2011) [4] S. H. Kim et al., IEDM, p.311 (2011) [5] K. Uchida et al., APL **82**, 17(2003) [6] K. Uchida et al., JAP **102**, 074510 (2007) [7] T. Skotnicki et al., IEEE TED **55**, p. 96 (2008)

WeD1-3 (Oral)
9:00 - 9:15

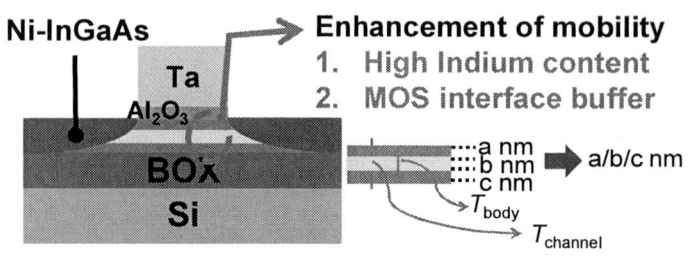

Fig. 1 Schematic cross-section of InGaAs-OI MOSFETs and key channel engineering in this work.

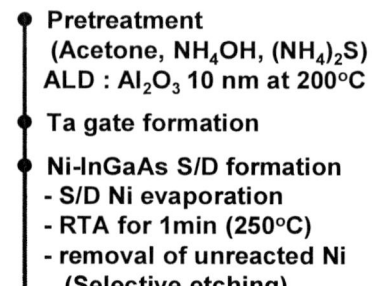

Fig. 2 Device fabrication process

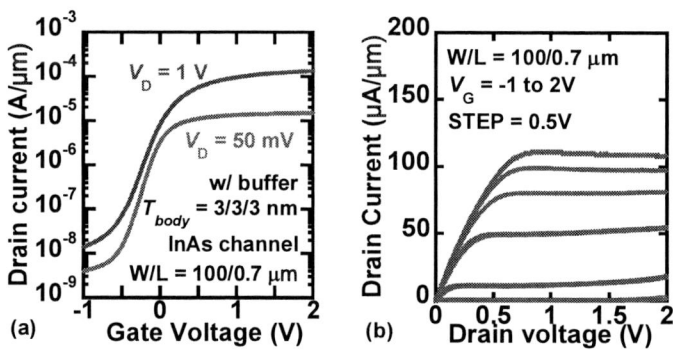

Fig. 3 (a) I_D-V_G and I_D-V_D characteristics of InAs-OI MOSFETs

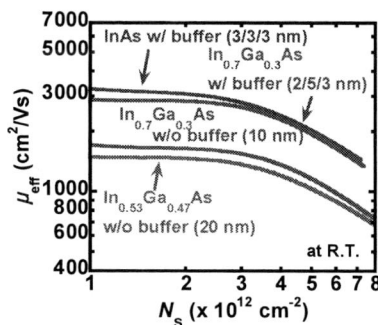

Fig. 4 The μ_{eff} characteristics of InGaAs-OI MOSFETs at R.T.

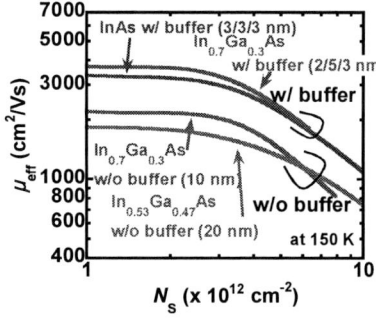

Fig. 5 The μ_{eff} characteristics of InGaAs-OI MOSFETs at 150 K

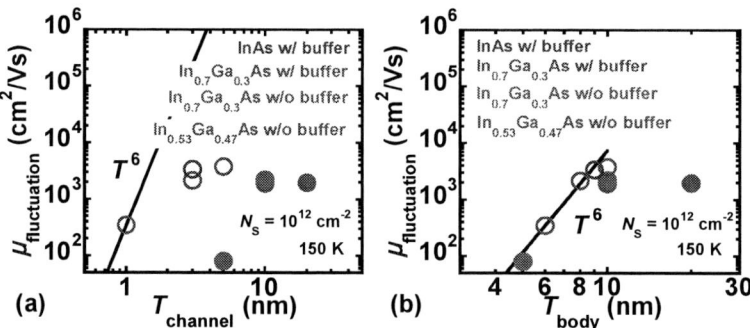

Fig. 6 $\mu_{fluctuation}$ of InGaAs-OI MOSFET with and without buffer as a function of (a) $T_{channel}$ and (b) T_{body}.

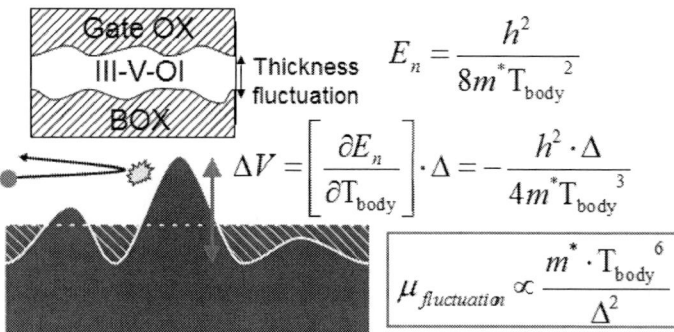

$$E_n = \frac{h^2}{8m^* T_{body}^2}$$

$$\Delta V = \left[\frac{\partial E_n}{\partial T_{body}} \right] \cdot \Delta = -\frac{h^2 \cdot \Delta}{4m^* T_{body}^3}$$

$$\mu_{fluctuation} \propto \frac{m^* \cdot T_{body}^6}{\Delta^2}$$

Fig. 7 Schematic illustration of thickness fluctuation scattering and related equations.

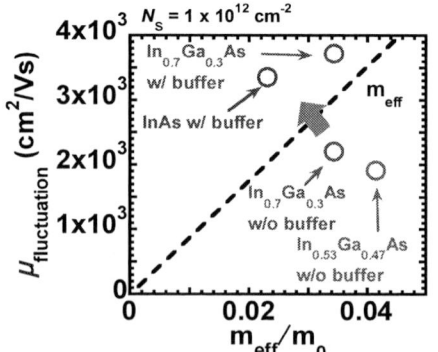

Fig. 8 $\mu_{fluctuation}$ characteristics of InGaAs-OI MOSFETs as a function of m_{eff}/m_0

978-1-4673-6130-9/13 $31.00 © 2013 IEEE

WeD1-4 (Oral)
9:15 - 9:30

Impact of Al₂O₃ ALD temperature on Al₂O₃/GaSb metal-oxide-semiconductor interface properties

Masafumi Yokoyama[1], Yuji Asakura[1], Haruki Yokoyama[2], Mitsuru Takenaka[1], and Shinichi Takagi[1]

[1]Department of Electrical Engineering and Information Systems (EEIS), Graduate School of Engineering,
The University of Tokyo
Yayoi 2-11-16, Bunkyo, Tokyo, 113-0032, Japan
yokoyama@mosfet.t.u-tokyo.ac.jp
[2]NTT Photonics Laboratories, NTT Corporation, Atsugi 243-0198, Japan

Abstract—**We have investigated the impact of the Al₂O₃ atomic-layer-deposition (ALD) temperature on Al₂O₃/GaSb metal-oxide-semiconductor (MOS) interface properties. We have found that the GaSb MOS interfaces are severely degraded with increasing the ALD temperature. X-ray photoelectron spectroscopy (XPS) measurements clarified that the main cause of the interface deterioration is the reduction of Sb oxides from GaSb surfaces.**

Keywords—*GaSb; MOS; interface; oxide; capacitor;*

I. INTRODUCTION

III-V compound semiconductor channel MOSFETs allow us to enhance the performance of CMOS devices in post scaling generation [1]-[4]. Particularly, Sb-based III-V compounds are promising channel materials for pMOSFETs [5]-[8]. Here, one of the most critical factors in Sb-based material MOSFETs is to realize the good MOS interfaces. It has been often pointed out that elimination of III-V native oxides is mandatory for providing good III-V MOS interfaces. It has been reported, on the other hand, that passivation using trivalent oxides such as Al₂O₃ and Ga₂O₃ could be a key to realize the superior MOS interface properties [9]-[11]. However, the influence of the interfacial III-V oxides on Sb-based MOS interfaces has not been fully understood yet.

In this paper, we have studied the impact of the Al₂O₃ ALD temperature on Al₂O₃/GaSb MOS interface properties with emphasis on the oxidation condition of GaSb interfacial layers. We have found that increase in the process temperature of gate-stack formation results in elimination of Sb oxides from GaSb surfaces and, as a result, that elimination of the Sb oxides deteriorates the GaSb MOS interfaces.

II. EXPERIMENTS

A. Prearing GaSb MOS capacitors

The experiments were carried out for p-type (100) GaSb substrates with carrier concentration (N_A) of 1.85×10^{16} cm⁻³. Figure 1 shows the process flow of GaSb MOS capacitors. In order to study the impact of ALD condition on Al₂O₃/GaSb MOS interface properties, we prepared the Al₂O₃/GaSb MOS capacitors with ~ 5 nm Al₂O₃ deposited at 150, 200, 250, and 300 °C by ALD after HCl cleaning. Here, HCl cleaning using

36% HCl solution and de-ionized water rinse was done repeatedly until achievement of the hydrophobic surfaces indicating elimination of most of the GaSb surface oxides. This cyclic HCl treatment allows us to achieve the surface oxide thickness of less than 1 nm and roughness down to less than ~ 0.2 nm due to removal of native oxides. Though HCl cleaning reduces most of the Sb and Ga oxides, the XPS observation revealed that there were small amounts of the Sb and Ga oxides on the GaSb surfaces. After formation of the gate insulators, Au gate electrodes and Al back contacts were formed by thermal evaporation deposition.

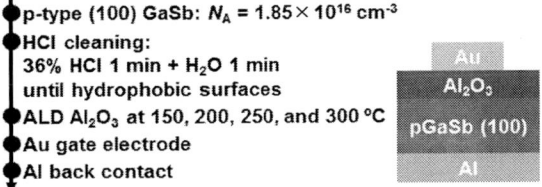

Fig. 1 Schematic illustration of GaSb MOS capacitor formation process flow.

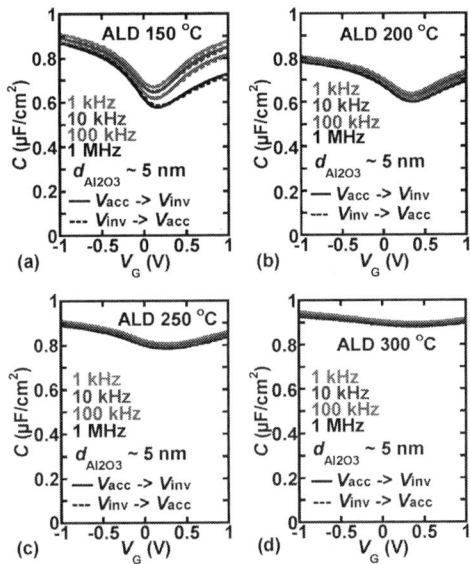

Fig. 2 Frequency dispersion with 1 kHz, 10 kHz, 100 kHz, and 1 MHz of $C - V$ curves of Al₂O₃/GaSb MOS capacitors with Al₂O₃ deposited at (a) 150, (b) 200, (c) 250, and (d) 300 °C, respectively. The solid and broken curves correspond to the measurements by sweeping the voltage from V_{acc} to V_{inv} and from V_{inv} to V_{acc}, respectively.

This work was supported by Development Program for Innovative Energy Efficiency Technology from NEDO.

WeD1-4 (Oral)
9:15 - 9:30

B. ALD temperatue dependence of capacitance versus gate voltage curves

The Al_2O_3 ALD temperature dependence of capacitance *versus* gate voltage ($C - V$) curves of Al_2O_3/GaSb MOS capacitors are evaluated. Figure 2 shows the frequency dispersion of the $C - V$ curves at room temperature of Al_2O_3/GaSb MOS capacitors fabricated at (a) 150, (b) 200, (c) 250, and (d) 300 °C, respectively. The red, blue, green, and black curves are for the $C - V$ curves with the frequency of 1 kHz, 10 kHz, 100 kHz, and 1 MHz, respectively. The solid and broken curves are measured by sweeping the voltage from V_{acc} to V_{inv} and from V_{inv} to V_{acc}, respectively. We found the deterioration of the $C - V$ curves with increasing the ALD temperature. It is difficult to achieve the capacitance modulation, when the Al_2O_3/GaSb MOS structures were fabricated at 250 °C and higher. These results indicate that the low temperature process is preferable in order to achieve the good Al_2O_3/GaSb MOS interfaces by *ex-situ* process.

C. Gate-stack formatin temperatue dependence of the residula Sb and Ga oxides

In order to understand the reason of the deterioration of the $C - V$ curves, we evaluated the Sb and Ga oxides at the GaSb MOS interfaces by XPS measurements. Figure 3 shows the Sb $4d$ and Ga $3d$ photoelectron spectra from the Al_2O_3/GaSb MOS interfaces. Here, the thickness of ALD Al_2O_3 is ~ 1 nm. The red, green, blue, and black curves are the results for 150, 200, 250, and 300 °C, respectively. As shown in Fig. 3(a), the peaks of Sb oxides are found in the energy range from 34 to 36 eV for the sample fabricated at 150 °C. However, the peaks of Sb oxides gradually reduced with increasing the ALD temperature. Finally, most of the Sb oxides were eliminated, when ALD was carried out at 300 °C. On the other hand, the peaks of Ga oxides are found in the energy range from 19 to 21 eV, even when ALD was done at 300 °C. The peaks of Ga oxides from the samples fabricated at 200, 250, and 300 °C are as high as those at 150 °C. These results indicate that the elimination of Sb oxides can deteriorate Al_2O_3/GaSb MOS interface properties and that the Ga oxides do not affect to the GaSb MOS interface properties. As a result, the XPS analyses suggest that the strong ALD temperature dependence of the Al_2O_3/GaSb MOS interface properties can be caused by the elimination of Sb oxides by higher ALD temperature.

In conclusion, we have found that the Al_2O_3 ALD temperature is important to fabricate Al_2O_3/GaSb MOS structures. The lower ALD temperature and resulting Sb oxides passivating GaSb MOS interfaces can lead to better Al_2O_3/GaSb MOS interfaces, suggesting a possible way to control the GaSb MOS interface properties.

REFERENCES

[1] S. Takagi *et al.*, "Device structures and carrier transport properties of advanced CMOS using high mobility channels," Solid. State. Electron. vol. 51, pp. 526 – 536, April 2007; "Carrier-Transport-Enhanced Channel CMOS for Improved Power Consumption and Performance," Trans. Electron Devices vol. 55, pp. 21 – 39, January 2008.

[2] Y. Xuan *et al.*, "High performance submicron inversion-type enhancement-mode InGaAs MOSFETs with ALD Al_2O_3, HfO_2 and HfAlO as gate dielectrics," IEDM Tech. Dig., pp. 637 – 640, Dec. 2007.

[3] M. Yokoyama *et al.*, "Thin Body III–V-Semiconductor-on-Insulator Metal–Oxide–Semiconductor Field-Effect Transistors on Si Fabricated Using Direct Wafer Bonding," Appl. Phys. Express vol. 2, p. 124501, Dec. 2009; "III-V-semiconductor-on-insulator n-channel metal-insulator-semiconductor field-effect transistors with buried Al_2O_3 layers and sulfur passivation: Reduction in carrier scattering at the bottom interface," Appl. Phys. Lette. vol. 96, p. 142106, April 2010; "Sub-10-nm Extremely Thin Body InGaAs-on-insulator MOSFETs on Si Wafers With Ultrathin Al_2O_3 Buried Oxide Layers," Electron Device Lett. vol. 32 pp. 1218 – 1220, Sept. 2011; "III–V/Ge High Mobility Channel Integration of InGaAs n-Channel and Ge p-Channel Metal–Oxide–Semiconductor Field-Effect Transistors with Self-Aligned Ni-Based Metal Source/Drain Using Direct Wafer Bonding," Appl. Phys. Express vol. 5, p. 076501, June 2012.

[4] M. Radosavljevic *et al.*, "Advanced High-K Gate Dielectric for High-Performance Short-Channel $In_{0.7}Ga_{0.3}As$ Quantum Well Field Effect Transistors on Silicon Substrate for Low Power Logic Applications," IEDM Tech. Dig., pp. 319 – 322, Dec. 2009; "Non-Planar, Multi-Gate InGaAs Quantum Well Field Effect Transistors with High-K Gate Dielectric and Ultra-Scaled Gate-to-Drain/Gate-to-Source Separation for Low Power Logic Applications," IEDM Tech. Dig., pp. 126 – 129, Dec. 2010; "Electrostatics Improvement in 3-D Tri-gate Over Ultra-Thin Body Planar InGaAs Quantum Well Field Effect Transistors with High-K Gate Dielectric and Scaled Gate-to-Drain/Gate-to-Source Separation," IEDM Tech. Dig., pp. 765 – 768, Dec. 2011.

[5] M. Radosavljevic *et al.*, "High-Performance 40nm Gate Length InSb P-Channel Compressively Strained Quantum Well Field Effect Transistors for Low-Power ($V_{CC} = 0.5$ V) Logic Applications," IEDM Tech. Dig. pp. 727 – 730, Dec. 2008.

[6] A. Ali *et al.*, "Advanced Composite High-k Gate Stack for Mixed Anion Arsenide-Antimonide Quantum Well Transistors," IEDM Tech. Dig. pp. 134 – 137, December 2010.

[7] A. Nainani *et al.*, "Engineering of Strained III-V Heterostructures for High Hole Mobility," IEDM Tech. Dig., pp. 857 – 860, Dec. 2009; "Development of High-k Dielectric for Antimonides and a Sub 350 °C III-V pMOSFET Outperforming Germanium," IEDM Tech. Dig., pp. 138-141, Dec. 2010.

[8] M. Xu, R. Wang, and P. D. Ye, "GaSb Inversion-Mode PMOSFETs With Atomic-Layer-Deposited Al_2O_3 as Gate Dielectric," Electron Device Lett. vol. 32, pp. 883 – 885, July 2011.

[9] L. Lin and J. Robertson, "Defect states at III-V semiconductor oxide interfaces," Appl. Phys. Lett. vol. 98, p. 082903, February 2011; "Passivation of interfacial defects at III-V oxide interfaces," J. Vac. Sci. Technol. B vol. 30 p. 04E101, May 2012.

[10] J. Robertson and L. Lin, "Bonding principles of passivation mechanism at III-V-oxide interfaces," Appl. Phys. Lett. vol. 99, p. 222906, November 2011.

[11] R. Suzuki *et al.*, "1-nm-capacitance-equivalent-thickness HfO_2/Al_2O_3/InGaAs metal-oxide-semiconductor structure with low interface trap density and low gate leakage current density," Appl. Phys. Lett. vol. 100, p. 132906, March 2012.

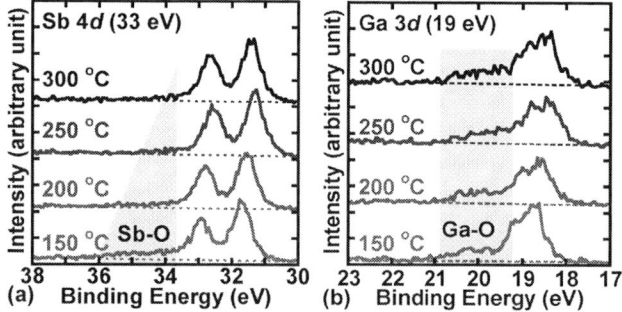

Fig. 3 XPS spectra of (a) Sb $4d$ and (b) Ga $3d$ from the Al_2O_3/GaSb MOS structures. The red, green, blue, and black curves are for ALD temperature of 150, 200, 250, and 300 °C, respectively.

WeD1-5 (Oral)
9:30 - 9:45

1/f-noise in Vertical InAs Nanowire Transistors

Karl-Magnus Persson, Martin Berg, Erik Lind, and Lars-Erik Wernersson
Department of Electrical- and Information Technology
Lund University
Lund, Sweden
karl-magnus.persson@eit.lth.se

Abstract—**The material quality at high-k interfaces are a major concern for FET devices. We study the effect on two types of InAs nanowire (NW) transistors and compare their characteristics. It is found that by introducing an inner layer of Al_2O_3 at the high-κ interface, the low frequency noise (LFN) performance regarding gate voltage noise spectral density, S_{Vg}, is improved by one order of magnitude per unit gate area.**

Keywords—*1/f-noise; high-κ; nanowire; InAs; FET*

I. INTRODUCTION

InAs nanowire (NW) FETs are a good candidate for future RF electronics, promising both low power dissipation and high speed operation [1][2][3]. To verify the feasibility in the technology, however, it is vital to fabricate high performance devices. By utilizing different evaluation methods, the delicate balance of different processing conditions can be evaluated; for a FET, good control over the interfaces is essential. Besides measurements of the *I-V* characteristics, it is of interest to determine the trap density and one common method is measurement of low frequency noise (LFN). In this abstract, we investigate the differences in LFN as well as I_{ON}, *SS, DIBL,* and other metrics for devices fabricated with two different high-κ films, HfO_2 and Al_2O_3/HfO_2, implemented for DC and RF operation, respectively [4][5].

II. FABRICATION

A. DC Devices

InAs NWs are grown on a doped InAs substrate. Seed Au particles defined with electron beam lithography (EBL) determines diameter, here 40 nm, and placement. The growth is made in a metal-organic-vapor-phase-epitaxy (MOVPE) growth chamber at 420 °C and with a Sn dopant molar fraction of $3.49 \cdot 10^{-8}$. The high-κ film is deposited both before and after the fabrication of the first separation layer. The two high-κ films, consisting of HfO_2, is deposited with atomic layer deposition (ALD) at 250 °C and is in total 7 nm, with an EOT of 1.5 nm. The source-gate separation layer consists of a 30 nm evaporated SiO_x film. During evaporation, the sample is tilted so that there will be a buildup of flakes on sides of the NWs, which can be removed in a wet-etch, leaving the lateral layer intact. This process is also used for the evaporated Ni gate and the film thickness sets $L_G = 35$ nm (Fig. 1a). The evaporated top metal is elevated with a spin-on polymer resist which is back etched.

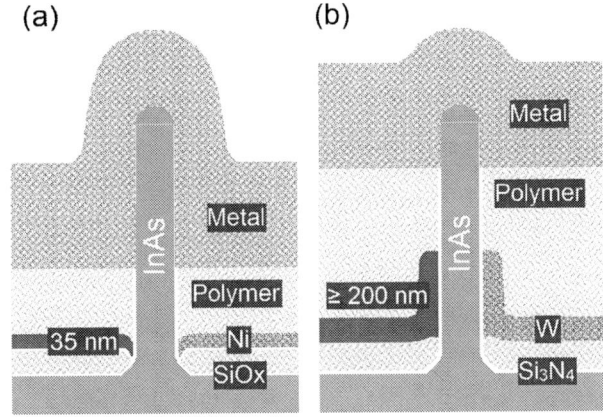

Fig. 1. (a) The DC device structure with specified materials and $L_G = 35$ nm. (b) The refined, RF device structure with specified materials and $L_G \geq 200$ nm.

B. RF Devices

To accommodate RF transistors and circuits, the DC device structure was refined; devices are placed on an isolating substrate. Also, targeting high yield, the gate process is exchanged. The processing flow chain begins with growth of a 300-nm-thick InAs layer on top of a Si substrate to mitigate growth of InAs NWs, and also, to act as contact mesa structure after being etched out. NWs with a diameter of 45 nm are grown at 420 °C and with a Sn dopant molar fraction of $1.07 \cdot 10^{-7}$. For both devices, the NWs are homogenously doped. The high-κ film is deposited with ALD and consists of 0.5 nm Al_2O_3 at 250 °C and 6.5 nm of HfO_2 at 100 °C, with a total EOT of 1.8 nm. The W-gate is sputtered with a layer thickness of 60 nm and is separated from the source mesa with a 60 nm Si_3N_4 film deposited with plasma enhanced chemical vapor deposition (PECVD) in a preceding step. The nitride film is removed from the sides of the NWs with a dry etch. The gate length is set by a defining polymer and a dry etch, giving $L_G \geq$ 200 nm (Fig. 1b). The top metal is sputtered and is elevated by a spin-on polymer resist that is back etched.

III. MEASUREMENT AND RESULTS

Devices are measured with a Keithley 4200 SCS and for the LFN characterization, a Lock-in amplifier and a LNA are added to the setup. The 1/f-noise is measured for $V_{DS} = 50$ mV and $f = 10$ Hz. Obtained data from measurements of normalized current noise spectral density, S_{Id}/I_{DS}^2, is shown in

---

This work has been supported by the Swedish Foundation for Strategic Research and VINNOVA.

WeD1-5 (Oral)
9:30 - 9:45

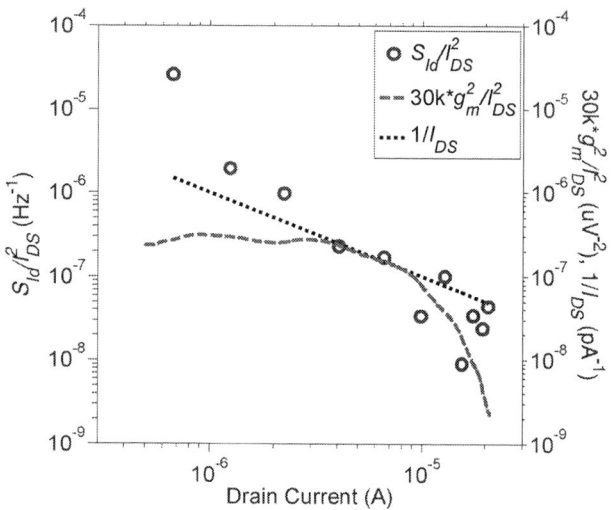

Fig. 2. S_{Id}/I_{DS}^2 plotted versus I_{DS} at V_{DS} = 50 mV and f = 10 Hz. Two curves have been added, g_m^2/I_{DS}^2 and $1/I_{DS}$, where the shape of the curves are indicative for LFN dominated by number fluctuations and mobility fluctuations, respectively. The factor of 30k was multiplied with the g_m^2/I_{DS}^2 curve to position it relevant to the S_{Id}/I_{DS}^2 data.

Fig. 2. Two curves have been added, g_m^2/I_{DS}^2 and $1/I_{DS}$, respectively. All measured performance metrics for the DC device and the refined RF device are benchmarked in Table I.

IV. DISCUSSION

Analyzing the data in Fig. 2, the S_{Id}/I_{DS}^2 seems to be dominated by mobility fluctuations in the subthreshold region, whereas number fluctuations better explain the data around threshold. The observed increase in S_{Id}/I_{DS}^2 for large I_{DS} could

be explained by noise generated by a combination of mobility fluctuations and series resistance. This interpretation differs from the analysis of data measured for the DC devices, where the entire I_{DS} range could be explained by number fluctuations stemming from oxide traps [4]. Still, the shown data in Fig. 2 is representative as other RF devices on the same batch sample show similar tendencies and have comparable gate voltage noise spectral density, S_{Vg}, ranging between 310 - 410 $\mu m^2 \mu V^2$/Hz. Comparing S_{Vg} and S_{Id}/I_{DS}^2 for the two different high-κ devices, the improvement is one order of magnitude for S_{Vg} and about two orders of magnitude for S_{Id}/I_{DS}^2. As the numbers given for the S_{Vg} are normalized to the gate area, the difference in the amount of charges for different channel lengths is accounted for. The data suggest that there is a lower interface trap concentration with the introduction of an Al_2O_3 film and the improvement can possibly be traced to a reduction in sub-oxides [6]. The reason for the SS degradation can be explained by the increase in doping and the inability to deplete the channel all the way through. This is partly the explanation for the increase in DIBL as well, where the undepleted carriers form a parallel parasitic resistance. The DIBL is also likely to be caused by other effects such as band-to-band tunneling. To accommodate the issue concerning the channel depletion, the wire diameter, D_{NW}, should be scaled and it has been shown that scaling the diameter to 28 nm improves the off characteristics (SS = 80/140 mV/V for V_{DS} = 0.05/0.5 V) keeping other parameters constant (although EOT = 1.3 nm) [5]. To maintain good on-performance when scaling, however, it is necessary to address the increase in series resistance; this could be done by, for example, introducing a heterogonous doping profile and/or thinning down the separation layers [3].

ACKNOWLEDGMENT

The authors would like to thank Mattias Borg, Jun Wu, and Johannes Svensson for growth of nanowires and materials.

TABLE I. PERFORMANCE METRICS

	HfO2	*Al₂O₃ / HfO₂*
D_{NW}	40 nm	45 nm
L_G	35 nm	200 nm
EOT	1.5 nm	1.8 nm
Doping	~$1 \cdot 10^{18}$ cm^{-3}	~$3 \cdot 10^{18}$ cm^{-3}
I_{ON}[a,b,c]	104 mA/mm	670 mA/mm
R_{ON}[b]	3.7 Ωmm	0.33 Ωmm
g_m[b,c]	227 mA/mm	1190 mA/mm
SS	110 mV/decade	500 mV/decade
DIBL[c]	100 mV/V	320 mV/V
S_{Vg}	5700 $\mu m^2 \mu V^2$/Hz	410 $\mu m^2 \mu V^2$/Hz
S_{Id} / I_D^2	$7.3 \cdot 10^{-7}$ Hz^{-1}	$8.9 \cdot 10^{-9}$ Hz^{-1}

a. $V_{DS} = V_{GS} - V_t = 0.5$ V

b. Values normalized to the NW circumference, $W = D_{NW} \cdot \pi$

c. Measured at $V_{DS} = 0.5$ V

REFERENCES

[1] C. P. Auth and J. D. Plummer , "Scaling theory for cylindrical, fully-depleted, surrounding-gate MOSFET's," *IEEE Electron Device Letters, IEEE* , vol.18, no.2, pp.74-76, Feb. 1997.

[2] D.-H. Kim, J. A. del Alamo, D. A. Antoniadis and B. Brar, "Extraction of Virtual-Source Injection Velocity in sub-100 nm III-V HFETs," *IEEE International Electron Devices Meeting (IEDM), 2009*, vol., no., pp. 35.4.1 - 35.4.4, 7-9 Dec. 2009.

[3] K. Jansson, E. Lind, and L.-E. Wernersson., "Performance Evaluation of III–V Nanowire Transistors," *IEEE Trans. Electron Devices.*, vol. 59, no. 9, pp. 2375–2382, Sep. 2012.

[4] K.-M. Persson, E. Lind, A. Dey, C. Thelander, H. Sjöland, and L.-E. Wernersson, "Low-Frequency Noise in Vertical InAs Nanowire FETs," *IEEE Electron Dev. Lett.*, vol. 31, no. 5, pp. 428-430, May 2010.

[5] K.-M. Persson, M. Berg, M. Borg, J. Wu, H. Sjoland, E. Lind and L.-E. Wernersson, "Vertical InAs Nanowire MOSFETs with I_{DS} = 1.34 mA/μm and g_m = 1.19 mS/μm at V_{DS} = 0.5 V," *IEEE Device Research Conference (DRC), 2012*, vol., no., pp. 195-196, 18-20 June 2012.

[6] J. Wu, E. Lind, R. Timm, M. Hjort, A. Mikkelsen, and L.-E. Wernersson, "Al₂O₃/InAs metal-oxide-semiconductor capacitors on (100) and (111)B substrates," *Appl. Phys. Lett.*, vol. 100, no. 13, pp. 132905, Mar. 2012.

WeD2-1 (Invited)
10:30 - 11:00

InP Based Photonic Integrated Circuits For DWDM Optical Communication

Beck Mason, Michael Larson, Yuliya Akulova, Srinath Kalluri

JDSU Transmission R&D
Milpitas, CA USA
beck.mason@jdsu.com

Abstract— **We review and discuss several new advances in InP based photonic integrated circuits and their application to 100Gb/s coherent transmission applications.**

Keywords—InP, Photonic Integration, Coherent

I. INTRODUCTION

As data rates in long haul DWDM optical communication links have progressed from 2.5 Gb/s to 10 Gb/s to 100 Gb/s and beyond it has become increasingly difficult to continue to scale solutions in cost, power and size. Early generation 100G coherent line side systems depend on discrete individual components based on multiple material systems with significant amounts of hybrid integration. These systems have demonstrated significant technical capability in terms of reach [1] and spectral efficiency [2] but in general have suffered from high power dissipation, high cost and large physical size.

An analogous situation existed in 10 Gb/s systems where the dominant transmission technology for many years was based on 300pin MSA transponders with separate tunable lasers and lithium Mach-Zehnder modulators. The development of more highly integrated InP based photonic components that combined the tunable laser and the Mach-Zehnder modulator into a single monolithic chip [3], along with the associated high density packaging technology, enabled a significant leap forward in density, cost and power dissipation. It also facilitated a rapid transition to pluggable tunable transceivers enabling increased flexibility in deployment and inventory management for the industry.

For 100Gb/s systems a similar transition is underway and there is a strong drive towards smaller more integrated components. The challenges inherent in this transition however are much greater due to the increased complexity associated with coherent Polarization Multiplexed Quadrature Phase Shift Keyed (PM-QPSK) transmission formats. In general these implementations rely on a larger number of photonic elements and place more stringent requirements on their performance in addition to the increased bandwidth requirements. These challenges are being met with a greater diversity of component technologies including Silicon [4] and Polymer [5] based components in addition to InP. In this paper we will discuss the development of key InP based photonic integrated components for supporting 100G PM-QPSK transmission.

II. NARROW LINEWIDTH LASER

The transmitter and local oscillator (LO) lasers (used for intradyne signal recovery) are critical components of coherent PM-QPSK transmission systems. Under conditions of high chromatic dispersion the linewidth of the LO laser can have a significant impact on the OSNR performance for the link. In order to maintain good OSNR performance LO linewidths below 500kHz are typically required (ref).

There are several types of tunable lasers that rely on carrier injection tuning for wavelength control. These have the benefit of being small low power and capable of rapid tuning. In most cases the devices employ a complex mirror structure that enables them to leverage a more modest index tuning range to cover a broader wavelength span an example of this is the Sampled Grating Distributed Bragg Reflector SGDBR laser [6]. The use of carrier injection tuning in SGDBR lasers causes them to suffer from additional tuning-induced loss. This excess loss is a challenge preventing them from achieving linewidths substantially below 1MHz across the C band.

Thermal tuning is an approach that can avoid tuning induced loss and has been used to demonstrate sub 400kHz linewidth in a super-structure grating (SSG) DBR laser using on-chip microheaters [7]. Unfortunately, the tuning power dissipation of these devices was prohibitively large (>1W) due to the high thermal conductance of the structure. In this paper, we discuss a thermally tuned sampled-grating (SG) DBR laser in which the tuning sections have been engineered for dramatically reduced thermal impedance. This has enabled a >40nm tuning range with total tuning heater power dissipation less than 120mW, linewidth below 300kHz, side mode suppression ratio (SMSR) above 40dB, and fiber coupled output power of +16dBm. Linewidth and SMSR versus ITU channel are shown in Figure 2. The low heat capacity of the tuning sections enables millisecond-timescale tuning speeds.

The device structure is shown schematically in Figure 1. It comprises sampled grating DBR front and back mirror tuning sections surrounding a phase tuning and gain section. Laser emission is further amplified by a semiconductor optical amplifier (SOA) prior to exiting the anti-reflection coated front facet. In the tuning sections below the lower cladding is an InGaAs layer which has been selectively removed through lateral undercut etching. In this way the tuning sections are thermally isolated from the underlying substrate. Each tuning section contains a thin film microheater with contact pads on both ends.

978-1-4673-6130-9/13 $31.00 © 2013 IEEE

WeD2-1 (Invited)
10:30 - 11:00

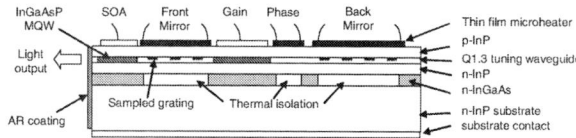

Figure 1. Schematic diagram of the thermally-tuned SGDBR laser

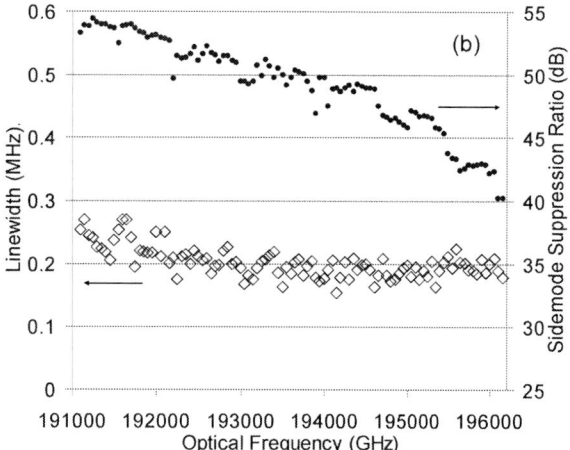

Figure 2. Linewidth and SMSR vs. ITU channel.

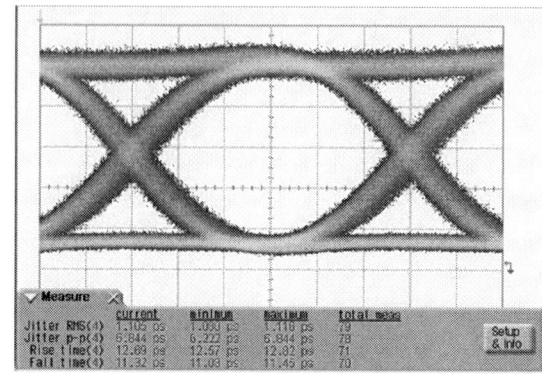

Figure 3. Transmitted eye diagram single InP MZ modulator at 30Gbaud

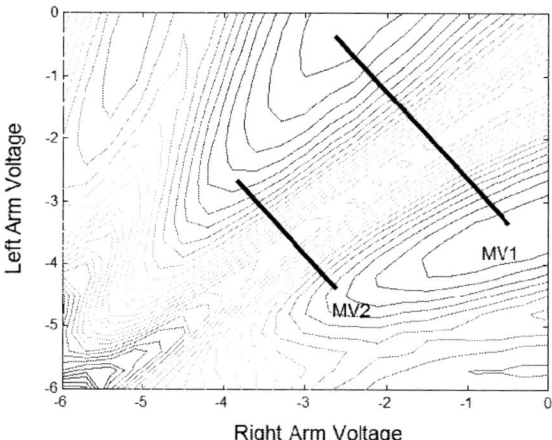

Figure 4. DC contour plot of intensity vs voltage

III. MACH-ZEHNDER MODULATOR

The other critical building block element required is a high bandwidth InP based MZ modulator. For a 100G PM-QPSK transmitter four of these are required in a dual nested configuration to enable In-phase and Quadrature modulation of the two polarization states. Modulators were fabricated based on a push-pull shallow ridge quasi-travelling wave design. Devices with phase modulator lengths of 800, 1000, 1200 and 1400 µm were fabricated yielding bandwidths in excess of 25GHz, low Vpi, high extinction ratio and good modulation capability at baud rates of 30 Gb/s and higher Figure 3.

The phase modulators employed in these devices were based on multi quantum well Quantum-confined Stark effect modulators. QCSE based phase modulators have a couple of disadvantages when compared to lithium niobate based devices. They have a nonlinear phase vs. voltage relationship, much greater wavelength dependence and exhibit both absorption and phase shift under bias.

The nonlinear nature of the phase vs voltage dependence leads to a more complicated optical transfer function which is shown in Figure 4. measured for a 1400 µm long device. This complex nature can be exploited to reduce the drive voltage by adjusting the DC bias on the modulator. Two different modulation vectors are shown on the plot. By choosing the appropriate bias we have been able to demonstrate operation at drive voltages as low as 1Vpp per side.

These results show the potential for InP based photonic integrated components to provide compact and efficient sources for 100G coherent PM-QPSK transmission.

[1] M. Salsi, C. Koebele, P. Tran, H. Mardoyan, E. Dutisseuil, J. Renaudier, M. Bigot-Astruc, L. Provost, S. Richard, P. Sillard, S. Bigo, and G. Charlet, "Transmission of 96×100Gb/s with 23% Super-FEC Overhead over 11,680km, using Optical Spectral Engineering", OMR2, OFC 2011, Los Angeles, CA, USA

[2] J.-X. Cai, Y. Cai, C. R. Davidson, D. G. Foursa, A. Lucero, O. Sinkin, W. Patterson, A. Pilipetskii, G. Mohs, and Neal S. Bergano, "Transmission of 96x100G pre-filtered PDM-RZ-QPSK channels with 300% spectral efficiency over 10,608km and 400% spectral efficiency over 4,368km", PDPB10, OFC 2010, San Diego, CA, USA

[3] Y. A. Akulova, G. A. Fish, P. Koh, P. Kozodoy, M. Larson, C. Schow, E. Hall, H. Marchaod, P. Abraham, L. A. Coldren, "10 Gb/s Mach-Zehnder modulator integrated with widely-tunable sampled grating DBR Laser," OFC 2004, Los Angeles, CA, USA

[4] Hyundai Park et al. "Device and Integration Technology for Silicon Photonic Transmitters", J. of Selected Topics in Quantum Electronics, Vol. 17, No. 3, May/June 2011

[5] Min-Cheol Oh et al., "Recent Advances in Electrooptic Polymer Modulators Incorporating Highly Nonlinear Chromophore ", J. of Selected Topics in Quantum Electronics, Vol. 7, No. 5, Spet./Oct 2001

[6] B. Mason, J. Barton, G.A. Fish, L.A. Coldren, S.P. DenBaars, "Design of sampled grating DBR lasers with integrated semiconductor optical amplifiers", Photon. Tech. Lett., Vol. 12 , Issue: 7

[7] H. Ishii et al., "Narrow Spectral Linewidth Under Wavelength Tuning in Thermally Tunable Super-Structure-Grating (SSG) DBR Lasers," IEEE J. Selected Topics in Quantum Electronics, vol. 1, no. 2, 401-407 (1995)

WeD2-2 (Oral)
11:00 - 11:15

17-Gb/s Direct Modulation of Lambda-scale Embedded Active Region Photonic Crystal Lasers

[1,3]Koji Takeda, [1,3]Tomonari Sato, [2,3]Akihiko Shinya, [2,3]Kengo Nozaki, [2,3]Hideaki Taniyama, [1,3]Koichi Hasebe, [1,3]Takaaki Kakitsuka, [2,3]Masaya Notomi, and [1,3]Shinji Matsuo

[1]NTT Photonics Labs., [2]NTT Basic Research Labs., [3]Nanophotonics Center
NTT Corporation
Kanagawa, Japan
e-mail address : takeda.koji@lab.ntt.co.jp

Abstract— **We demonstrated the direct modulation of photonic-crystal nanocavity lasers to realize on-chip optical interconnects. A maximum 3-dB bandwidth of 16.2 GHz was obtained. We achieved a 17-Gb/s eye opening with a 35.3-fJ/bit energy cost.**

Keywords—photonic crystal laser; nanocavity laser; optical interconnects; direct modulation;

I. INTRODUCTION

On-chip and off-chip optical interconnects have attracted great interest as regards reducing the total power consumption of CMOS chips, and various types of light sources have been proposed. According to the total power limitation of CMOS chips, an energy of no more than 10 fJ is allowed for an on-chip light source when transmitting a single bit signal [1]. In addition, high-speed direct modulation of the light sources is desirable to enable us to take advantage of wide-bandwidth optical interconnects. To meet the energy requirement, we have to reduce the volume of the laser active region to less than a cubic micron. The small active volume is advantageous as regards low power consumption; only a small amount of current is required to obtain a sufficient carrier concentration for the high-speed direct modulation. In this context, we have developed a photonic-crystal (PhC) nanocavity laser that we call a lambda-scale embedded active region PhC (LEAP) laser. We have already demonstrated the continuous-wave operation of a current-driven LEAP laser at room temperature [2]. Although LEAP lasers can be directly modulated, the bit rate has been limited to 10 Gb/s [3]. Recently, we have employed an InAlAs sacrificial layer and an InGaAlAs active region to reduce leakage current and to obtain a larger gain at higher temperature, respectively [4]. Since a LEAP laser with an InGaAlAs active region has a differential gain higher than one with an InGaAsP active region, direct modulation with a bit rate higher than 10 Gb/s has been expected.

In this paper, we describe the high-speed direct modulation of a LEAP laser with an InGaAlAs active region. Small signal responses exhibited a maximum 3-dB bandwidth of 16.2 GHz; we obtained eye opening at a maximum bit rate of 17 Gb/s. The energy required to transmit a single bit signal was 35.3 fJ. High-speed direct modulation allows us to realize wide-bandwidth on-chip optical interconnects.

II. DEVICE STRUCTURE AND FABRICATION

A scanning electron microscopic (SEM) image of the LEAP laser is shown in Fig. 1. The structure and fabrication procedure were similar to those described in our previous report [3]. The size of the active region was $2.5 \times 0.3 \times 0.15$ μm^3 and it consisted of three InGaAlAs quantum wells. The lattice constant of the PhC was 420 nm. We used an InAlAs sacrificial layer to block leakage current.

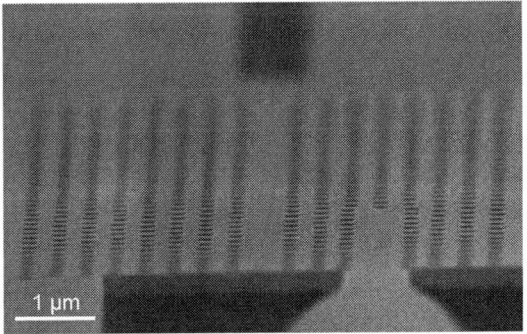

Figure 1. Scanning electron microscopic image of the LEAP laser. We used an embedded active region with a volume of $2.5 \times 0.3 \times 0.15$ μm^3 in the center line-defect waveguide. The lattice constant of the PhC was 420 nm.

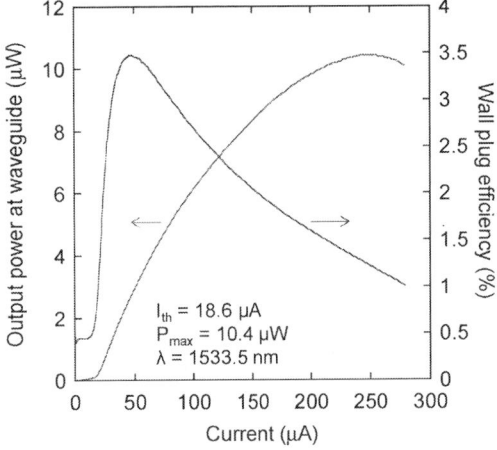

Figure 2. Output power of the device at the waveguide versus injected current. The right axis shows the wall plug efficiency of the LEAP laser.

Part of this work was supported by the New Energy and Industrial Technology Development Organization (NEDO).

WeD2-2 (Oral)
11:00 - 11:15

$$\text{Energy cost} = \frac{V_{bias} \cdot I_{bias}}{1.3 f_{3dB}} \quad (1)$$

where I_{bias} is the injected current, V_{bias} is the applied voltage, and f_{3dB} is the measured 3-dB bandwidth. We assumed that the maximum bit rate was 1.3 times larger than the 3-dB bandwidth [5] when estimating the energy cost. The calculated energy cost is shown as the plot (blue circles) in Fig. 3(b). At a low current injection up to 80 μA, we achieved an energy cost of less than 10 fJ/bit.

To confirm the high-speed modulation ability of the LEAP laser, we applied a large signal and observed the eye diagram using a pulse-pattern generator and a digital sampling oscilloscope. As in the previous measurement, we used an EDFA and a BPF. In the eye-diagram measurement, we injected a 200-μA current and used the pattern-lock mode of the oscilloscope. The inset in Fig. 3(b) shows the obtained eye diagram, which has a bit rate of 17 Gb/s and a signal amplitude of 1.2 V_{pp}. The applied signal was a 2^7-1 pseudo-random bit sequence. We achieved an eye opening with an injected power of 600 μW and a bit rate of 17 Gb/s, resulting in an energy cost of 35.3 fJ/bit. The operating condition of the eye-diagram measurement is shown by the open symbols in Fig. 3(b). The actual bit rate was slower than that expected from the small signal measurement because we used the probe de-embedding function of the LCA to extract the frequency response of the air coplanar probe. However, the impedance mismatch and the frequency response of the probe were included in the eye-diagram acquisition, which reduced the operating speed. This is the first demonstration of a PhC laser with a bit rate above 15 Gb/s.

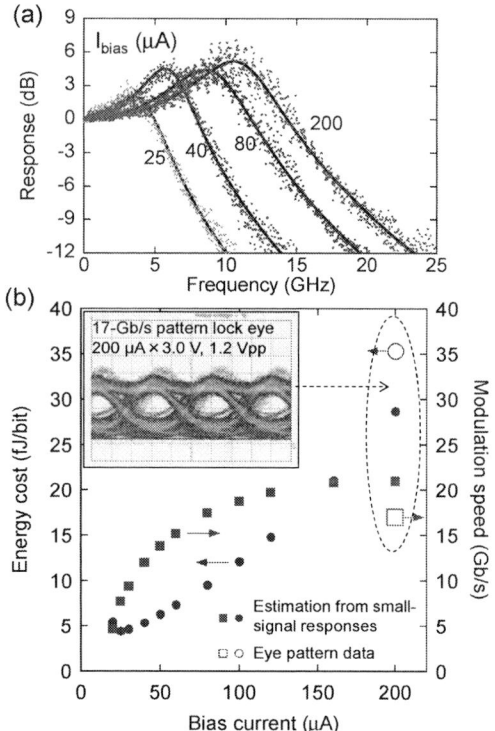

Figure 3. (a) Small signal responses of the LEAP laser with an injected current ranging from 25 to 200 μA. (b) Estimated modulation speed (blue circles) and energy required to transmit a single bit signal (red symbols). The inset shows an eye diagram with a bit rate of 17 Gb/s and an injected current of 200 μA. The open symbols correspond to the operating condition used to acquire the eye diagram.

III. EXPERIMENTAL RESULTS

We measured the output power from the device versus the injected current (L-I characteristic) using a tapered fiber. The result is shown in Fig. 2. To calculate the output power at the waveguide, we assumed a coupling loss of 6.68 dB between the tapered fiber and the waveguide. We obtained a threshold current of 18.6 μA and a maximum output power of 10.4 μW. The lasing wavelength was approximately 1533.5 nm. The wall plug efficiency, which is the optical output power divided by the applied electric power, was derived from the L-I characteristic. We obtained a maximum efficiency of 3.48% at an injected current of 48 μA.

We then characterized the dynamic behavior of the LEAP laser using a 40-GHz air-coplanar probe and an Agilent Lightwave Component Analyzer (LCA). The small signal responses are shown in Fig. 3(a) with an injected current ranging from 25 to 200 μA. Due to the large coupling loss between the tapered fiber and the waveguide, we used an erbium-doped fiber amplifier (EDFA) and an optical band-pass filter (BPF) to amplify the output light at the fiber to be received by the LCA. We obtained a maximum 3-dB bandwidth of 16.2 GHz.

We estimated the energy required to transmit a single bit signal from the measured 3-dB bandwidth as

IV. CONCLUSION

We have developed a current-driven LEAP laser with an InGaAlAs active region that has a gain higher than InGaAsP, and reported its high-speed direct modulation. The small signal responses had a maximum 3-dB bandwidth of 16.2 GHz. We achieved an eye opening at a bit rate of 17 Gb/s, where we injected a 200-μA current with an applied voltage of 3.0 V. As a result, an energy of 35.3 fJ was needed to generate a single bit signal. We believe that the 17-Gb/s direct modulation of the LEAP laser means that CMOS chips have the great advantage of being used for wide-bandwidth optical communication.

REFERENCES

[1] D. A. B. Miller, "Device requirements for optical interconnects to silicon chips", Proc. IEEE, vol. 97, pp. 1166—1185 (2009).

[2] S. Matsuo et al., "Room-temperature continuous-wave operation of lateral current injection wavelength-scale embedded active-region photonic-crystal laser," Optics Express, vol. 20, pp.3773—3780 (2012).

[3] S. Matsuo et al., "10-Gbit/s direct modulation of electrically driven photonic crystal nanocavity laser," Optical fiber communication (OFC), PDP5A.7 (2012).

[4] T. Sato et al., "95°C CW operation of InGaAlAs multiple-quantum-well photonic-crystal nanocavity laser with ultra-low threshold current," IEEE Photonics Conference (IPC), WF2 (2012).

[5] R. S. Tucker, J. M. Wiesenfeld, P. M. Downey, and J. E. Bowers, "Propagation delays and transition times in pulsemodulated semiconductor lasers," Appl. Phys. Lett., vol. 48, pp. 1707—1709, June 1986.

WeD2-3 (Oral)
11:15 - 11:30

Room-temperature Continuous-wave Operation of Lateral Current Injection Membrane Laser

Kyohei Doi[1], Takahiko Shindo[2], Mitsuaki Futami[1], Jieun Lee[1], Takuo Hiratani[1],
Daisuke Inoue[1], Shu Yang[1], Tomohiro Amemiya[2], Nobuhiko Nishiyama[1], Shigehisa Arai[1,2]

[1]*Department of Electrical and Electronic Engineering, Tokyo Institute of Technology*
[2]*Quantum Nanoelectronics Research Center, Tokyo Institute of Technology*
2-12-1-S9-5 O-okayama, Meguro-ku, Tokyo 152-8552, Japan
E-Mail: doi.m.ac@m.titech.ac.jp

Abstract—**Toward realization of an ultralow-power-consumption semiconductor light source for optical interconnection, we have been investigating the lateral current injection (LCI) membrane distributed feedback (DFB) laser. This time, we realized membrane Fabry-Perot (FP) laser with 220 nm core thickness and demonstrated room-temperature continuous-wave (CW) operation with a threshold current of 3.5 mA for the cavity length of 700 μm and the stripe width of 1.0 μm, which is almost the same as the theoretical value.**

Keywords— *membrane laser; lateral current injection; strong optical confinement; optical interconnects*

I. INTRODUCTION

It is predicted that the progress of the processing speed and integration of large scale integrated circuits (LSIs) will soon confront limitation associated with RC delay and large power dissipation in the electrical interconnection in the global wiring. As one of the promising approaches for solving this problem, an introduction of optical interconnection instead of the electrical interconnection has been extensively studied [1]. An ultralow-power-consumption semiconductor laser is strongly required for such optical interconnections, and we demonstrated a GaInAsP/InP membrane distributed feedback (DFB) laser consisting of a thin semiconductor core layer sandwiched by low refractive-index claddings such as air, benzocyclobutene (BCB), and SiO₂. Since the membrane structure produces a large refractive-index difference between the core layer and the cladding layers and supports strong optical confinement into the active region, it leads to ultralow power consumption operation [2]. Previously, an optically pumped membrane laser with low threshold pump power 0.34 mW under room temperature continuous wave (RT-CW) was demonstrated [3]. Toward an injection-type membrane laser, a lateral current injection (LCI) structure [4] was introduced and an injection-type GaInAsP/InP membrane DFB laser was demonstrated [5]. Recently, RT-pulsed operation with a threshold current of 3.8 mA was realized for LCI membrane DFB lasers with the core thickness of 450 nm where the active region consists of 5 quantum-wells (5QWs) and the optical confinement factor in each quantum-well was ξ = 2.1%/well [6].

This time, we would like to report a RT-CW operation of a LCI GaInAsP membrane laser with the core thickness of 220 nm by introducing highly Be-doped p-GaInAs contact layer for the first time. A threshold current of 3.5 mA and a differential quantum efficiency of 11% from the front facet were obtained.

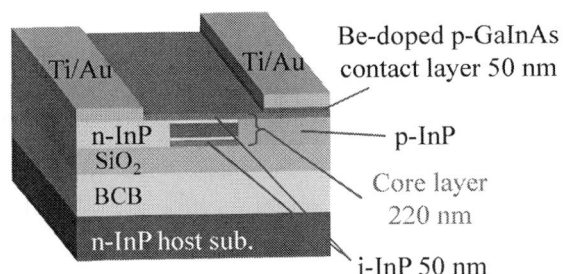

Fig. 1 Schematic structure of the membrane laser with Be-doped contact layer.

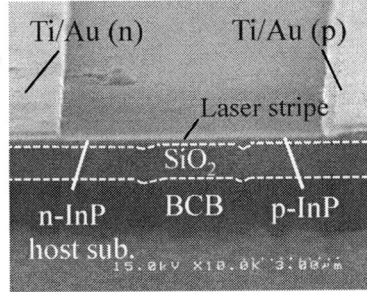

Fig. 2 Cross sectional SEM view of the fabricated device.

II. DESIGN AND FABRICATION

Figure 1 shows the schematic structure of LCI membrane laser with a Be-doped contact layer. Top and bottom cladding layers were composed of air (n = 1) and SiO₂ (n = 1.45), respectively. The device was fabricated as follows. Firstly, an initial wafer with a Be-doped p-GaInAs (N_A = 8 × 10¹⁸ /cm³) contact layer was prepared on an n-InP substrate by gas-source molecular-beam-epitaxy (GSMBE) method; the core layer consists of five 1% compressively-strained (CS) Ga₀.₂₂In₀.₇₈As₀.₈₁P₀.₁₉ 5QWs (6 nm thick each) with -0.15% tensile-strained (TS) Ga₀.₂₆In₀.₇₄As₀.₄₉P₀.₅₁ barriers (10 nm thick each), and an undoped-GaInAsP optical confinement layers (OCLs, λ_g = 1.2 μm, 15 nm thick each). The total thickness of the core layer including 50-nm-thick undoped-InP cap layers is 220 nm. The introduction of the Be-doped contact underneath the QWs had an advantage of low dopant diffusion. The optical confinement factor ξ is 3.2%/well which is 1.5 times larger than that of the previously reported device with 450 nm core thickness [5],[6]. Secondly, the LCI structure was fabricated by two-step organo-metallic-vapour-

978-1-4673-6130-9/13 $31.00 © 2013 IEEE

WeD2-3 (Oral)
11:15 - 11:30

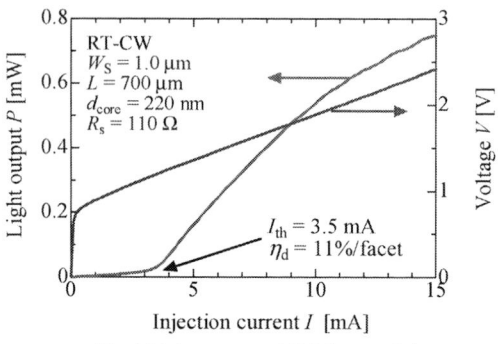

Fig. 3 Light output and *V-I* characteristics.

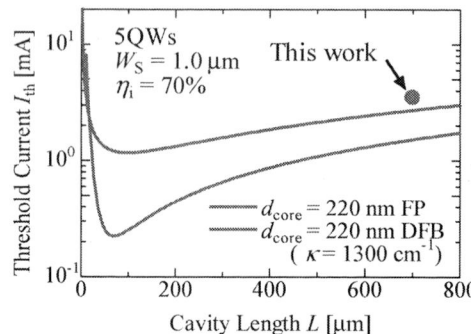

Fig. 4 Calculated threshold current as a function of cavity length.

phase-epitaxy (OMVPE) selective area regrowth. A 7-µm wide mesa was formed by CH_4/H_2 reactive ion etching (RIE) and an n-InP ($N_D = 4 \times 10^{18}$ /cm³) was selectively regrown at both sides of the mesa as a cladding layer with a SiO_2 mask. After etching the part of the mesa and the one side of the n-type cladding layer, a p-InP ($N_A = 4 \times 10^{18}$ /cm³) was regrown in the same way. Then, 1-µm-thick SiO_2 bottom cladding layer was deposited on the wafer. Meanwhile, 2-µm-thick BCB was spin-coated onto the InP host substrate and it was thermally pre-cured for its polymerization in N_2 environment at 210°C. We used the InP substrate just because of easiness of cleavage, but it can be any substrate such as Si or SOI. The wafer and host substrate were bonded with a bonding pressure of about 25 kPa at 130°C, and completely solidified by hard-curing at 250°C under N_2 atmosphere. Subsequently, the InP host substrate and etch-stop layers were removed by polishing and wet etching. A part of the top Be-doped p-GaInAs contact layer near the stripe edge was removed by wet chemical etching. Finally, Ti/Au electrodes were evaporated on the n-InP and p-InP regions. Figure 2 shows a cross-sectional SEM view of the fabricated device around the core layer.

III. EXPERIMENTAL RESULTS

The light output and voltage–current ($V–I$) characteristics of the LCI-membrane laser with the core thickness of 220 nm were shown in Fig. 3. The cavity length and the stripe width were 700 µm and 1.0 µm, respectively. As can be seen, the threshold current of 3.5 mA (a threshold current density of 100 A/cm²/well) and the differential quantum efficiency of 11%/facet were obtained under a RT-CW condition. The rise-up voltage, differential series resistance, and the voltage at the threshold were 0.8 V, around 110 Ω, and 1.2 V, respectively. Further reduction of the series resistance will be required for ultralow threshold operation of a LCI membrane DFB laser with very short cavity structure.

Figure 4 shows the calculated threshold current as a function of the cavity length for a LCI membrane FP laser and a LCI membrane DFB laser with surface grating, where the depth of surface grating on the top of the laser stripe was assumed to be 30 nm and the index-coupling coefficient κ was estimated to be 1300 cm⁻¹. As can be seen, the threshold current of the fabricated device is almost the same as the calculated value ($I_{th} = 2.7$ mA @ $L = 700$ µm, $W_S = 1.0$ µm) by assuming

an internal quantum efficiency of $\eta_i = 70\%$ and the reflectivity of $R = 20\%$ for both facets (It is much smaller than that in conventional lasers in the thin core membrane structure). As the next step, we try to realize a LCI membrane DFB laser with very short cavity for ultralow threshold operation.

IV. CONCLUSION

The LCI membrane laser with 220 nm core thickness was realized and a RT-CW operation with threshold current of 3.5 mA and external quantum efficiency of 11%/facet was achieved for the cavity length of 700 µm and the stripe width of 1.0 µm. These results indicate ultralow threshold operation of the LCI membrane DFB laser with very short cavity.

ACKNOWLEDGMENT

This work was supported by JSPS KAKENHI (Grant numbers 24246061, 24656046, 22360138, 21226010, 23760305, and 10J08973), the Council for Science and Technology Policy (CSTP) under the FIRST program, and the Ministry of Internal Affairs and Communications under the SCOPE program.

REFERENCES

[1] D. A. B. Miller, "Device requirements of optical interconnects to silicon chips," *Proc. IEEE*, Vol. 97, No. 7, pp. 1166-1185, July 2009.

[2] S. Sakamoto, H. Naitoh, M. Ohtake, Y. Nishimoto, T. Maruyama, N. Nishiyama, and S. Arai, "85°C Continuous-Wave Operation of GaInAsP/InP-Membrane Buried Heterostructure Distributed Feedback Lasers with Polymer Cladding Layer," *Jpn. J. Appl. Phys.*, Vol. 46, No. 47, pp. L1155-L1157, Nov. 2007.

[3] S. Sakamoto, H. Naitoh, M. Ohtake, Y. Nishimoto, S. Tamura, T. Maruyama, N. Nishiyama, and S. Arai, "Strongly index-coupled membrane BH-DFB lasers with surface corrugation grating," *IEEE J. Sel. Top. in Quantum Electron.*, Vol. 13, No. 5, pp. 1135-1141, Sept./Oct. 2007.

[4] K. Oe, Y. Noguchi, and C. Caneau, "GaInAsP lateral current injection lasers on semi-insulating substrates," *IEEE Photon. Technol. Lett.*, Vol. 6, No. 4, pp. 479-481, Apr. 1994.

[5] T. Shindo, M. Futami, T. Okumura, R. Osabe, T. Koguchi, T. Amemiya, N. Nishiyama, and S. Arai, "Lasing Operation of Lateral-Current-Injection Membrane DFB Laser with Surface Grating," *The 16th Opto-Electronics and Communications Conference (OECC2011)*, 6D3-7, July 2011.

[6] M. Futami, T. Shindo, K. Doi, T. Amemiya, N. Nishiyama, and S. Arai, "Low-Threshold Operation of LCI-Membrane-DFB Lasers with Be-doped GaInAs Contact Layer," *International Conference on Indium Phosphide and Related Materials (IPRM2012)*, Th-2C, Aug. 2012.

WeD2-4 (Oral)
11:30 - 11:45

Mode Locked InAs/InP Quantum dash based DBR Laser monolithically integrated with a semiconductor optical amplifier

Siddharth Joshi[1], Nicolas Chimot[1], Ricardo Rosales[2], Sophie Barbet[1], Alain Accard[1], Abderrahim Ramdane[2] & François Lelarge[1]

3-5 Lab, a joint laboratory Alcatel-Lucent, Bell Labs France, Thales Research and Technology and CEA Leti.
Laboratoire de Photonique et de Nanostructures, CNRS, France
Marcoussis, France
siddharth.joshi@3-5lab.fr

Abstract—In this paper we present the first demonstration of a InAs/InP Quantum Dash based mode Locked Laser (MLL) compatible with uncooled operation. For integration purpose, we designed a Distributed Bragg Reflector (DBR) mirror in order to close the cavity without disturbing the mode-locking efficiency. As a demonstration of integration, we fabricated such DBR monolithically integrated with a semiconductor optical amplifier. This opens the way to the integration e.g. of frequency comb generators in photonic integrated circuits.

Keywords— Semiconductor Mode Locked Laser, Distributed Bragg Reflector, Photonic Integration

I. INTRODUCTION

Mode Locked Lasers based on InAs/InP quantum dash material have been well explored in the recent years by various groups and the mode locking characteristics of such lasers have been studied in both single section and two section device formats [1],[2],[3]. In order to fully exploit the unique performances of such Qdash lasers, it is of paramount importance to control accurately the frequency spacing required in many applications. In addition, it would be necessary to monolithically integrate such frequency comb generator in photonic integrated circuits. In this paper we present an effective DBR design, which was used to efficiently close the cavity while still maintaining mode-locking efficiency, equivalent to that of discrete Fabry Pérot lasers.

II. MATERIAL QUALIFICATION

The precise control over material quality and expertise in the growth of Qdash layers had allowed various demonstrations on mode locked lasers. In this section we present some results from Fabry Pérot cavities of the material used for DBR lasers, to qualify the quality of the material that was used for the tests on DBR lasers. Fig.1, shows the frequency comb generated by QDash material in a one-section device showing a -3dB bandwidth of 12 nm. This material was used to validate the persistence of efficient mode locking, up to 90°C, demonstrating the potential of Qdash MLL for uncooled operation as shown in Fig.2.

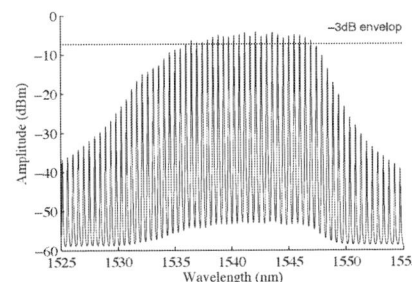

Figure 1: -3dB Envelop of Qdash MLL frequency comb.

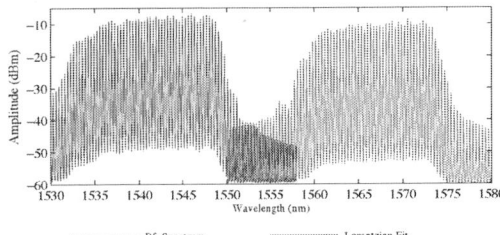

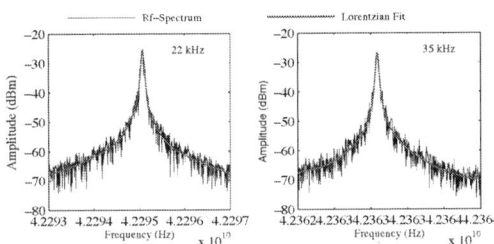

Figure 2: shows the optical and RF mode beating spectra of these lasers at 20°C (left) and 90°C (right). The RF-line-width obtained at 90°C is about 35 kHz against 22 kHz obtained at 20°C.

III. DBR CAVITY MODE LOCKED LASERS

A typical problem that is usually encountered with the Fabry Pérot cavity is the uncertainty in its length, which induces an error of ~3GHz in the free spectral range, corresponding to facet cleavage precision. It is important to lock the generated frequency comb on to the ITU-T recommendation for DWDM. This error can be significantly reduced if the cavity can be closed internally using two DBR mirrors as the positioning and the gain section length is determined with high precision using standard lithographic

---

This work had been sponsored by ITN-PROPHET & ANR-TELDOT. The authors would like to acknowledge the Marie Curie-Initial Training Networks, European Commission and Agence Nationale de la Recherche, France for the financial support.

WeD2-4 (Oral)
11:30 - 11:45

techniques. This further opens the way for integrating such frequency comb with photonic integrated circuits for increased functionality. In both cases, it is of a great importance to develop a DBR building block that does not degrade the unique performances of Qdash based mode locked lasers lasers.

Using the same material as that of the Fabry Pérot lasers described above a DBR mirror can instead be used to efficiently close the cavity. The DBR length and coupling coefficient can be designed in such a way to provide the same reflectivity as that of an as-cleaved Fabry Pérot cavity and a broad pass band allowing the entire 3dB envelop to fit in.

A. Design of DBR mirror for Mode Locking

Increasing the length of a grating in a DBR increases the mirror reflectivity while increasing the refractive index contrast between the materials in the Bragg pairs increases both the reflectivity and the bandwidth [4]. Thus a short Bragg section with high coupling co-efficient can be used to fabricate a DBR with the same characteristics as that of a Fabry-Pérot cavity. With this idea, several designs with varying coupling coefficients κ and grating lengths are tested and are presented in Fig.3.

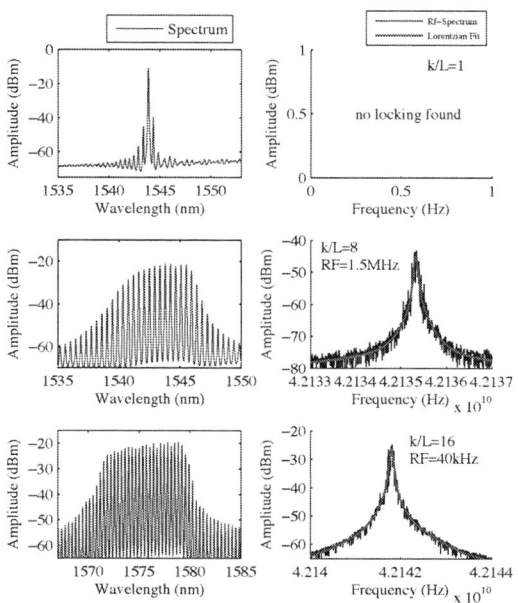

Figure 3: Optical Spectra and corresponding RF Linewidth of lasers with different DBR designs, the value of κ/L for case a=1, case b= 8 and case c=16.

It is observed that when the Bragg section length is large with low value of κ, the mode locking efficiency is not good and a value as high as 7MHz for radio-frequency (RF) linewidth is obtained as most of the modes are filtered off from the spectrum internally (Fig.3a). In this case no self-pulsation is observed. On the contrary, when a small grating length with large coupling coefficient is used, RF-linewidth as low as 40kHz is achieved (Fig.3c). For further demonstration, the mode-locking efficiency was assessed using an auto-correlator. As shown in Fig.4, the auto-correlation trace shows pulses of 2ps obtained with a DBR laser having a RF-linewidth of 60 kHz.

B. Integration to Semiconductor Optical Amplifier(SOA)

These DBR cavities offer a building block for integration to various active devices like an SOA or a modulator section to achieve an integrated functional device on a single chip. We successfully integrated an SOA in a simple way by using the same active medium as that of the laser, which resulted in an overall gain of about 10dB for an amplifier length of 1mm. Such

integration had been explored previously on Quantum- well lasers for other applications by [5].

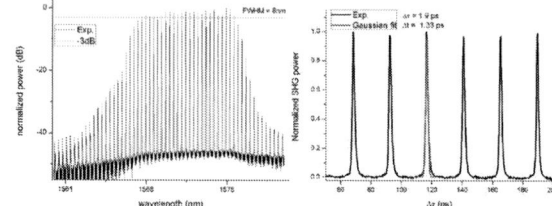

Figure 4: Optical spectrum and auto-correlation trace of DBR lasers, showing 2 ps wide optical pulses.

Fig.5(a) shows the laser integrated to a SOA and Fig.5(b) shows the optical spectra recorded after the SOA section. This result demonstrates the potential integration of a QDash-MLL with SOA or modulator section. The mode locking efficiency results with this three-section device will be presented at the conference.

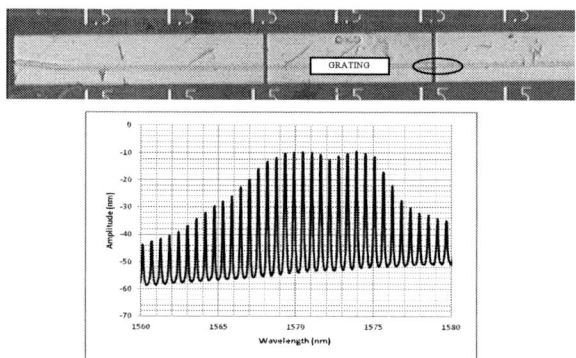

Figure 5: Optical Image of the Device (top), the Optical Spectrum obtained after amplification through SOA (bottom).

IV. CONCLUSION

We present the first demonstration of a primary building block for monolithic integration of Q-Dash based mode locked lasers for InP photonic integrated circuits. We also demonstrate an effective approach of locking the frequency comb generated by a Q-Dash based laser on to the selected ITU-Grid by closing the cavity using DBR-mirrors.

REFERENCES

[1] F. Lelarge, B. Dagens, J. Renaudier, R. Brenot, A. Accard, F. Van Dijk, D. Make, O. LeGouezigou, J. G. Provost, F. Poingt. J. Landreau, O. Drisse, E. Derouin, B. Rousseau, F. Pommereau, and G. H. Duan, ―Recent advances on InAs/InP quantum dosh based semiconductor lasers and optical amplifiers operating at 1.55 μm,‖ IEEE J. Sel. Top. Quantum. Electron. 13, pp. 111–124, 2007.

[2] R. Rosales, S. Murdoch, R. T. Watts, K. Merghem, A. Martinez, F. Lelarge, A. Accard, L. P. Barry, and A. Ramdane, ―High performance mode locking characteristics of single section quantum dash lasers,‖ Opt. Express 20(8), pp. 8649–8657, 2012.

[3] K. Merghem, R. Rosales, S. Azouigui, A. Akrout, A. Martinez, F. Lelarge, G.-H. Duan, G. Aubin, A. Ramdane, Low Noise Performance of Passively Mode-Locked Quantum-Dash-based Laser Under External Optical Feedback Appl. Phys. Lett. 95, 131111 (2009)

[4] Ajoy Ghatak, K. Thyagarajan, "An Introduction to Fiber Optics," Cambridge University Press, 1998.

[5] Lianping Hou; Haji, M.; Dylewicz, R.; Bocang Qiu; Bryce, A.C.; , "Monolithic 45-GHz Mode-Locked Surface-Etched DBR Laser Using Quantum-Well Intermixing Technology," Photonics Technology Letters, IEEE , vol.22, no.14, pp.1039-1041, July15, 2010 doi: 10.1109/LPT.2010.2049566

978-1-4673-6130-9/13 $31.00 © 2013 IEEE

ThD1-1 (Invited)
8:30 - 9:00

Asymmetric Dual-Grating Gate InGaAs/InAlAs/InP HEMTs for Ultrafast and Ultrahigh Sensitive Terahertz Detection

Taiichi Otsuji, Takayuki Watanabe, Stephane Boubanga-Tombet, Tetsuya Suemitsu
Research Institute of Electrical Communication, Tohoku University
Sendai, Japan
otsuji@riec.tohoku.ac.jp

Dominique Coquillat, Wojciech Knap
LC2 Laboratory. University of Montpellier and CNRS
Montpellier, France

Denis Fateev, Vyacheslav Popov
Kotelnikov Institute of Radio Engineering and Electronics (Saratov Branch), RAS
Saratov, Russia

Abstract—**This paper reviews recent advances in ultrafast and ultrahigh sensitive broadband terahertz detection utilizsing asymmetric double-grating-gate InP-based high-electron-mobility transistors, demonstrating a record responsivity of 2.2 kV/W at 1 THz under drain-unbiased conditions with a superior low noise equivalent power of 15 pW/√Hz and 6.4 kV/W even at 1.5 THz under drain-biased conditions.**

Keywords—*terahertz; plasmon; detector; HEMT; grating*

I. INTRODUCTION

Hydrodynamic nonlinearities of two-dimensional (2D) plasmons in high-electron-mobility transistors (HEMTs) are promising for fast and sensitive rectification/detection of terahertz (THz) radiation [1], which can be applied to real-time THz sensing and future THz wireless communications [2]. When a terahertz radiation excites the 2D plasmons in a HEMT, local carrier densities as well as local carrier drift velocity in the channel are modulated by the THz radiation, resulting in generation of quadratic plasma-wave current with a rectified DC component, giving rise to a photovoltaic effect at the high-impedance drain terminal under a common-source asymmetric boundary condition. Recently, InP- and GaN-based HEMTs as well as Si-MOSFETs have demonstrated improved responsivities [3,4], approaching 1 kV/W at 1 THz by introducing narrowband dipole antenna merged with the gate electrode [3]. A metal grating gate is an efficient broadband coupler between plasmons and THz radiation, but suffers from poor responsivity [5]. This paper reviews recent advances in ultrahigh sensitive, broadband THz detection based on InP-based HEMTs incorporating authors' original asymmetric double-grating-gate (A-DGG) structure.

II. DEVICE STRUCTURE AND ANALYSIS

Typical A-DGG HEMT structure of the authors' proposal [6] is depicted in Fig. 1. The asymmetry of the cavity boundaries is greatly enhanced by applying different gate voltages V_{g1} and V_{g2} to the two different sub-gratings of the A-DGG. Terahertz electric current distribution and resultant photovoltaic response were numerically simulated using a self-consistent electromagnetic approach combined with the perturbation theory for the hydrodynamic equations for 2D plasmons in HEMTs [6]. Figure 2a shows the electric-current distribution, underneath an A-DGG unit cell. It is clearly seen that the photocurrent (thus the responsivity) dramatically increases in the undepleted regions, as the portions of the channel under the second sub-grating gates is strongly depleted. One can see a giant enhancement of responsivity by four orders of magnitude (Fig.2a) in asymmetric structures ($d_1 \neq d_2$) compared to conventional symmetric ones ($d_1 = d_2$). Two different processes may be considered to analyze the enhancement of the photovoltage in our A-DGG-FET structures: i) resonant excitation of plasmon modes in undepleted portions of the 2D electron channel that generate strong THz photocurrent due to the nonlinear behavior of these plasmons, and ii) non resonant excitation of overdamped plasmons in depleted regions. The strong depleting of those

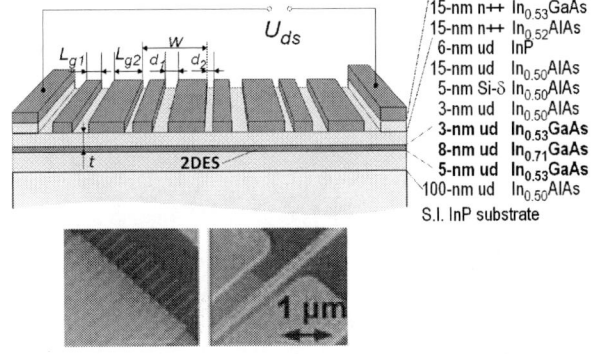

Figure 1. Fig. 1. Schematic view and SEM images of an A-DGG HEMT THz detector. L_{g1} = 200 nm, L_{g2} = 1.6 μm, d_1 = 200 nm, d_2 = 400 nm.

ThD1-1 (Invited)
8:30 - 9:00

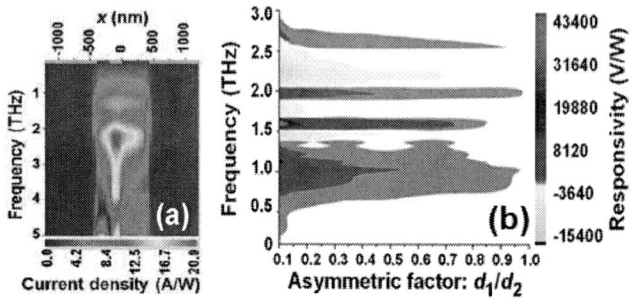

Figure 4. Simulated current distribution (a) and responsivity (b) for an A-DGG HEMT. L_{g1} = 200 nm, L_{g2} = 800 nm, d_1 = 200 nm, d_2 = 400 nm.

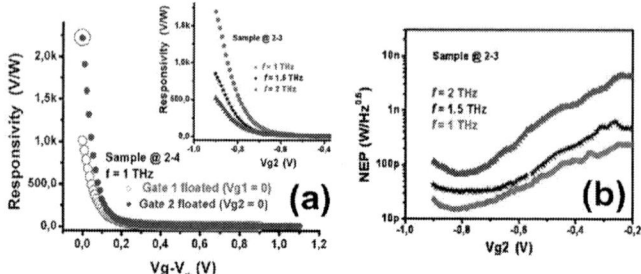

Figure 2. Measured responsivity (a) and NEP (b) under drain-unbiased conditions at 1, 1.5, and 2 THz at 300K.

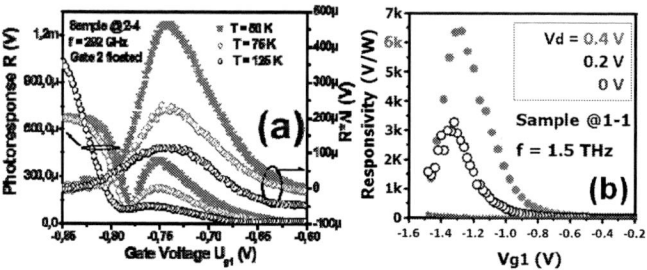

Figure 3. Measured responsivity in resonant mode at 292 GHz at low temperatures under drain-unbiased conditions (a) and in nonresonant mode at 1.5 THz at 300K under drain-biased conditions (b).

regions greatly enhances the channel resistance, which leads to enormous enhancement of the photovoltage induced between the source and drain contacts of the entire A-DGG-FET structure. Therefore these structures combine the advantages of both, resonant and non-resonant plasmonic THz detectors.

III. EXPERIMENTS

A-DGG HEMTs have been designed and fabricated using InAlAs/InGaAs/InP heterostructure material systems (Fig. 1). The asymmetric factor, d_1/d_2, was fixed to be 0.5. A chirped grating structure for the G1 finger L_{g1} was introduced, which can compensate the electron density chirp along the channel under a specific drain-source bias V_{ds} condition [7]. Figure 3 shows measured responsivities at 300 K (a) and related noise equivalent power (NEP) (b) under 1-THz radiation. Responsivities of as high as 2.2 kV/W and NEP below 15 pW/√Hz have been obtained [7]. These values are lower than those of commercial room temperature THz detectors such as Golay cells (200 - 400 pW/√Hz) or SBDs (100 pW/√Hz). Figure 4(a) shows measured photovoltaic response under 292-GHz radiation at low temperatures (50 K, 75 K and 125 K) [8]. By decreasing the temperature, one can see a peak evolving as a shoulder and narrowing on the detection curves. This peak is identified as gated plasmon resonance under the unbiased portions of the channel. The frequency width of this resonance can be estimated as $\delta f = (f/2)*(\delta U_{g1}/(U_{g1} - U_{th1}))$ where f is the incident frequency. This width varies from 279 GHz to 100 GHz when the temperature decreases. This corresponds to the plasma waves damping time of the order of 20 ps and an intrinsic fast response speed of the same order. Figure 4(b) shows measured responsivities at 300 K when constant source to drain voltages V_{ds} are applied under illumination at 1, 1.5 and 2 THz. One can see marked increase in responsivity from 0.4 to 6.4 kV/W at 1.5 THz when V_{ds} reaches 0.4 V and the responsivity showing a monotonic decrease with increase in radiation frequency [8]. The obtained responsivity in the THz range is more than one order of magnitude higher than any existing SBD-based detectors.

IV. CONCLUSION

InP-based A-DGG HEMTs demonstrated an ultrahigh sensitive and fast THz detection with superior low noise power. Further enhancement on the THz detection responsivity is expected by increasing the asymmetry of the A-DGG.

ACKNOWLEDGMENT

This work is supported by the JST-ANR WITH, Japan.

REFERENCES

[1] M. Dyakonov, and M. Shur, "Detection, mixing, and frequency multiplication of terahertz radiation by two-dimensional electronic fluid," *IEEE Trans. Electron. Dev.*, vol. 43, pp. 380-387, 1996.

[2] M. Tonouchi," Cutting-edge terahertz technology," *Nature Photon.*, vol. 14, pp. 97-105, 2007.

[3] T. Tanigawa, T. Onishi, S. Takigawa and T. Otsuji, "Enhanced responsivity in a novel AlGaN/GaN plasmon-resonant terahertz detector using gate-dipole antenna with parasitic elements," *68th Device Research Conf. Dig.*, pp. 167-168, Notre Dame, IN, June 2010.

[4] F. Schuster, D. Coquillat, H. Videlier, M. Sakowicz, F. Teppe, L. Dussopt, B. Giffard, T. Skotnicki, and W. Knap, "Broadband terahertz imaging with highly sensitive silicon CMOS detectors," *Opt. Express*, vol. 19, pp. 7827-7832, 2011.

[5] D. Coquillat, S. Nadar, F. Teppe, N. Dyakonova, S. Boubanga-Tombet, W. Knap, T. Nishimura, T. Otsuji, Y. M. Meziani, G. M. Tsymbalov, and V. V. Popov, "Room temperature detection of sub-terahertz radiation in double-grating-gate transistors," *Opt. Express*, vol. 18, pp. 6024–6032, 2010.

[6] V. V. Popov, D. V. Fateev, T. Otsuji, Y. M. Meziani, D. Coquillat, W. Knap, "Plasmonic terahertz detection by a double-grating-gate field-effect transistor structure with an asymmetric unit cell," *Appl. Phys. Lett.*, vol. 99, pp. 243504-1-4, 2011.

[7] T. Watanabe, S. Boubanba Tombet, Y. Tanimoto, Y. Wang, H. Minamide, H. Ito, D. Fateev, V. Popov, D. Coquillat, W. Knap,Y. Meziani, and T. Otsuji, "Ultrahigh sensitive plasmonic terahertz detector based on an asymmetric dual-grating gate HEMT structure," *Solid State Electron.*, vol. 78, pp. 109-114, 2012..

[8] S. Boubanga-Tombet, Y. Tanimoto, T. Watanabe, T. Suemitsu, W. Yuye, H. Minamidev, H. Ito, V. Popov, and T. Otsuji, "Asymmetric dual-grating gate InGaAs/InAlAs/InP HEMTs for ultrafast and ultrahigh sensitive terahertz detection," *70th Device Research Conf. Dig.*, pp. 169-170, Penn-State, PA, USA, June 19, 2012.

ThD1-2 (Oral)
9:00 - 9:15

Improvement in Nonlinear Characteristics of Zero Bias GaAsSb-based Backward Diodes

Tsuyoshi Takahashi[1,2], Masaru Sato[1,2], Yasuhiro Nakasha[1,2], and Naoki Hara[1,2]

[1]Fujitsu Laboratories Ltd., [2]Fujitsu Limited
10-1 Morinosato-Wakamiya, Atsugi 243-0197, Japan
e-mail: takahashi.tsuyo@jp.fujitsu.com

Abstract— A high curvature coefficient (γ) of 49.4 V^{-1} was achieved at zero bias using p-GaAsSb/i-InAlAs/n-InGaAs backward diodes that were lattice-matched to an InP substrate. γ indicates a higher value than that of ideal Schottky diodes (39.6 V^{-1}). Backward diodes with such a high γ are applicable in mixers. The doping concentration in the p-GaAsSb layer was optimized to obtain an ideal energy band structure for a backward diode. The impedance-matched voltage sensitivity ($\beta_{v,opt}$) at 94 GHz was estimated to be 6,069 V/W using a diode of 3.0 μm diameter.

Keywords—backward diode; GaAsSb; zero bias; non-linearity; millimeter-wave; lattice match; sensitivity

I. INTRODUCTION

Sb-based backward diodes are attractive because of their high sensitivity in the millimeter-wave region [1]. Recently, we have developed p$^+$-GaAsSb/i-InAlAs/n-InGaAs backward diodes that were lattice-matched to an InP substrate [2]. The GaAsSb-based backward diode indicated a voltage sensitivity ($\beta_{v,opt}$) as high as 20,000 V/W when an input impedance was matched with the diode at 94 GHz. However, they have a curvature coefficient (γ) of 35 V^{-1} which was smaller than 39.6 V^{-1}, i.e. the ideal value of Schottky diodes. In this study, we report an achievement of a high γ value exceeding that of the Schottky diodes at zero bias. By controlling doping concentration in an p-GaAsSb layer, a high γ of 49.4 V^{-1} was obtained. This property is applicable in millimeter-wave mixer ICs with low conversion loss.

II. EXPERIMANTAL

A. Curvature coefficient

To improve GaAsSb-based backward diode characteristics, the effect of doping concentration was investigated. For p-type semiconductor of ideal backward diodes, the energy difference between a valence band and Fermi level must be decreased. Diode characteristics, such as detector sensitivity, strongly depend on a curvature coefficient (γ) that indicates a non-linearity characteristic of the diodes. The γ, defined by $(\partial^2 I / \partial V^2)/(\partial I / \partial V)$, is expressed as

$$\gamma = 4/(V_p + V_n) + 2/\hbar\sqrt{2\varepsilon_s m^* / N^*}, \qquad (1)$$

where V_p and V_n are degeneracies on the p-side and n-side; e.g. V_p is the energy difference from the Fermi level to the valence

This work was supported in part by the research and development program of the Ministry of Internal Affairs and Communications, Japan.

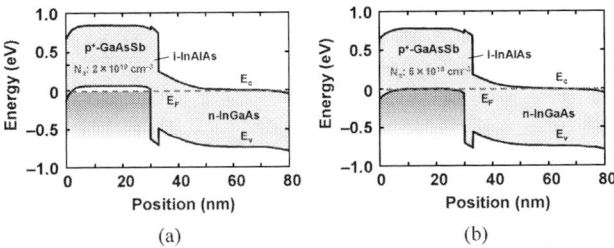

Fig. 1. Energy band diagrams of GaAsSb-based backward diode. (a) $N_a = 2 \times 10^{19}$ cm^{-3}, (b) $N_a = 5 \times 10^{18}$ cm^{-3}.

band of p-GaAsSb. $\hbar$ is the reduced Plank constant, ε_s is a semiconductor permittivity, m^* is the average effective mass, and N^* is the effective doping concentration [3]. From (1), V_p and V_n must be reduced to improve γ. Control of the doping concentration in diodes is one of the methods to decrease V_p and V_n. Fig. 1 shows the calculated energy band diagrams of GaAsSb-based backward diodes in thermal equilibrium. In the figure, the band diagram at p-GaAsSb layer was compared when the case of acceptor doping concentration (N_a) was 2 × 10^{19} and 5 × 10^{18} cm^{-3}. In contrast, the donor concentration (N_d) of an n-InGaAs was kept constant at 1 × 10^{18} cm^{-3}, which indicates a small V_n of about zero. From the calculated band structures, the γ value is expected to be improved by decreasing N_a in the GaAsSb layer.

B. Device fabrication

An epitaxial layer structure of a GaAsSb-based backward diode was grown by metal organic chemical vapor deposition (MOCVD) on a semi-insulating InP substrate. The diode layer consisted of p$^+$-GaAs$_{0.51}$Sb$_{0.49}$/i-In$_{0.52}$Al$_{0.48}$As/n-In$_{0.53}$Ga$_{0.47}$As (30/3/50 nm) heterojunctions that was lattice-matched to the InP substrate. Easy integration should be expected with low noise amplifiers (LNA) of InP-based HEMTs. It is also effective in improving the surface morphology and dislocations in the epitaxial wafers. N_a in the p$^+$-GaAsSb layer was 5 × 10^{18}, 1 × 10^{19}, and 2 × 10^{19} cm^{-3}. In contrast, N_d in the n-In$_{0.53}$Ga$_{0.47}$As was 1 × 10^{18} cm^{-3}. An i-InAlAs layer was inserted as a barrier layer which was thin enough to cause electron tunneling. Epitaxial layers consisting of n$^+$-In$_{0.53}$Ga$_{0.47}$As/n-In$_{0.8}$Ga$_{0.2}$As were grown on p$^+$-GaAsSb layer to form an ohmic contact by interband tunneling. An anode electrode was formed on the n$^+$-In$_{0.53}$Ga$_{0.47}$As layer using TiW metal. Ti/Pt/Au was evaporated to form a self-aligned cathode electrode [4].

ThD1-2 (Oral)
9:00 - 9:15

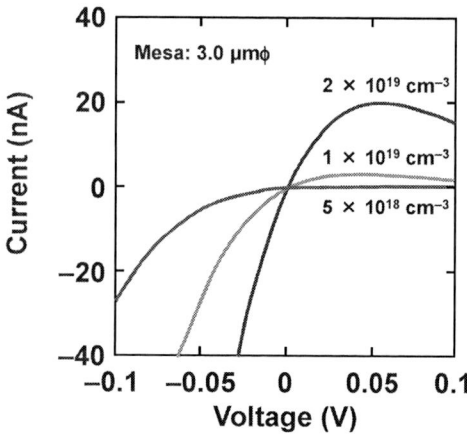

Fig. 2. *I–V* characteristic of GaAsSb-based backward diodes with different doping concentration in the p-GaAsSb layer. The circular mesa diameter is 3.0 μm.

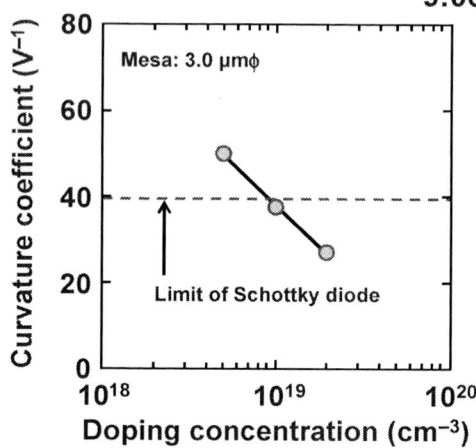

Fig. 3. Dependence of curvature coefficient as a function of acceptor doping concentration in the p-GaAsSb layer.

III. DEVICE CHARACTERISTICS

A. I–V Characteristic

The *I–V* characteristics of the fabricated GaAsSb-based diodes are shown in Fig. 2. A typical curve of backward diodes was obtained for each diode which has the same mesa diameter of 3.0 μm. The current was reduced in accordance with a decrease of acceptor doping concentration in the p-GaAsSb layer. The absolute values of the obtained curvature coefficient γ at zero bias were 49.4, 37.5, and 26.8 V^{-1} for doping concentration in a p^+-GaAsSb layer of 5×10^{18}, 1×10^{19}, and 2×10^{19} cm^{-3}. Fig. 3 shows the curvature coefficient dependence as a function of acceptor doping concentration in the p-GaAsSb layer. When the doping concentration was reduced from 1×10^{19} to 5×10^{18} cm^{-3}, the γ exceeded 39.6 V^{-1} which indicated an ideal value of Schottky diodes ($\gamma_{Schottky}$) given by q/nkT. The high γ of the backward diode indicates a superior parabolic characteristic and is thought to be helpful for mixer applications. Furthermore, it is also attractive for millimeter-wave detection because of the high sensitivity of the GaAsSb-based backward diode.

B. Sensitivity

Voltage sensitivity ($\beta_{v,opt}$) measurements were carried out under 94 GHz and zero bias conditions. The GaAsSb-based backward diode, which had an acceptor concentration of 5×10^{18} cm^{-3}, indicated $\beta_{v,opt}$ of 6,069 V/W, when the mesa diameter was 3.0 μm. Similarly, $\beta_{v,opt}$ of 4,722 and 3,616 V/W were obtained for the GaAsSb-based diodes with N_a of 1×10^{19} and 2×10^{19} cm^{-3}. These results indicate that large $\beta_{v,opt}$ at zero bias will be led by large γ value at zero bias.

C. Device parameters

Device parameters were extracted from an equivalent circuit model. The model includes a junction resistance (R_j), a junction capacitance (C_j), a series resistance (R_s), a parasitic capacitance (C_p), an interconnection inductance (L_p), and a pad capacitance (C_{pad}). R_j was determined to be 22.1 MΩ at zero bias when the backward diode has N_a of 5×10^{18} cm^{-3} and the mesa diameter was 3.0 μm. The extracted C_j and R_s were 13.1

fF and 24.1 Ω, respectively. Thus, the cutoff frequency (f_c) was estimated to be 504 GHz from an expression of $1/(2\pi\,C_j \cdot R_s)$. The f_c improved from the GaAsSb-based backward diode we reported previously [2]. Further improvement of device characteristics is needed for high frequency applications such as a terahertz region.

IV. SUMMARY

A high curvature coefficient (γ) of 49.4 V^{-1} was achieved at zero bias using lattice-matched p-GaAsSb/i-InAlAs/n-InGaAs backward diodes. γ indicated higher a value than that of ideal Schottky diodes. The GaAsSb-based backward diodes with such a high γ are applicable in mixers and other millimeter-wave applications.

ACKNOWLEDGMENT

The authors would like to thank T. Ohki for discussing the device process and T. Ito for his technical assistance in fabricating the devices. The authors would also like to thank Professor O. Ueda of Kanazawa Institute of Technology for his helpful discussions and encouragement.

REFERENCES

[1] Z. Zhang, R. Rajavel, P. Deelman, and P. Fay, "Sub-micron area heterojunction backward diode millimeter-wave detectors with 0.18 pW/Hz$^{1/2}$ noise equivalent power," *IEEE Microw. Wireless Components Lett.*, vol. 21, pp. 267–269, May 2011.

[2] T. Takahashi, M. Sato, Y. Nakasha, and N. Hara, "Lattice-matched p$^+$-GaAsSb/i-InAlAs/n-InGaAs zero-Bias backward diodes for millimeter-wave detectors and mixers," in *Proc. IEEE 25nd Int. Conf. IPRM, 2012*, Tu-1E.2.

[3] S. M. Sze and K. K. Ng, *Physics of semiconductor devices*, Third edition. New Jersey: Wiley, 2007, p. 436.

[4] T. Takahashi, M. Sato, T. Hirose, and N. Hara, "Energy band control of GaAsSb-based backward diodes to improve sensitivity of millimeter-wave detection," *Jpn. J. Appl. Phys.*, vol. 49, p. 104101, 2010.

978-1-4673-6130-9/13 $31.00 © 2013 IEEE

ThD1-3 (Oral)
9:15 - 9:30

Characterization and Modeling of Zero Bias rf-Detection Diodes based on Triple Barrier Resonant Tunneling Structures

Gregor Keller*, Anselme Tchegho*, Benjamin Münstermann*, Werner Prost*, Franz-Josef Tegude*, and Michihiko Suhara**

*Center for Semiconductor Technology and Optoelectronics, University of Duisburg Essen,
Lotharstr. 55, D-47057 Duisburg, Germany
**Electrical and Electronic Engineering, Graduate School of Science and Engineering,
Tokyo Metropolitan University, 1-1 Minami-ohsawa, Hachioji, Tokyo, 192-0397, Japan

Abstract— **InP-based resonant tunneling diodes with symmetrical I/V-characteristics have shown their excellent high frequency performance for THz signal generation. For signal detection we present a device with an additional third barrier to create an unsymmetrical I/V-characteristic. Sensitivity measurements are performed and further improvements by scaling of the active device area are discussed. To allow SPICE based circuit simulation an approach for a large signal model is presented.**

Keywords—RTD, Triple Barrier RTD, rectification, InP, tunneling diode

I. INTRODUCTION

In the field of THz frequency signal generation resonant tunneling diodes (RTD) have proven their excellent high frequency performance [1]. The devices benefit from their large current densities beyond 1000 kA/cm^2 at moderate parasitic capacitances [2]. These features can be used to design quantum structures with an additional third barrier [3]. By asymmetric design of the second well these elements provide nonlinearity in the I/V-characteristic, so that these elements are promising candidates for rf signal detection.

II. DEVICES

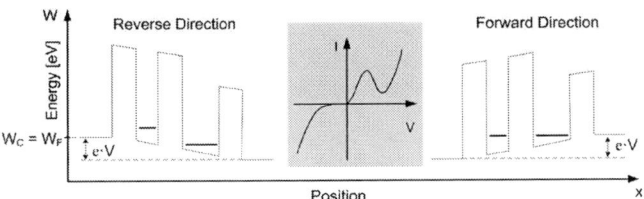

Figure 1. Conduction band edge for reverse and forward operation

The principle of function of the investigated device is shown in fig 1. The wells are designed to have a misalignment for the different discrete energy levels. In forward direction tunnel current flow is possible because the energy levels in both wells are in resonance. This is not possible in reverse direction so that the current flow is blocked [4].

The investigated structure is grown by MBE on semi-insulating InP. The quantum structure is embedded into heavily Si-doped contact layer with doping concentrations up to

$3.7e19$ cm^{-3}, to provide good ohmic contact to the intrinsic structure. The quantum structure is formed with one 1.7 nm InAlAs and two 1.7 nm AlAs barriers. The quantum wells are formed by an InGaAs/InAs/InGaAs layer stack. The 1.17 nm thin InGaAs layers are used to smooth the interfaces, because AlAs and InAs are not lattice matched to the InP substrate. The InAs used for the two wells has a thickness of 2.42 nm in the first and 1.21 nm for the second well.

III. CHARACTERISATION

The dc current voltage characteristics of the fabricated devices were measured on-wafer with a semiconductor parameter analyzer at room temperature.

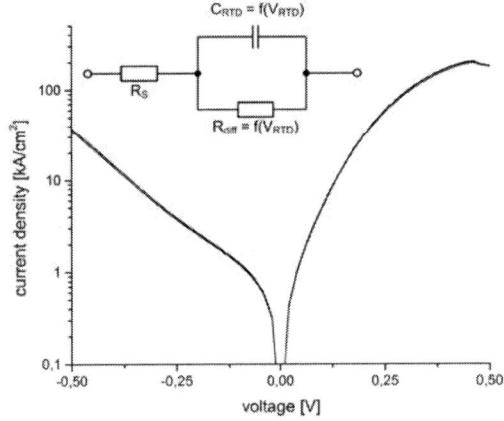

Figure 2. Current voltage characteristic of a 1 μm^2 device at room temperature

The measurements show high current densities, up to 230 kA/cm^2 at peak voltage of 0.45 V. With an s-parameter analyzer sensitivity measurements are performed on-wafer, with coplanar contacted diodes. Zero bias sensitivity measurements of the detected output voltage (V_{det}) are performed at a frequency of 30 GHz with 1 MΩ resistive load. For this the mismatch between the rf-source and the device under test is unaccounted. The input power is measured at the end of the cable, so the power loss at the coplanar probe tip is not taken into account, too. The device shows no compression for power levels up to -10 dBm. Due to the low thermionic current for higher voltages also higher input levels should be possible. To calculate the available sensitivity for impedance matched devices the following equation is used [5].

$$\beta_{v,opt} = \frac{V_{Det}}{P_{in}(1-|\Gamma|^2)} \qquad (1)$$

Here Γ is the return loss caused by the impedance mismatch between source and diode. The result for a 1 µm^2 device, in dependence of the input power, is presented in the following figure.

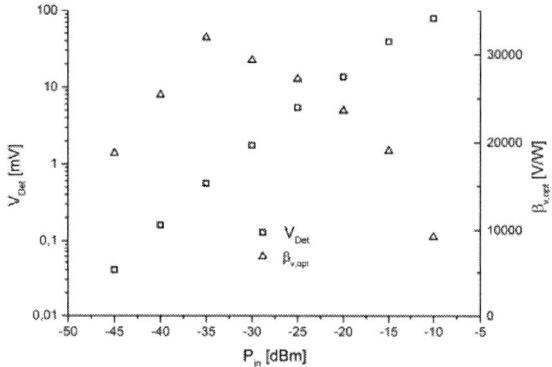

Figure 3. Detected voltage and matched sensitivity of a 1 µm^2 device at 30 GHz

By down scaling to devices with smaller active area the matched sensitivity can be improved due to smaller intrinsic capacitance. The different lines represent structures with a different number of contact fingers but the same active area for each finger.

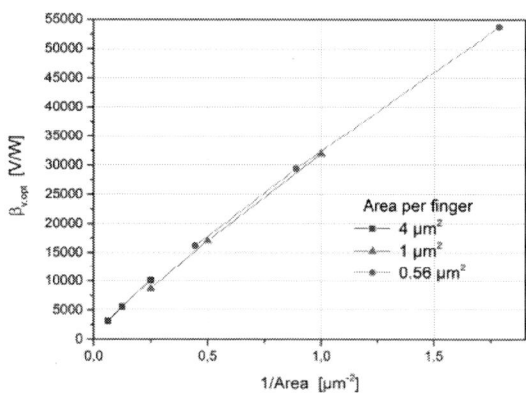

Figure 4. Scaling the active area of the devices at 30 GHz with -35 dBm input power

IV. MODELLING FOR CIRCUIT SIMULATION

Based on RTD large signal models an approach for TBRTD can be achieved. For this, the forward and backward characteristic is modeled with the same equations but a different set of parameters like presented in the following equation:

$$I(V_{RTD}) = \left. \begin{matrix} I_{TP,neg} + I_{TN,neg} + I_{Therm,neg} \\ I_{TP,pos} + I_{TN,pos} + I_{Therm,pos} \end{matrix} \right\} \begin{matrix} V_{RTD} \le 0 \\ V_{RTD} > 0 \end{matrix} \qquad (2)$$

The current in backward direction is basically dominated by the thermionic component. For low voltages a current component from tunneling electrons, but with small peak current density and low peak voltage can also be observed. In forward direction the characteristic is mainly given by the tunnel current, with low influence of the thermionic component, especially for low voltages. The equations used for the modeling of the tunnel current are based on the publication by Yan [6]. In our case in the negative differential region only the Gaussian component of the tunnel current is used. To describe the thermionic current component in both directions, the leakage current is calculated with the equation published by Broekaert [7]. To perform first simulations a constant capacitance scaling with the active area has been used. The voltage depending modeling of this intrinsic capacitance is in progress so that further detailed results can be expected. With large signal S-Parameter simulations, the matched sensitivity up to 1.2 THz is analyzed for diodes with different active area.

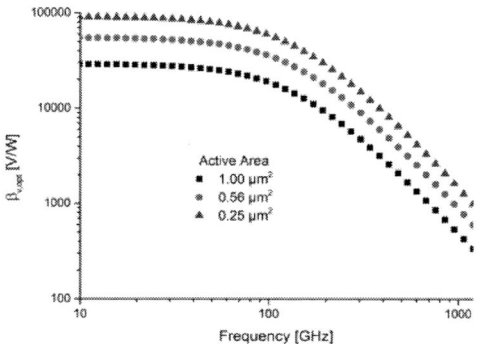

Figure 5. Simulated matched sensitivity for -35 dBm input power

The simulations show matched sensitivities above 1 kV/W up to 1.1 THz for a device with 0.25 µm^2 active area.

REFERENCES

[1] H. Kanaya, H. Shibayama, R. Sogabe, S. Suzuki and M.Asada, „Fundamental oscillation up to 1.31 THz in resonant tunneling diodes with thin well and barriers", Applied Physics Express, vol. 5, no. 12, December 2012

[2] M. Feiginov, C. Sydlo, O. Cojocari, and P. Meissner, "Resonant-tunneling-diode oscillators operating at frequencies above 1.1 THz", Applied Physics Letters, vol 99, no. 23, December 2011

[3] R. Sekiguchi, Y. Koyama, and T. Ouchi, „Subterahertz oscillations from triple-barrier resonant tunneling diodes with integrated patch antennas", Applied Physics Letters, vol. 96, Issue 6, February 2010

[4] G. Keller, A. Tchegho, B. Muenstermann, W. Prost, and F.J. Tegude, "Sensitive high frequency envelope detectors based on triple barrier resonant tunneling diodes", IPRM 2012

[5] T. Takahashi, M. Sato, Y. Nahasha, and N. Hara, "Lattice-matched p+-GaAsSb/i-InAlAs/n-InGaAs zero-bias backward diodes for millimeter-wave detectors and mixer", IPRM 2012

[6] Z. Yan, and M.J. Deen, "New RTD large-signal DC model suitable for PSPICE", IEEE Transaction on Computer-Aided Design of Integrated Circuits and Systems, vol. 14, no. 2, pp.167-172, February 1995

[7] T. Broekaert, B. Brar, J. Wagt, A. Seabaugh, F. Morris, T. Moise, E. Beam, and G. Frazier, "A monolithic 4-bit 2-Gsps resonant tunneling analog-to-digital converter" IEEE Journal of solid state circuits", vol. 33, no. 9, September 1998

Extremely-High Sensitive Terahertz Detector based on Dual-Grating Gate InP-HEMTs

Yuki Kurita, Kengo Kobayashi, and Taiichi Otsuji

Research Institute of Electrical Communication, Tohoku University
2-1-1 Katahira, Aoba-Ku, Sendai 980-8577, Japan
Email: kurita@riec.tohoku.ac.jp

Guillaume Ducournau

Institut d'Electronique, de Microélectronique et de Nanotechnologie (IEMN), 59562 Villeneuve d'Ascq Cedex, France

Yahya M. Meziani

Universidad de Salamanca, Salamanca 37008, Spain

Vyacheslav V. Popov

Kotelnikov Institute of Radio Engineering and Electronics RAS, 410019 Saratov, Russia

Wojciech Knap

LC2 Lab., Univ. Montpellier 2 & CNRS, F-34095 Montpellier, France

Abstract—We report on an extremely-high sensitive terahertz (THz) detector based on our original asymmetric dual-grating gate high electron mobility transistors (A-DGG HEMTs) designed and fabricated using InAlAs/InGaAs/InP material systems. The obtained responsivity is 22.7 kV/W at 200 GHz. To the best of our knowledge, this value is the record responsivity ever reported for this frequency range at room temperature.

Keywords—terahertz, detection, two-dimensional plasmon, high electron mobility transistor.

I. INTRODUCTION

In the modern terahertz (THz) science and technologies, verious kinds of THz detectors, for example, Golay cells [1], pyroelectric detectors [2], bolometers [3] as well as Schottky barrier diodes (SBDs) [4] are used. These detectors except SBDs are thermoelectric types so that they exhibit slow response speed although having rather excellent detection sensitivity. The SBDs can serve fast response speed but suffer from poor sensitivity at higher frequencies due to their electron transit-type mechanism. In such a situation hydrodynamic nonlinearities of two-dimensional (2D) plasmons in high electron mobility transistors (HEMTs) have attracted attention due to their potentiality of fast and sensitive rectification and detection for THz radiation [5, 6]. Under the source-terminated and drain-opened asymmetric boundary conditions, the rectified dc photocurrent in the HEMT channel gives rise to a dc photovoltage at the drain terminal. So far by utilizing such a photovoltaic effect of 2D plasmons excellent THz detection sensitivities have been reported including 5 kV/W at 290 GHz from Si-nMOSFETs [7] and 2.2 kV/W at 1 THz from InGaAs/InAlAs/InP HEMTs [8] at room temperature. In this paper, we demonstrate an extremely high responsivity of 22.7 kV/W at 200 GHz from our original asymmetric dual-grating-gate (A-DGG) HEMT [8, 9].

II. DEVICE DESIGN AND FABRICATION

We designed and fabricated an A-DGG HEMT as shown in Fig. 1 [8, 9]. The device structure consists of an $In_{0.52}AlAs/In_{0.53/0.70/0.53}GaAs/InP$ heterostructure with selective doping in the InAlAs layer (the electron mobility is 11,000 $cm^2/(Vs)$). The spacing between metal fingers of the two sub-grating gates is set to be d_1 and d_2 with $d_1/d_2 = 0.5$. This A-DGG scheme breaks the mirror symmetry of the internal electric field distribution of the 2D plasmon cavities in the HEMT channel, resulting in enormous increase in responsivity by orders of magnitude compared with those of symmetric DGG structures [9]. The two sub-gratings: G1 (with length L_{g1}) and G2 (with length L_{g2}) were formed with 65 nm thick Ti/Au/Ti. The designed parameters of the sample we used for the experiment are shown in table 1.

III. EXPERIMENTS AND DISCUSSIONS

We estimated the responsivity R_v using the equation $R_v = \Delta U \times S_t / (P_t \times S_d)$, where ΔU is the drain-source voltage induced by the THz radiation, S_t is radiation beam spot size, P_t is the total source power of the THz radiation, and S_d is the active area of the detector. Figure 2 shows an experimental setup for THz detection. We used a 200-GHz source and two TPX lenses to focus the THz beam at the surface of the device. The responsivity ΔU was observed by a lock-in technique. All the following measurements were performed under zero drain bias condition, which gave rise to low noise figure of the detector.

Figure 3 shows the responsivity at 200 GHz as a function of the gate bias V_{g2} (whereas the gate bias V_{g1} for G1 was fixed at 0 V) for sub-grating G_2 using sample #1-1. We obtained the highest responsivity of 22.7 kV/W at $V_{g2} = -0.9$ V, close to the threshold voltage. To the best of our knowledge, this is the highest value ever reported.

We also conducted an experiment using a 600-GHz source. Figure 4 shows the comparison of the responsivity using sample #1-2. The magnitude of both frequencies is not so high

ThD1-4 (Oral)
9:30 - 9:45

due to the gate leakage current, but we can see the shift of these peaks (at V_{g2} = -0.7 V for 200 GHz and -0.8 V for 600 GHz). Considering the quality factor (given by the product of angular frequency and electron momentum relaxation time) of a 2D plasmon cavity (defined by each sub-grating finger of G1) at measured frequencies, the detection are made in the non-resonant mode. For this case the peak responsivity should be obtained at the threshold bias point independent of the radiation frequencies. Some new aspects are expected from the measured results. Furthers experiments and device modeling are needed, which will be a future work.

Acknowledgment

The authors thank NTT-AT Corp. for the cooperation in processing the sample fabrications. This work has been supported in part by the JST/ANR "WITH", a Japan-France strategic collaborative research project. The experiment was conducted in IEMN, Lille, France.

References

[1] M. J. E. Golay, "Theoretical consideration in heat and infra-red detection, with particular reference to the pneumatic detector," Rev. Sci. Instrum., **18**, 5 (1947).

[2] R. W. Whatmore, "Pyroelectric devices and materials," Rep. Prog. Phys., **49**, 12 (1986).

[3] P. L. Richard, "Bolometers for Infrared and milimater waves," J. Appl. Phys., **76**, 1 (1994).

[4] E. R. Brown, A. C. Young, J. E. Bjarnason, J. D. Zimmerman, A. C. Gossard, and H. Kazemi, Int. J. High Speed Electron. Sys., **17**, 383 (2007).

[5] M. Dyakonov and M. Shur, "Detection, Mixing, and Frequency Multiplication of Terahertz Radiation by Two-Dimensional Electronic Fluid," IEEE Trans. Electron Devices, **43**, 380-387 (1996).

[6] W. Knap, M. Dyakonov, D. Coquillat, F. Teppe, N. Dyakonova, J. Łusakowski, K. Karpierz, M. Sakowicz, G. Valusis, D. Seliuta, I. Kasalynas, A. El Fatimy, Y.M. Meziani, T. Otsuji, "Field effect transistors for terahertz detection: physics and first imaging applications," Int. J. Infr. Milli. Terhrz. Waves, **30**, 1243 (2009).

[7] F. Schuster, D. Coquillat, H. Videlier, M. Sakowicz, F. Teppe, L. Dussopt, B. Giffard, T. Skotnicki, W. Knap, "Broadband terahertz imaging with highly sensitive silicon CMOS detectors," Opt. Express, **18**, 7827 (2011).

[8] T. Watanabe, S. Boubanga Tombet, Yudai Tanimoto, Yuye Wang, Hiroaki Minamide, Hiromasa Ito et al., "Ultrahigh sensitive plasmonic terahertz detector based on an asymmetric dual-grating gate HEMT structure," Solid State Electron., **78**, 109 (2012).

[9] V. V. Popov, D. V. Fateev, T. Otsuji, Y. M. Meziani, D. Coquillat, and W. Knap, "Plasmonic terahertz detection by a double-grating gate field-effect transistor structure with an asymmetric unit cell," Appl Phys Lett, **99**, 243504 (2011).

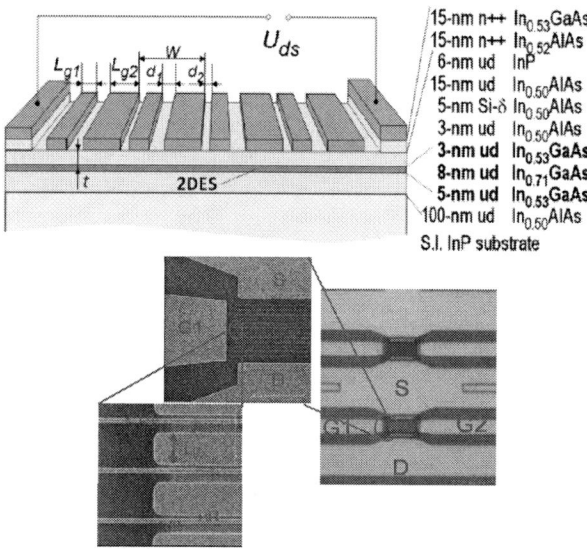

Fig. 1. Schematic (upper) and SEM images (lower) of the A-DGG HEMT.

Table 1. Design parameters of the fabricated samples.

Sample #	1-1	1-2
L_{g1} (nm)	200	400
d1/d2 (nm)	200/400	400/800
L_{g2} (nm)	1600	1600
# of fingers: G1/G2	8/9	6/7
Active area (μm^2)	20·20	20·20

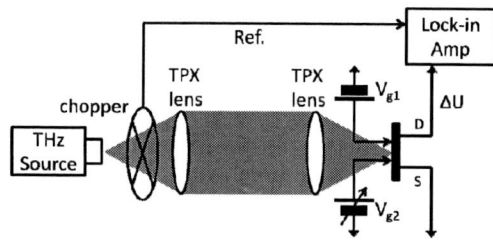

Fig. 2. Experimental setup of THz detection with 200/600 GHz sources and two TPX lenses.

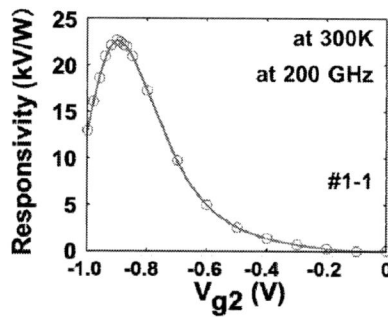

Fig. 3. Responsivity of sample #1-1 as a function of V_{g2} (V_{g1} = 0 V) at 200GHz.

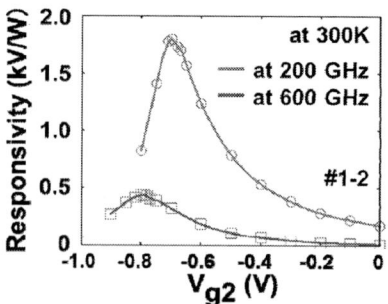

Fig. 4. Measured responsivity at 200 and 600 GHz (Vg1 = 0V) for sample #1-2.

ThD1-5 (Oral)
9:45 - 10:00

High Performance Modulation Doped AlGaAs/InGaAs Thermopiles (H-PILEs) for Uncooled IR FPA Utilizing Integrated HEMT-MEMS Technology

Masayuki Abe[1], Kian Siong Ang[2], René Hofstetter[2], Hong Wang[2], and Geok Ing Ng[2]

[1]3D-bio Co., Ltd., 3-10-6 Shibusawa, Hadano, Kanagawa 259-1322, Japan.
[2]Nanyang Technnological University, NOVITAS-Nanoelectronics Centre of Excellence,
50 Nanyang Avenue, Singapore 639798.
E-mail: abe@ba2.so-net.ne.jp

Abstract-**Novel thermopile based on modulation doped AlGaAs/InGaAs heterostructures is proposed and developed for uncooled infrared FPA (Focal Plane Array) image sensor application. The high sensitivity performance is designed to be the responsivity R of 33,000 V/W with the response time τ of 8 ms, and the high speed performance is designed to be R of 4,900 V/W with τ of 110 μs, under the 2 μm design rule. Based on integrated HEMT-MEMS technology, the 32x32 matrix FPA is fabricated to demonstrate its enhanced performances by black body measurement. The technology presented here demonstrates the potential of this approach for low-cost uncooled infrared FPA application.**

Keywords- Seebeck effect, heterostructure-thermopile, H-PILE, AlGaAs/InGaAs, AlGaN/GaN, HEMT, MEMS, FPA, infrared image sensor

I. INTRODUCTION

Infrared image sensors are essential for automotive vehicle night-vision, rescue robot-eye vision, thermal imaging in biology and medicine, remote sensing in security surveillance systems, and THz electronics applications. The uncooled broad band thermal detectors, either of pyroelectric, bolometric VO_x [1] and a-Si [2], or thermopile types [3], [4], for detection of far-infrared radiation of around 10 μm wavelength, still require high speed operation as well as high responsivity performances with low cost. In this paper, a novel heterostructure thermopile (H-PILE) based on modulation-doped AlGaAs/InGaAs structures is developed [5], and thermopile design, fabrication and evaluated performances are presented and discussed for uncooled infrared FPA (Focal Plane Array) image sensor applications.

II. DEVICE DESIGN AND FABRICATION

The schematic cross sectional structure of H-PILE based on modulation-doped $Al_{0.25}Ga_{0.75}As/In_{0.25}Ga_{0.75}As$ on a GaAs substrate grown by MOCVD, and the energy band diagram is shown in Fig. 1 [6].The conventional HEMT LSI fabrication process [7] is applied to process steps, and the MEMS process is also applied with selective surface-etching to form suspended diaphragm of pixel area and deposition process of absorber. Figure 2 shows the pixel area for infrared detection, for high sensitivity (Type-A) and high speed (Type-C) designs.

For high sensitivity TypeA design, the responsivity R is calculated by $R=\alpha\Sigma_N(S_p R_{thp} - S_n R_{thn})$, with the absorption coefficient $\alpha=1$ and the number of pile-couples N=2. $S_{p,n}$ is the Seebeck coefficient of p- and n-piles. The Seebeck effect consideration has been theoretically discussed in details [5], [6]. The Seebeck coefficient $S=S_p-S_n=2,120$ μV/K, where $S_n=S_{0n}$(diffusion component)+S_{phn} (phonon-drag component) $=-(850+20)=-870$ μV/K, and $S_p=S_{0p}$(diffusion component) +S_{php}(phonon-drag component)$= 1,200+50=1,250$ μV/K. The thermal resistance $R_{thp,n}$ of each pile (p and n) shown in Fig. 2, are calculated to be $R_{thp}=WL/wt=4.9x10^6$, and $R_{thn}=1.2x10^7$ K/W, respectively, where the thermal resistivity W=9.2 for AlGaAs and 2.3 Kcm/W for GaAs, beam length L=137 μm, beam width w=3 μm, the thickness of InGaAs channel layer d=10 nm, the beam thickness t=0.34 for n-type and 0.55 μm for p-type pile, respectively [5].The R is calculated to 33,000 V/W, one order of magnitude higher than that of poly-Si [3]. The detectivity D* is calculated by $D^*=R(A\Delta f/(4k_B T R_{el}))^{1/2}$. We use the detection area A=50x50 μm^2, the frequency band $\Delta f=1$ Hz, k_B the Boltzmann constant, the absolute temperature T=300K. The total electric resistance $R_{el}=\rho L/wt$, $\rho=(q\mu N_s/d)^{-1}$ for series pile-couples of N=2 is 240 kΩ. The resistivity of InGaAs layer, $\rho_e=6.5x10^{-4}$ Ωcm and $\rho_h=2.0x10^{-3}$ Ωcm are calculated for electron and hole, respectively, where q is the elementary charge, μ the mobilities of 8,000 for electron and 320 cm^2/Vs for hole, the sheet carrier concentration is $1.2x10^{12}$ cm^{-2} for the 10 nm thick of 2DEG and/or 2DHG. The D* is calculated to be $5.6x10^9$ cmHz$^{1/2}$/W, which is one order of magnitude higher than that of bolometer [1]. The response time is calculated to be 8 ms using $\tau=C_{heat} R_{th,total}$, with the heat capacitance of the sensor $C_{heat}=4.2x10^{-9}$ J/K and the total thermal resistance of parallel pile-couples (N=2) of $R_{th,total}=1.8x10^4$ K/W.

For high speed Type-C design, the R is calculated by the same step as Type-A, with $\alpha=1$ and N=8. The $R_{thp,n}$ of each pile (p_1, p_2, n_1, and n_2) shown in Fig. 2, are calculated to be $R_{thp1}=1.4x10^5$, $R_{thp2}=2.1x10^5$, $R_{thn1}=5.4x10^5$, and $R_{thn2}=3.6x10^5$ K/W, respectively. The R is calculated to 4,900 V/W. The R_{el} for N=8 is 35 kΩ. The D* is calculated to $2.0x10^9$ cmHz$^{1/2}$/W, which is one order of magnitude higher than that of bolometer [1]. The τ is calculated to 110 μs, two orders of magnitude higher speed than 10 ms of VO_x-bolometer [1] and 25 ms of poly Si thermopile [3], using with the $C_{heat} = 7.0 \times 10^{-9}$ J/K and the total $R_{th,total} = 1.5 \times 10^4$ K/W for N=8. The performance territory in responsivity versus response time is in prospect for Type-A and Type-C, respectively, as shown in Fig. 3.

III. MEASURED RESULTS AND DISCUSSION

Black body measurements have been carried out directly without any filter (▼, ♦, and ■), and with the band pass filter (▲, 14>λ>8 μm), under normal atmospheric pressure. The measured data (▼) for Type-A are plotted at T_b=900-1,200K. The measured data (♦, ■, and ▲) for Type-C are plotted at T_b=800-1,200K. The calibrated data (○, □ and △) are derived from the measured plotted data (■) for Type-C multiplied by λT_b–integral rate [8], i.e., integrated over wavelength λ for hemispherical radiation intensity based on Planck's equation for λ>3 μm, λ>5 μm (Ge-window) and λ>8 μm, respectively. It is noted that the calibrated data (△) for λ>8 μm are close to the measured data (▲) with filter (14>λ>8 μm). The T_b dependence of V_{out} is also calculated from $V_{out}=RP_{in}$, where P_{in} is the incident radiated power, R is assumed to be from 1,000 to 50,000 V/W, as a parameter. The calibrated data for λ>5 μm are close to R of 3,000 V/W. The measured value for R_{el} is 2.8 MΩ, larger than the designed value of 35 kΩ. This would be due to process induced large contact resistance of p- and n-electrodes. The detectivity D* might be roughly estimated to be 1.4×10^8 cmHz$^{1/2}$/W, although D* has to be analyzed based on frequency dependence of noise spectrum measurement.

In conclusion, these devices are expected to realize high performances due to superior Seebeck coefficient and the excellently high mobility of 2DEG and 2DHG in high purity channel layers at the heterojunction interface, and this technology demonstrates the low-cost potentiality for uncooled infrared FPA application.

REFERENCES

[1] InfraVision LSI,Inc.,Uncooled Bolometer FPA SpecSheet, Canada, 1998.
[2] A.Durand, C. Minassian, J. L. Tissort, M. Vilain, P. Robert, A. Touvignon, J. M. Chiappa, and Pistre, "Uncooled amorphous silicon TEC-less 1/4 VGA IRFPA with 25μm pixel-pitch for high volume application", Infrared Technology and Applications XXXV, proc. of SPIE, 7298, 2009.
[3] M. Hirota, Y. Nakajima, Y. Hirose, M. Saito, F. Satou, and M. Uchiyama, "120x90 element thermoelectric infrared focal plane array with precisely patterned Au-black absorber", Proc. 19th Sens.Symp. 2002, pp. 117-121.
[4] A. Dehe, D. Pavlidis, K. Hong, and H. L. Hartnagel, "InGaAs/InP thermoelectric infrared sensor utilizing surface bulk micromachining technology", IEEE Trans. Electron Devices, 44, Jul.1997, pp. 1052-1059.
[5] M. Abe, "Modulation doped heterostructure thermopiles for uncooled infrared image-sensor application", IEICE Trans. Electron., Vol. E93-C, Aug. 2010, pp.1302-1308.
[6] M. Abe, N. Kogushi, K. S. Ang, R. Hofstetter, K. Manoj, L. N. Retnam, H. Wang, G. I. Ng, C. Jin and D. Pavlidis, "High-performance modulation-doped heterostructure-thermopiles for uncooled infrared image-sensor application", IEICE Trans. Electron. E95-C, Aug. 2012, pp.1354-1362.
[7] M. Abe, T. Mimura, N. Kobayashi, M. Suzuki, M. Kosugi, M,Nakayama, K. Odani, and I. Hanyu, "Recent advances in ultrahigh-speed HEMT LSI technology", IEEE Trans. on Electron Dev. 36, Oct. 1989, pp. 2012-2031.
[8] R. Siegel and J. R. Howell, Thermal Radiation Heat Transfer (2nd Ed.), Hemisphere Publishing Corp., McGraw-Hill, 1981.

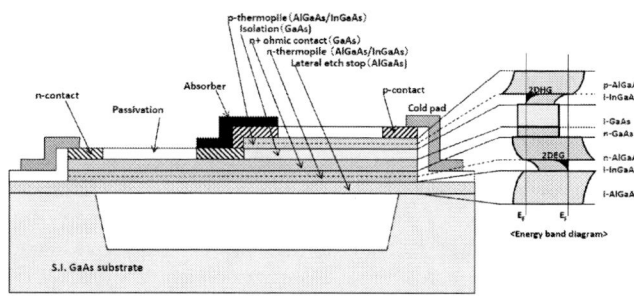

Fig. 1 Cross sectional structure of AlGaAs/InGaAs H-PILE.

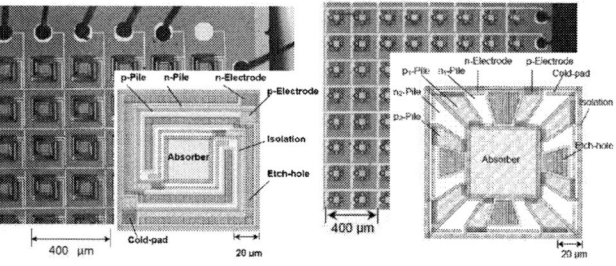

(a) Type-A   (b) Type-C

Fig. 2 Microphotograph of pixel area, for 32x32 matrix array.

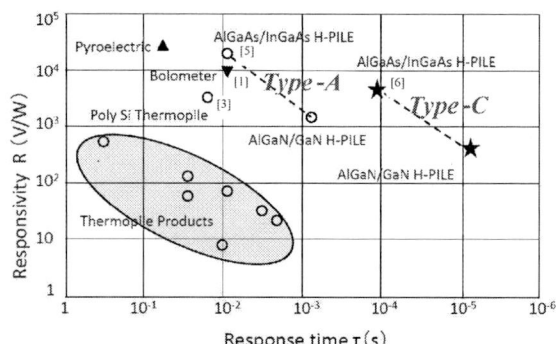

Fig. 3 Performance territory for R vs. τ, compared with alternative uncooled infrared sensor.

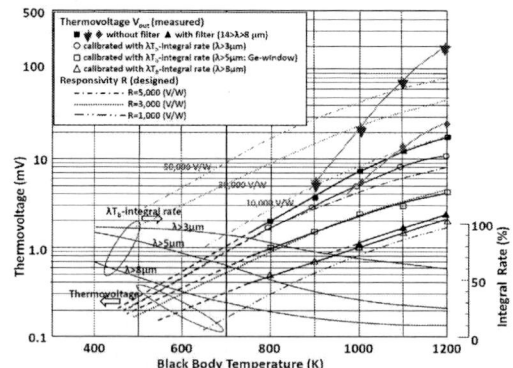

Fig. 4 Thermovoltage and λT_b–integral rate, vs. black body radiation temperature.

978-1-4673-6130-9/13 $31.00 © 2013 IEEE

ThD1-6 (Oral)
10:00 - 10:15

Frequency Modulation in mm-Wave InGaAs MOSFET/RTD Wavelet Generators

Mikael Egard[1,2], Mats Ärlelid[1,2], Lars Ohlsson[1], B. Mattias Borg[1], Erik Lind[1], and Lars-Erik Wernersson[1]

[1]Electrical and Information Technology, Lund University, Lund Sweden
[2] Present address: Acconeer, Lund, Sweden
lars-erik.wernersson@eit.lth.se

Abstract—Co-integration of an InGaAs MOSFET and an RTD is performed to realize a wavelet generator. The large transconductance of the MOSFET (1.9 mS/μm) is used to switch the current in an oscillator circuit and coherent wavelets down to 41 ps are generated in the frequency domain of 50 to 100 GHz. The lowest power consumption measured is 1.9 pJ/pulse. It is found that the supply bias can be used to modulate the center frequency of the wavelets.

Keywords—InGaAS MOSFET; RTD; wavelet generator; mm-Wave circuits;

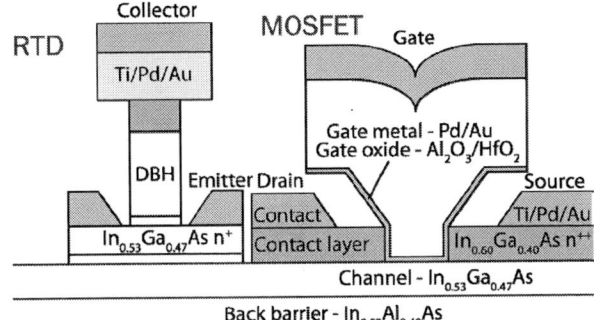

Fig. 1 Schematic layout of RTD and InGaAs MOSFET.

I. INTRODUCTION

InGaAs-based MOSFETs are currently being developed with the aim to increase the drive current in MOS transistors [1]. This integration of III-V channels in the MOS-technology furthermore allows for the introduction of more advanced heterostrcutures in the transistor technology, an approach that has been the back-bone for the III-V industry for several decades.

The speed of III-V transistors has in the past been utilized for the generation of short pulses, so called wavelets [2,3,4]. The generation and detection of short pulses find applications in communication, radar and imaging. Desired key aspects of these wavelets include a well-defined and sharp rise-time of the pulse, as well as the possibility to avoid ringing once the system is turned off. These factors determine the obtainable bit-rates for communication systems and the accuracy in radar measurements. Besides, low power consumption is yet another key aspect.

In this paper, we have fabricated mm-wave wavelet generators by integrating an InGaAs-based MOSFET into a RTD-based oscillator circuit. The MOSFET is used as a switch in the circuit and effectively turns the oscillator on and off. The MOSFET has been designed with low on-resistance using regrowth technology to provide maximum output power. E-mode operation and the isolated gate allows the use of a large gate-swing to control the oscillator. In particular, we show how the center frequency of the wavelet can be modulated by the DC bias applied to the circuit.

II. FABRICATION

The fabrication is based on a MBE-grown heterostructure, where an AlAs/InGaAs/InAs/InGaAs/AlAs double barrier (1.4/0.9/2.0/0.9/1.4 nm) is grown above a 10-nm-thick InGaAs channel deposited on an 190-nm-thick InAlAs back-barrier on

This work is supported by the Swedish Research Council and the Swedish Foundation for Strategic Research.

a SI-InP wafer. The MOSFETs are first processed after selectively etching the RTD layers outside the effective RTD area, Fig. 1. The MOSFET is formed using an HSQ dummy gate and regrowth of InGaAs/InP source-drain regions[5]. After release of the dummy gate, an Al_2O_3/HfO_2 bi-layer ALD film is deposited as gate isolation and a Pd/Au gate contact is evaporated. The RTD is defined by selective etching and the MOSFET and RTD is integrated in a coplanar waveguide, Fig. 2. This stub acts as a resonant tank, but also provides reactive bias stabilization via a large MIM capacitor. Hence, no additional DC path is required in the generator circuit. The details of the processing can be found elsewhere [4].

III. DEVICE AND CIRCUIT CHARACTERIZATION

The InGaAs surface-channel MOSFETs have been individually tested and they show an extrinsic transconductance of 1.9 mS/μm and a drive current of 2.0 mA/μm at 55 nm gate length (V_{ds}=0.5V) [5]. The on-

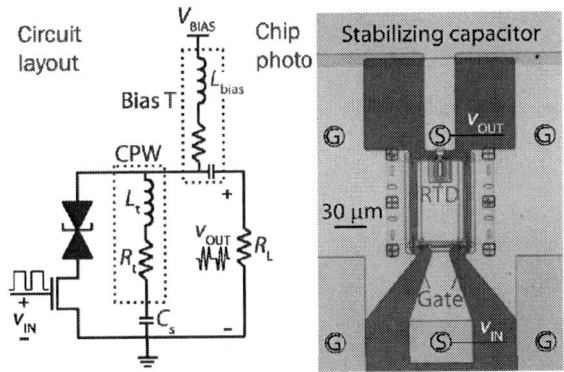

Fig. 2 Schematic circuit layout (left) and chip-photo (right).

ThD1-6 (Oral)
10:00 - 10:15

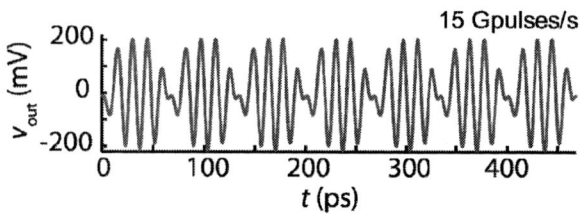

Fig. 3 Measured Wavelets at 15Gpulses/s.

resistance was evaluated to be 199 $\Omega\mu m$. Detailed RF-characterization has furthermore been performed, demonstrating a f_t of 292 GHz and f_{max} of 244 GHz [5]. From small signal-modelling, an instrinsic transconductance of 3.0 mS/μm is determined and that the RF-properties are limited by parasitic capacitance around the T-gate. It is also demonstrated that the tranconductance shows a frequency dependency that is attributed to the frequency response of traps within the high-k film. Individual measurement of the RTDs showed a peak current density of 122 kA/cm^2 and a PVR of 12.

The circuits have been tested on-wafer using an Agilent N4906B to provide input signals to the gate and using a 100 GHz Lecroy sample oscilloscope to evaluate the wavelets generated. Typical pulses show a peak output power of 7 dBm with repetition rates up to 15 Gpulses/s, Fig. 3. The shortest pulses measured were 41 ps at 70 GHz. Based on the short wavelet duration, the energy required to generate one single pulse was evaluated to be 1.9 pJ/pulse.

The frequency performance of these wavelets has been evaluated by fourier transforming the signal from the time-domain to the frequency-domain. Typical 10-dBc bandwidths of 20.6 GHz was evaluated. In an earlier wavelet generator implementation based on a gated tunnel diode, we used the DC bias at the gate to modulate the center frequency of the wavelet by approximately 5 GHz [6]. Using the MOSFET/RTD design, it was observed that the gate bias is less effective to tune the center frequency, whereas the bias voltage V_{BIAS} effectively may be utilized to modulate the center frequency, Fig. 4. Depending on the circuit design, a modulation of 10 to 20 GHz was observed as V_{BIAS} was varied between 1 and 2 V.

REFERENCES

[1] M. Radosavljevic, B. Chu-Kung, S. Corcoran, G. Dewey, M. Hudait, J. Fastenau, J. Kavalieros, W. Liu, D. Lubyshev, M. Metz, K. Millard, N. Mukherjee, W. Rachmady, U. Shah, and R. Chau, "Advanced high-k gate dielectric for high-performance short-channel In0.7Ga0.3As quantum well field effect transistors on silicon substrate for low power logic applications," in Proc. IEEE IEDM, Dec. 2009, pp. 1–4.

[2] Y. Nakasha, Y. Kawano, T. Suzuki, T. Ohki, T. Takahashi, K. Makiyama, T. Hirose, and N. Hara, "A _-band wavelet generator using 0.13-m InP HEMTs for multi-gigabit communications based on ultra-wideband impulse radio," in *IEEE MTT-S Int. Microw. Symp. Dig.*, Jun. 2008, pp. 109–112.

[3] M. Ärlelid, M. Egard, L. Ohlsson, E. Lind and L.-E. Wernersson Impulse-based 4 Gbps radio link at 60 GHz Electronics Letters 47, 467 (2011)

[4] M. Egard, M. Arlelid, L. Ohlsson, B. M. Borg, E. Lind, and L.-E. Wernersson In0.53Ga0.47As RTD-MOSFET mm-Wave Wavelet Generator IEEE Electron Dev. Lett., 33, 970 (2012)

[5] M. Egard, L. Ohlsson, M. Arlelid, K,-M. Persson, M. Borg, F. Lenrick, R. Wallenberg, E. Lind, L.-E. Wernersson High-Frequency Performance of Self-Aligned Gate-Last Surface Channel In0.53Ga0.47As MOSFET IEEE Electron Dev. Lett., 33, 369 (2012)

[6] M. Ärlelid, M. Egard, E. Lind, and L.-E. Wernersson, *Coherent V-Band Pulse Generator for Impulse Radio BPSK* IEEE Microwave and Wireless Components Letters 20, 414 (2010)

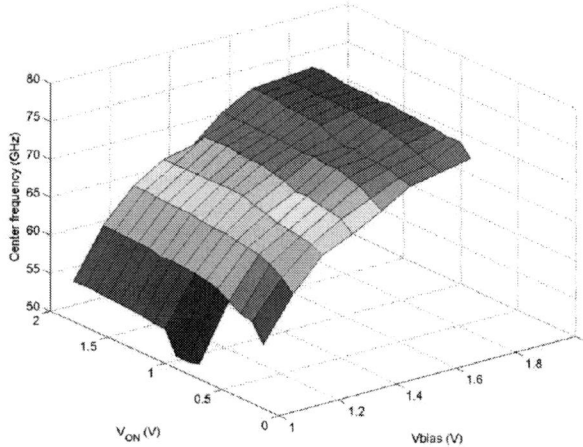

Fig 4. Measured frequency performance for a wavelet generator.

978-1-4673-6130-9/13 $31.00 © 2013 IEEE

ThD1-7 (Oral)
10:15 - 10:30

Ultrashort pulse generators using resonant tunneling diodes with improved power performance

Dongpo Wu, Jie Pan, Katsutaro Mizumaki, Masayuki Mori, and Koichi Maezawa
Graduate School of Science and Engineering, University of Toyama
3190 Gofuku, Toyama 930-8555, Japan
maezawa@ieee.org

Abstract— **RTD pulse generators were designed and fabricated for improved power performance. 10-ps class pulse width with higher peak power was demonstrated for the circuit having an exponentially tapered transmission line impedance converter.**

Keywords—resonant tunneling, pulse generator, tapered transmission line

I. INTRODUCTION

A resonant tunneling diode (RTD) is one of the most promising THz devices. Oscillations higher than 1 THz have been already reported for RTD oscillators [1]. The negative differential resistance (NDR) of the RTDs is also the basis for generating ultrashort pulses. We have reported 6-ps full-width of half maximum (FWHM) pulses can be generated with simple RTD circuits [2]. Such ultrashort pulses can be used for various applications, such as ultra wide band (UWB) communications, and signal sampling. However, there is a crucial problem that the RTD pulse generators can output only a limited power. This is due the fact that the voltages characteristic to the RTD *I-V* curve, such as the peak and valley voltages, are fixed to the small values, and it is difficult to increase them. This leads to the fixed and small voltage swing. In this paper, we will discuss how to improve the output power of the RTD pulse generators and report its experimental demonstration.

II. RTD PULSE GENERATORS

A. Operation Principle

Figure 1 shows the circuit configuration of the basic pulse generator circuit using an RTD. Load line diagram of the series circuit of the resistor and RTD is shown in Fig. 2. When the input voltage exceeds the V_{th1}, the output voltage jumps from stable point 1 to stable point 2. Similarly, when the input voltage decreases to V_{th2}, a transition occurs from stable point 3 to stable point 4. When the input voltage is swept in a range including V_{th1} and V_{th2}, the transition occurs at rising and falling edges of the input voltages. Since these transitions take an extremely short time owing to the nature of RTDs, one can easily obtain sub 10-ps pulses using a high-pass filter.

B. Limitation and Method to Improve the Power

As described in the Introduction, the RTD pulse generators

This work was supported by Grant in Aid for Challenging Research 22656080 from MEXT, Research Institute of Electrical Communication, Tohoku University, .VDEC in collaboration with Agilent Technologies Japan.

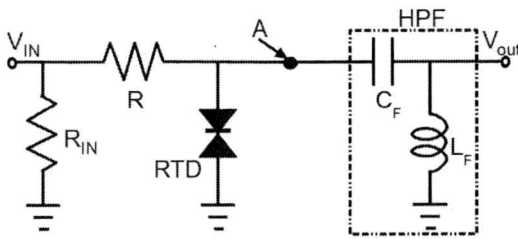

Fig. 1 Basic circuit configuration of the RTD pulse generator.

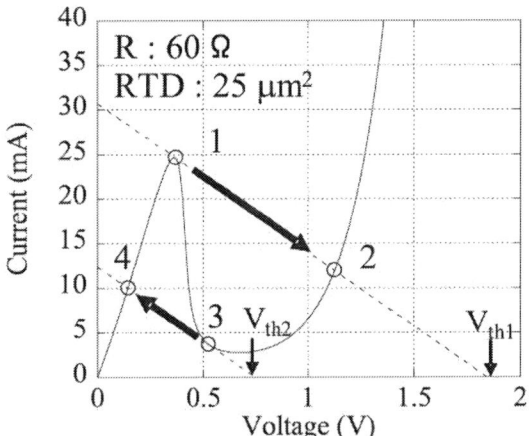

Fig. 2 Load line diagram of the series circuit consisting of an RTD and a resistor.

can output only a limited power due to the fixed peak, valley, and forward voltages. This means voltage swing can not be increased even if the RTD area is increased. To supply maximum power to the standard 50 Ω load, the load impedance should be converted to much smaller values. However, it is difficult to use conventional impedance conversion circuits consisting of inductors and capacitors, since the ultrashort pulses contain very wide and high frequency components. In this paper, we apply the exponentially tapered transmission line for impedance conversion [3]. A tapered transmission line has an impedance that slowly varies spatially from source end to load end, which has superior wide band characteristics. We inserted the exponentially tapered transmission line into the node A in the Fig. 1 to convert the standard 50 Ω load impedance to small values.

ThD1-7 (Oral)
10:15 - 10:30

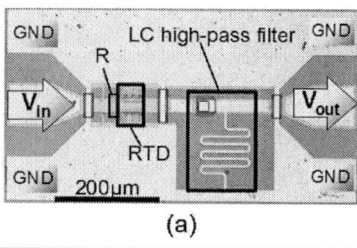

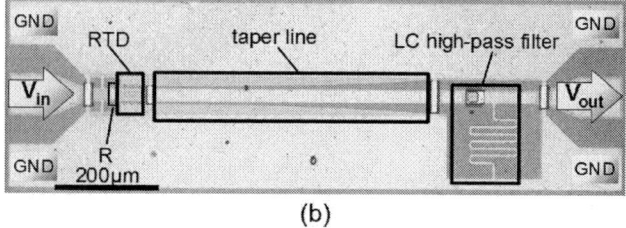

Fig. 3 Fabricated RTD pulse generator circuits. (a)without tapered transmission line, (b) with tapered transmission line.

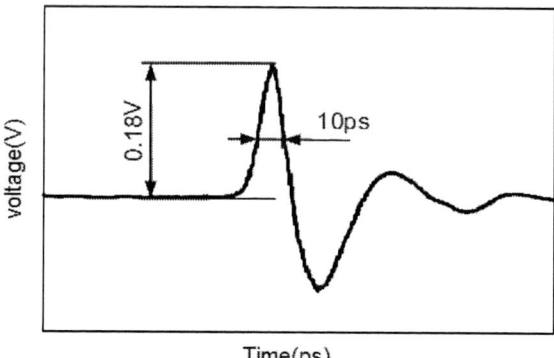

Fig. 4 Measured output wave form of the circuit having the tapered transmission line impedance converter.

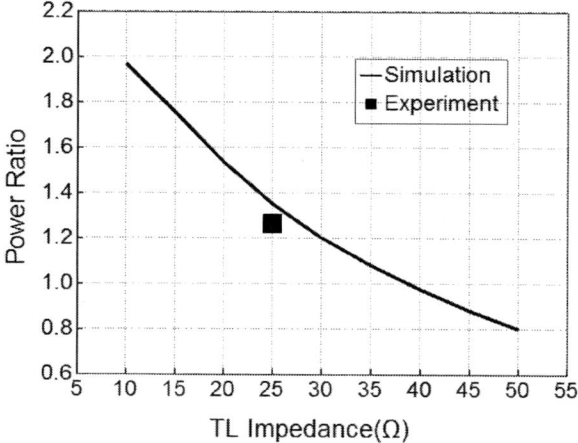

Fig. 5 Peak power improvement ratio as a function of the converted impedance.

III. FABRICATION PROCESS

The circuits were fabricated on the RTD/HEMT epitaxial layers on an InP substrate using conventional photo-lithography and lift-off process [2]. Microphotographs of the fabricated circuits with and without the tapered transmission line are shown in Fig. 3. The circuits consist of 6 RTDs, and the size of each RTD was 17.5 μm^2. The peak current density was 1.5×10^5 A/cm^2. The series resistor was fabricated using HEMT layers grown under the RTD layer. The circuits were designed with coplanar transmission lines and the MIM capacitors. The tapered transmission line was designed to convert the 50 Ω load impedance to 25 Ω.

IV. RESULTS AND DISCUSSION

The circuits were tested on the wafer using RF probes (dc to 110 GHz). The output signal was observed using a Lecroy SDA100 sampling oscilloscope with a 100 GHz sampling module. A 1 GHz sine wave was used for the input signal with a dc offset of 1.0 V. An example of the obtained output signal is shown in Fig. 4. Ultrashort pulses were observed. The FWHM of the pulses was about 10 ps, which was restricted by the bandwidth of the measurement system. The peak height and FWHM of these short pulses strongly depend on the frequency components higher than 100 GHz, which were out of the bandwidth of our measurement system. Therefore, the internal widths of the pulses are estimated to be much smaller than the obtained widths.

It is demonstrated that the peak power was increased by about 30% when we compared it to the circuit without the tapered transmission line (Fig. 3.(a)). This power performance can be much improved if we use smaller impedances than 25 Ω. Figure 5 shows the simulation result showing the peak power improvement ratio as a function of the converted impedance. The experimental results were also plotted. As shown in the figure, the output power can be much improved if one use the

conversion impedance of 10 Ω. Such small impedance can be easily designed if one use microstrip configuration instead of coplanar configuration.

In summary, we have designed and fabricated RTD pulse generators having an exponentially tapered transmission line impedance converter. Improved power performance was demonstrated experimentally with these circuits.

REFERENCES

[1] H. Kanaya, H. Shibayama, R. Sogabe, S. Suzuki, and M. Asada, "Fundamental Oscillations up to 1.31 THz in Resonant Tunneling Diodes with Thin Well and Barriers," Appl. Phys. Express 5, 2012, 124101

[2] N. Kamegai, S. Kishimoto, K. Maezawa, T. Mizutani, H. Andoh, K. Akamatsu, and H. Nakata, "Ultrashort Pulse Generators Using Resonant Tunneling Diodes and Their Integration with Antennas on Ceramic Substrates," Jpn. J. Appl. Phys. 47, 2008, pp. 2833-2837.

[3] F. Okada, *Principles & Applications of Microwave Engineering*, 2004, Sankaido, Tokyo, Japan (in Japanese)

ThD2-1 (Invited)
11:00 - 11:30

Sub-50nm Indium Phosphide High Electron Mobility Transistor Technology for Terahertz Monolithic Microwave Integrated Circuits and Systems

Invited Paper

Stephen Sarkozy, Xiaobing Mei, Wayne Yoshida, Po-Hsin Liu, Ling-Shine Lee, Joe Zhou, Kevin Leong, Vesna Radisic, William Deal, and Richard Lai

Aerospace Systems
Northrop Grumman Corporation
Redondo Beach, USA
stephen.sarkozy@ngc.com

Abstract—**This paper reports on the process and technology of the sub-50nm InP HEMT MMIC process which has enabled signal amplification up to 670 GHz. In particular, considerations not commonly addressed such as the related processing requirements and uniformity of transistors to establish working chipsets are discussed. Finally, initial burn in data is presented as the technology evolves from a research and development process to production**

Keywords—InP; HEMT; THz, fmax, Transistor, mmW; LNA; PA

I. INTRODUCTION

Recent years have seen a significant increase the in the development of systems [1-4] and monolithic microwave integrated circuits (MMICs) and chipsets [5-7] operating in the deep millimeter wave and sub-millimeter wave regimes, and even approaching terahertz (Thz) frequency operation. There are several motivations for utilizing these portions of the electromagnetic spectrum. Firstly, they are relatively untapped and as such offer large amounts of available bandwidth for short distance, narrow beam links, either for high data rate commercial communication (such as personal area networks or cellular backhaul applications), or covert links for military purposes. Several medical demonstrations such as dental and skin screening are available using wavelengths near Terahertz, and solid states methods of generating and detecting signals can produce significant improvements in resolution and system size, weight, and power, giving rise to portable system. Lastly, scientific missions such as radio astronomy and planetary monitoring, which currently use mixer front ends for extreme frequency operation, often operate at cryogenic temperatures and lack sensitivity without ultra long integration times. Solid state amplifiers can provide a medley of benefits, increase noise figure in the front end, increase output power for active systems, and significantly reduced d.c. power consumption for driving local oscillator chains.

Indium Phosphide (InP) materials have been responsible for the operation of active circuits at these frequencies, both through High Electron Mobility Transistors (HEMTs) [8] and also Heterjunction Bipolar Transistors (HBTs). While

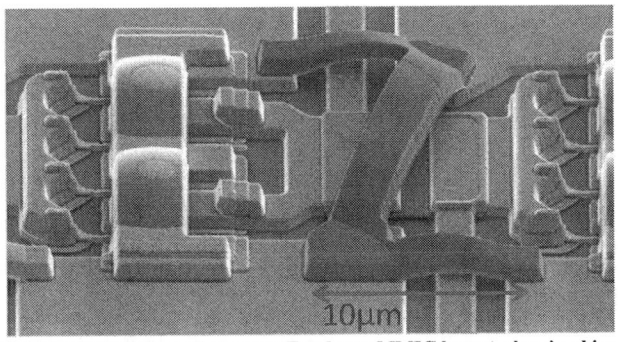

Figure 1 SEM image of compact Terahertz MMIC layout, showing bias lines (orange), blocking capacitors (green) and anairbridge (red)

numerous publications have detailed the circuit records and system demonstrations made with these technologies, comparatively little has been said about the foundation technology itself. This paper focuses on the Northrop Grumman Aerospace System (NGAS) sub-50nm InP HEMT transistor which has demonstrated record frequency and noise figure performance from d.c. to 670 GHz. Notably performance of the transistor itself shall be supplemented with description of the scaled parameters of the overall process, focusing on uniformity requirements need to realize 10 stage amplifiers. Additionally, design rules such as passive current handling capabilities are discussed. Lastly, initial transistor temperature dependence data is presented.

II. THZ FRONTSIDE PROCESSING

THz ICs require ever more compaction in order to keep electrical dimensions constant with the ever shrinking wavelength of higher frequency electromagnetic propagation. While with modern i-line photolithography techniques this has in general proven straightforward even in the deep millimeter and even sub-millimeter regimes, approaching THz frequencies introduces new challenges, chief among which are the ratios between the height of metal stacks and the gap spacing in coplanar waveguide transmission lines, and the necessity to

ThD2-1 (Invited)
11:00 - 11:30

provide uniform RF grounding by connecting ground planes on

Figure 2 SEM image of non-rectangular MMICs, with the dotted line depicting the dicing path.

both sides of the signal via multiple very thin airbridges. Some of the miniaturized structures needed to support operation above 500 GHz are seen in Figure 1.

III. • THz BACKSIDE PROCESSING

The challenges in backside processing for sub-millimeter wave and THz frequency circuits are made manifest in multiple ways. Most obviously, the handling of substrates which are reduced to ever thinner dimensions with decreasing wavelength is physically challenging. From a design motivation, thinner substrates are required to suppress undesired propagation modes. In addition to handling, thinner substrates provide a uniformity challenge, as each micron in Total Thickness Variation (TTV) becomes a significant percentage of the wafer thickness. With increasing frequency, a small window of TTW is tolerable.

A further challenge to the backside process is illustrated by Figure 2, where circuits are seen to be realized in a non-rectangular shape. Creating these chips depends on etching away large portions of the substrate in what is popularly known as the "cookie-cutter" process, which uses Reactive Ion Etching, similar to the process used for through vias. The driving need behind these oddly shaped die are a balance between an electrically narrow chip near the electromagnetic dipole antenna, and a physically wide chip to enable sufficient space to design functional circuitry.

IV. MMIC UNIFORMITY AND RELIABILITY

With the dearth of available gain at high frequencies and the compact MMIC topologies discussed early, it becomes necessary for several gain stages (as many as 10) to be biased from a single gate and single drain voltage. With bond pads not scaling with frequency, MMICs lack sufficient real estate to allow for individual bias for each transistor. Thus should transistors behave in a non-uniform fashion, a significant reduction in performance will occur. Additionally, relatively high current densities are needed to obtain sufficiently high transconductance for signal amplification. Initial chip and

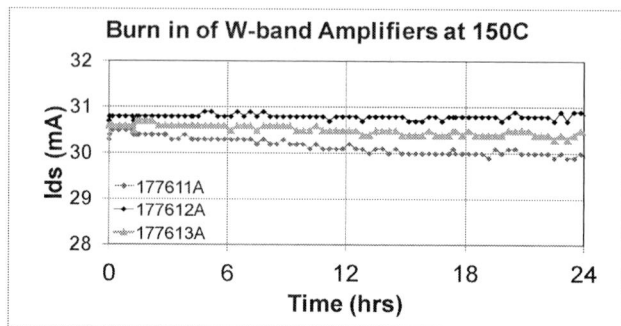

Figure 3 Current evolution of a W-band Amplifier burned in at 150C for 24 hours.

module level burn in test have been performed. The evolution of current at 150C is shown in Figure 3 for a 3-stage W-band amplifier.

ACKNOWLEDGMENT

The authors acknowledge the many contributions of the Northrop Grumman Aerospace Systems fabrication facility, include frontside, backside, and dice-sort-visual technicians, as well as Molecular Beam Epitaxy and Electron Beam Lithography support.

REFERENCES

[1] A. Hirata, R. Yamaguchi, T. Kosugi, H. Takahashi, K. Murata, T. Nagatsuma, N. Kukutsu, Y. Kado, N. Iai, S. Okabe, S. Kimura, H. Ikegawa, H. Nishikawa, T. Nakayama, and T. Inada, "10-Gbit/s wireless link using InP HEMT MMICs for generating 120-GHz-band millimeter-wave signal," IEEE Trans. Microwave Theory Tech., vol. 57, no. 5, pp. 1102-1109, 2009

[2] I. Kallfass, J. Antes, T. Schneider, F. Kurz, D. Lopez-Diaz, S. Diebold, H. Massler, A. Leuther, and A. Tessmann, "All Active MMIC-Based Wireless Communication at 220 GHz, " IEEE Trans. THz Sci. Technol., vol. 1, no. 2, pp. 477-487, Nov. 2011

[3] S. Sarkozy, J. Drewes, K. M. K. H. Leong, R. Lai, X. B. Mei, W. Yoshida, M. Lange, J. Lee, and W. R. Deal, "Amplifier Based Broadband Pixel For Sub-Millimeter Wave Imaging", Optical Engineering, vol. 51 no. 9 Sep. 2012.

[4] S.T. Brown, B. Lambrigtsen, R.F. Denning, T. Gaier, P. Kanaslahti, B.H. Lim, J.M. Tanabe, A.B. Tanner, "The High-Altitude MMIC Sounding Radiometer for the Global Hawk Unmanned Aerial Vehicle: Instrument Description and Performance," IEEE Transactions on Geoscience and Remote Sensing, vol. 49, no. 9, pp. 3291-3301, Sep. 2011.

[5] W. R. Deal, X. B. Mei, K. M. K. H Leong, V. Radisic, S. Sarkozy, and R. Lai, "THz Monolithic Integrated Circuits Using InP High Electron Mobility Transistors," IEEE Trans. THz Sci. Technol., vol. 1, no. 1, pp. 25-32, Sep. 2011

[6] L. Samoska, "An Overview of Solid-State Integrated Circuit Amplifiers in the Submillimeter-Wave and THz Regime," IEEE Trans. THz Sci. Technol., vol. 1, no. 1, pp. 9-24, Sep. 2011.

[7] V. Radisic, K. M. K. H. Leong, S. Sarkozy, X. B. Mei, W. Yoshida, P-H Liu, W. R. Deal, and R. Lai, "220-GHz Solid-State Power Amplifier Modules," IEEE Journal of Solid State Circuits, vol. 47, no. 10, Oct. 2012.

[8] R. Lai et al., "Sub 50 nm InP HEMT device with Fmax greater than1 THz," IEEE 2007 IEDM Conf. Dig., 609–611, IEEE, New York, USA (2007)..

ThD2-2 (Oral)
11:30 - 11:45

35 nm mHEMT Technology for THz and ultra low noise applications

Arnulf Leuther, Axel Tessmann, Michael Dammann, Hermann Massler, Michael Schlechtweg, Oliver Ambacher

Fraunhofer Institut for Applied Solid State Physics (IAF), Freiburg, Germany

Abstract—**In this paper we present a very compact 0.28×0.55 mm^2 six-stage terahertz monolithic integrated circuit (TMIC) using 35 nm gate length metamorphic high electron mobility transistors (mHEMTs). A linear gain of 20.3 dB at 610 GHz and more than 18 dB over the bandwidth from 557 to 616 GHz was achieved for a drain voltage V_d of only 0.6 V. The noise performance of the 35 nm mHEMT was investigated on the basis of a packaged H-band amplifier which achieved a small-signal gain of more than 20 dB between 220 and 300 GHz. The averaged measured noise figure was 6.1 dB which is to our knowledge the lowest published value of any MMIC technology in this frequency range. To determine the transistor reliability accelerated lifetime tests in air were done. Based on a 20 % g_{m_max} degradation failure criterion a median time to failure of 1.8×10^5 h at a channel temperature of 75 °C and V_{DS} = 0.8 V was extrapolated .**

I. INTRODUCTION

The frequency range above 100 GHz has attracted increasing interest in recent years for applications in the fields of radar, radiometry, spectroscopy and communication. This interest is motivated by the advantages of large band width and small wave length on system level. In the past the access to this frequency range was mainly limited to Schottky diode multipliers and mixers. In the meantime with the improvement of transistor technologies MMIC amplifiers up to an operation frequency of 670 GHz [1] were presented. The advantages of MMIC technology compared to Schottky diodes are the potential of signal amplification, higher functionality combined with lower system costs.

In this paper we present for the first time a TMIC with operating frequency above 600 GHz based on mHEMT technology. The fabricated high-gain six-stage amplifier uses 35 nm gate length transistors and utilizes grounded coplanar waveguide topology. For the evaluation of the noise performance of the technology hot-cold measurements of a packaged 230 to 290 GHz broad band amplifier are presented. With progressive device scaling the gate-to-channel separation is shrinking and device reliability is getting an issue. The median time to failure (MTTF) of the 35 nm mHEMTs was investigated with the help of accelerated life time tests.

II. TECHNOLOGY

The molecular beam epitaxy (MBE) growth of the metamorphic heterostructures is done on 100 mm semi-insulating GaAs substrate wafers. A 1.1 µm thick graded InAlGaAs buffer layer transforms the lattice constant from the GaAs to the InP value. For the electron transport a $In_{0.8}Ga_{0.2}As$ channel layer is used with Si δ-doped $In_{0.52}Al_{0.48}As$ barriers on both sides of the channel. The layer sequence is capped with a highly doped $In_{0.53}Ga_{0.47}As$ layer to achieve low source resistance.

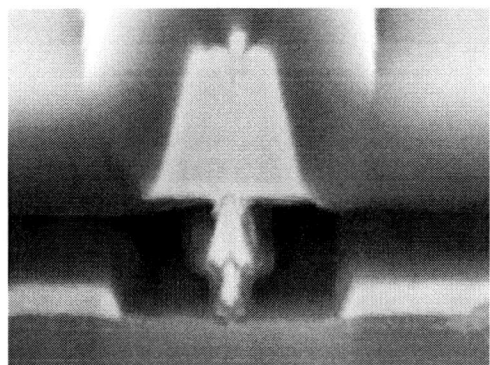

Figure. 1: SEM cross section of the gate foot.

The gate is defined with two separate electron beam lithography layers. In the first step a 35 nm PMMA 950K resist opening is transferred in a SiN layer by dry etching [2]. In the second lithography a 100 nm T-Gate is fabricated on top of the first opening with the help of a 3-layer PMMA resist. During the gate evaporation after recess etching the SiN opening acts as a shadow mask for the metal. The gate is encapsulated in BCB to reduce parasitic capacitances. On top of the e-beam written gate an additional gate head is created by using the first metal interconnect layer. This additional gate head reduces the gate line resistance which is important for high f_{max} and low noise figure. The electrical transistor parameters are listed in table 1. To suppress substrate modes the wafers are thinned down to 50 µm, through substrate vias are dry etched and the wafers are gold plated on the back side.

TABLE I

Electrical parameters of the 35 nm mHEMT ($w_g = 2 \times 10$ µm)

R_c	0.03 $\Omega \cdot$mm
R_s	0.1 $\Omega \cdot$mm
$I_{d, max}$	1600 mA/mm
V_{th}	-0.3 V
$BV_{off\text{-}state}$	2.0 V
$BV_{on\text{-}state}$	1.5 V
$g_{m, max}$	2500 mS/mm
f_T	515 GHz
f_{max}	> 1000 GHz

ThD2-2 (Oral)
11:30 - 11:45

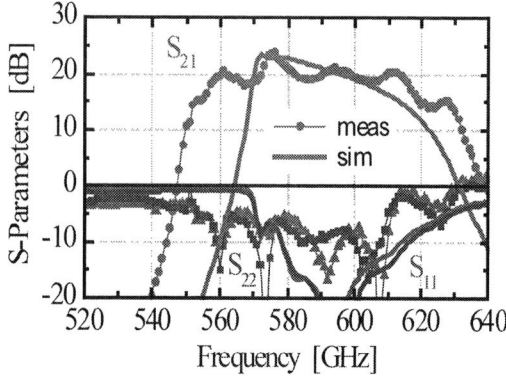

Figure 2. On-wafer measured and simulated S-parameters of the six-stage mHEMT amplifier TMIC from 520 to 640 GHz.

III. CIRCUIT RESULTS

To achieve reasonable gain in the WR-1.5 waveguide band, the gate width of the used common-source devices was chosen to be $w_g = 2 \times 4$ µm. The total chip size of the amplifier circuit is due to the high frequency only 0.28×0.50 mm^2.

The measured and simulated S-parameters of the six-stage metamorphic amplifier circuit are depicted in Fig. 2, over the frequency range from 520 to 640 GHz. A linear gain of 20.3 dB was achieved at 610 GHz, by applying a drain voltage of $V_d = 0.6$ V and a gate voltage of $V_g = 0.1$ V [3]. The drain current at this bias point was $I_d = 24$ mA. Between 557 and 616 GHz a small-signal gain of more than 18 dB was obtained.

The noise performance of the 35 nm mHEMT technology was investigated with the help of a packaged 230 to 290 GHz amplifier. The three-stage H-band amplifier MMIC was designed to achieve high gain and large bandwidth in combination with low noise figure. Therefore, a cascode configuration, consisting of a series connection of one HEMT in common source and one in common gate configuration was utilized. The gate width in each stage is 2×10 µm. As shown in Fig. 3, the amplifier achieved a small-signal gain of more than 20 dB between 220 and 300 GHz, when applying a drain voltage of $V_d = 1.4$ V, a second gate voltage of $V_{g2} = 0.7$ V, a gate voltage of $V_g = 0.1$ V and a drain current of $I_d = 28$ mA. An average noise figure of 6.1 dB was achieved for the

measurement range from 230 to 300 GHz. Which to our knowledge is the lowest published noise figure of any amplifier MMIC in this frequency range.

IV. RELIABILITY

There exists only few reliability data for sub 50 nm HEMT transistors [4]. In order to determine the device lifetime of the 35 nm mHEMT, accelerated lifetime tests were performed in air at $V_{DS} = 0.8$ V for three different channel temperatures. Based on a 20 % g_{m_max} degradation failure criterion and the log-normal distribution, we determine an activation energy of 1.1 eV and a median time to failure of 1.8×10^5 h at a channel temperature of 75 °C. The corresponding Arrhenius plot is shown in Fig. 4.

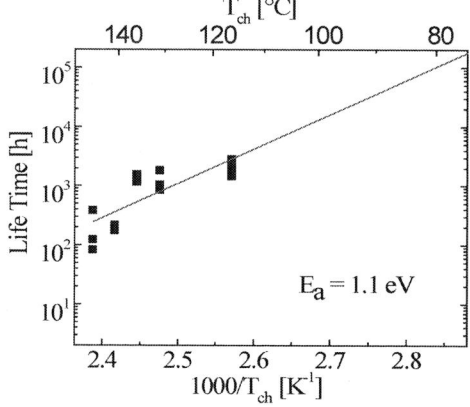

Figure 4: Arrhenius plot of 35 nm mHEMTs measured in air at $V_{DS} = 0.8$ V.

V. CONCLUSION

With the help of the developed 35 nm mHEMT technology it was possible to fabricate a TMIC amplifier with 20 dB gain at 600 GHz. Hot-cold measurements of a packaged H-band amplifier reveal a record noise figure of 6.1 dB in the 230 to 300 GHz frequency range. For the transistors a median time to failure of 1.8×10^5 h at a channel temperature of 75 °C and $V_{DS} = 0.8$ V was extrapolated. These results impressively demonstrate that mHEMT technology is highly suitable for next-generation TMIC applications.

REFERENCES

[1] William Deal, X. B. Mei, Kevin M. K. H. Leong, Vesna Radisic, S. Sarkozy, and Richard Lai, "THz Monolithic Integrated Circuits Using InP High Electron Mobility Transistors", *IEEE Tans. Terahertz Science and Technology*, vol. 1, no. 1, p. 25, September

[2] A. Leuther, A. Tessmann, H. Massler, R. Lösch, M. Schlechtweg, M. Mikulla, O. Ambacher, "35 nm Metamorphic HEMT MMIC Technology," 20th International Conference on Indium Phosphide and Related Materials, MoA3.3, May 2008.

[3] A. Tessmann, A. Leuther, H. Massler, M. Seelmann-Eggebert, " A high gain 600 GHz amplifier TMIC using 35 nm metamorphic HEMT technology", IEEE Compound Semiconductor Integrated Circuit Symposium, CSICS 2012, pp.14-17

[4] Vesna Radisic, Kevin M. K. H. Leong, Stephen Sarkozy, Xiaobing Mei, Wayne Yoshida, Po-Hsin Liu, William R. Deal, Richard Lai, "220 GHz Solid-State Power Amplifier Modules", IEEE J. Solid-State Circuits, vol.47, no. 10, 2012, pp.2291-229

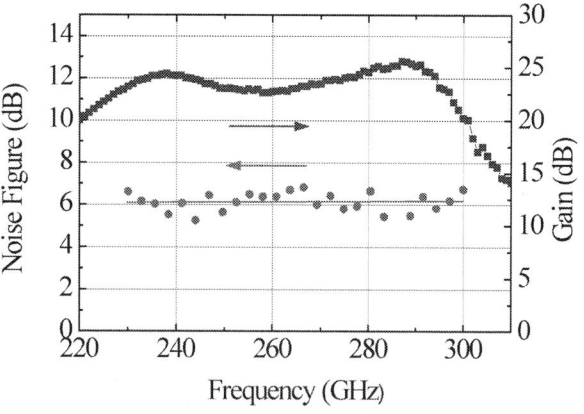

Figure 3: Small signal gain and noise figure of the H-band amplifier.

978-1-4673-6130-9/13 $31.00 © 2013 IEEE

ThD2-3 (Oral)
11:45 - 12:00

250-290 GHz Amplifier
in 75-nm InP HEMT Technology
Using Inverted Microstrip Transmission Line

Hiroshi Matsumura[1,2], Shoichi Shiba[1,2], Masaru Sato[1,2], Tsuyoshi Takahashi[1,2],
Toshihide Suzuki[1,2], Yasuhiro Nakasha[1,2] and Naoki Hara[1,2]
[1]Fujitsu Laboratories Ltd., [2]Fujitsu Limited
Atsugi, Japan
Matsumura.hiro@jp.fujitsu.com

Abstract— **In this paper, we present the development of J-band amplifier in 75-nm InP HEMT technology. The circuit utilizes a six-stage common-source amplifier. An inverted microstrip line (IMSL) structure is employed for matching networks of the amplifier. The developed amplifier realizes a small signal gain of 17.3 dB and a 3-dB band width of 35 GHz from 254 GHz to 289 GHz.**

Keywords— *common-source amplifier, inverted microstrip transmission line, J-band, HEMT.*

I. INTRODUCTION

Submillimeter-wave RF circuits are key components for radio astronomy, imaging sensor, or ultrahigh-speed wireless communication systems. During the last decade, advances in the process technology of InP-based high electron mobility transistors (HEMTs) have opened the fabrication of submillimeter-wave monolithic integrated circuits. Several amplifiers have been reported with operating frequencies around 300 GHz in 35-nm InP HEMT technology [1-5]. Furthermore, the 300-GHz amplifier modules with implementing MMIC amplifier are reported. According to these studies, the substrate of MMICs was thinned to 50 um in order to suppress unwanted substrate modes [2, 4]. As the operation frequency is higher, the distance between grounds in MMIC and mounting board is more considerable. Two large ground planes, which distance is larger than a quarter wavelength, may induce the propagation of unwanted wave and the oscillation in amplifier. So as a 300-GHz MMIC amplifier, the InP substrate must be thinned to smaller than 70 um and mounted many through substrate vias (TSVs) for connecting two ground planes as shown in Fig. 1. Then it reduces mechanical strength of MMIC.

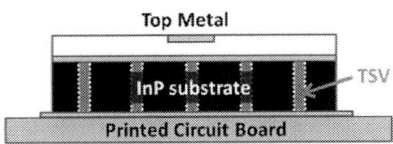

Fig. 1   Conventional implementation for 300-GHz MMIC

In this study, a MMIC implementation with flip-chip technology is planned. The inverted microstrip line (IMSL)

structure is employed for transmission line, and its ground is connected to an outer ground by metal bumps instead of TSV. And furthermore, this implementation realizes suppressing the substrate modes without thinning substrate, because the ground planes of top metal connects to an outer ground shortly by bumps. The developed J-band amplifier MMIC in 75-nm InP HEMT technology with a gain of 17.3 dB and a bandwidth of 35-GHz using IMSL is presented.

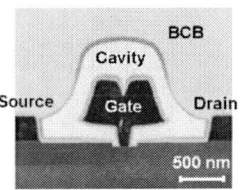

Fig. 2   75 nm InP HEMT technology

II. DEVICE TECHNOLOGY

The HEMT wafer employs 75-nm gate length indium phosphide (InP) technology and grown by metal-organic chemical vapor deposition on a 3-in semi-insulating InP substrate. A SEM micrograph of the gate cross-section is shown in Fig. 2. An electron-beam lithography technique was used to fabricate both the gate electrode and gate recess. The gate electrode, consisting of Ti/Pt/Au (5/30/500 nm), was evaporated on an InP etching-stopper layer and lifted off. Cavity structure was introduced in order to decrease parasitic capacitances such as C_{gs} and C_{gd} [6]. The maximum oscillation frequency (f_{max}) of the 75-nm InP HEMT is 660 GHz.

The conductor layer for double-layer routing was made of Au. A benzocyclobutene (BCB), having a relative permittivity of 2.8, was used. The thickness of BCB was 4.6 μm.

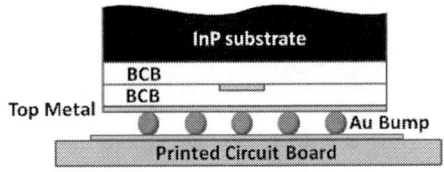

Fig. 3   Target structure of MMIC implementation

978-1-4673-6130-9/13 $31.00 © 2013 IEEE

III. DESIGN

A. Transmission Line

The IMSL structure is employed for all transmission line in design of circuits. The target implementation is showed in Fig. 3. The top metal is used for ground and connected to an outer ground with Au bumps by flip-chip bonding technology.

Besides, the chip size of MMIC using IMSL is able to be small. As the InP substrate with its high relative permittivity of 11.9 is located just above the signal line, the effective dielectric constant for the transmission line is higher than that of interlayer dielectrics. In order to design matching network accurately, the transmission characteristic of IMSL was computed by electro-magnetic field analysis.

B. Common-source Amplifier

The six-stage J-band amplifier was designed to achieve high gain and large bandwidth. A common-source topology was utilized. The schematic of a single amplifier stage is shown in Fig. 4. Complex conjugate matching network is utilized for inter-stage matching with a series capacitor, C_1, and transmission lines, TL_1-TL_5. IMSL structure is employed for all transmission lines. The DC bias is supplied by feed line independent of the matching network. The utilized transistors have a gate width of 2×20 µm.

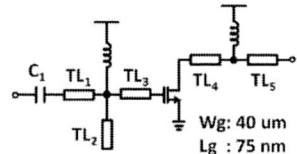

Fig. 4   Schematic of common-source amplifier (one stage)

IV. MEASURED RESULTS

A die photograph of the six-stage common-source amplifier is shown in Fig. 5. The circuit is entirely covered by top metal for ground plane of IMSL. And many bump pads are placed at the top metal for flip-chip bonding. The size of the chip is 0.76×1.94 mm² including pads. The amplifier draws 58.0 mA from a 1.2-V supply voltage. Total power consumption was 69.6 mW.

On-wafer S-parameter measurements of the amplifier was performed using an Agilent PNA with VDI J-band frequency extender modules and WR-3.4 on-wafer probes developed at Cascade Microtech. Fig. 6 shows the measured S-parameters of J-band amplifier. A small signal gain was 17.3 dB. A 3-dB bandwidth was obtained from 254 to 289 GHz. S_{11} and S_{22} were less than −10 dB. And in-band group delay indicated flat characteristic as shown in Fig. 7.

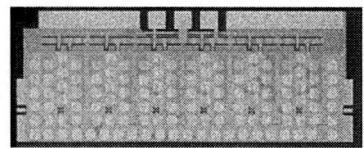

Fig. 5   Die photograph of 6-stage amplifier

V. CONCLUSION

This paper has presented high frequency amplifier using six-stage common-source amplifier with inverted microstrip transmission line structure for matching circuits. The structure realizes the implementation without reducing mechanical strength of MMIC. We designed and fabricated six-stage amplifier for submillimeter-wave amplifier in 75-nm InP HEMT technology. Measured results show a small signal gain of 17.3 dB and a 3-dB bandwidth of 35 GHz from 254 to 289 GHz. The power consumption was 69.6 mW from 1.2 V supply.

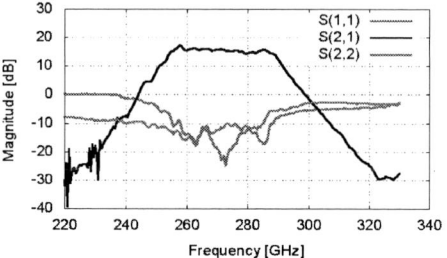

Fig. 6   Measured S-parameters

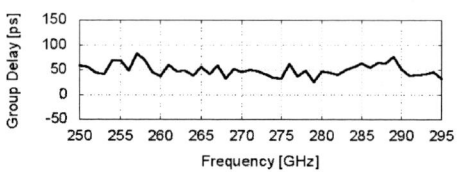

Fig. 7   Measured group delay

ACKNOWLEDGMENT

The authors would like to thank the device processing group in Fujitsu Laboratories Ltd. This work was supported in part by the research and development program on Multi-tens gigabit wireless communication technology at subterahertz frequencies of the Ministry of Internal Affairs and Communications, Japan.

REFERENCES

[1] W. R. Deal, X.B. Mei, V. Radisic, W. Yoshida, P.H. Liu, J. Uyeda, M. Barsky, T. Gaier, A. Fung and R. Lai, "Demonstration of a S-MMIC LNA with 16-dB Gain at 340-GHz," Compound Semiconductor Integrated Circuit Symposium, 2007. CSIC 2007. IEEE

[2] D. Pukala, L. Samoska, T. Gaier, A. Fung, X. B. Mei, W. Yoshida, J. Lee, J. Uyeda, P. H. Liu, W. R. Deal, V. Radisic, and R. Lai, "Submillimeter-Wave InP MMIC Amplifiers From 300–345 GHz," Microwave and Wireless Components Letters, IEEE, Jan. 2008

[3] A. Tessmann, A. Leuther, H. Massler, M. Kuri and R. Loesch, "A Metamorphic 220-320 GHz HEMT Amplifier MMIC," Compound Semiconductor Integrated Circuits Symposium, 2008, IEEE

[4] A. Tessmann, A. Leuther, V. Hurm, H. Massler, M. Zink, M. Riessle, R. Loesch, M. Schlechtweg and O. Ambacher, "A 300 GHz mHEMT amplifier module," Indium Phosphide & Related Materials 2009, IEEE

[5] A. Tessmann, A. Leuther, H. Massler, V. Hurm, M. Kuri, M. Zink, M. Riessle and R. Loesch, "High-gain submillimeter-wave mHEMT amplifier MMICs," Microwave Symposium Digest (MTT), 2010 IEEE MTT-S International

[6] T. Takahashi, M. Sato, Y. Nakasha, T. Hirose, and N. Hara, "Improvement of RF and noise characteristics using a cavity structure in InAlAs/InGaAs HEMTs," IEEE Trans. Electron Devices, August 2012

ThD2-4 (Oral)
12:00 - 12:15

Comparative Study on Frequency Limits of Nanoscale HEMTs with Various Channel Materials

Yutaro Nagai[1], Shohei Nagai[1], Jun Sato[1], Shinsuke Hara[1], Hiroki I. Fujishiro[1],
Akira Endoh[2], Issei Watanabe[2], and Akifumi Kasamatsu[2]

[1]Tokyo University of Science, [2]National institute of Information and Communication Technology

[1]2641 Yamazaki, Noda, Chiba 278-8510, Japan [2]4-2-1 Nukui-Kitamachi, Koganei, Tokyo 184-8795, Japan

j8112633@ed.tus.ac.jp

Abstract— The frequency limits in intrinsic f_T of the nano-scale HEMTs with various channel materials are investigated by using the quantum-corrected Monte Carlo simulation. The device C with the InSb channel exhibits the higher intrinsic f_T from the lower V_{ds} because of the higher electron velocity. Owing to the shorter limit in the L_g scaling at the lower V_{ds}, the device C also exhibits the higher frequency limit in the intrinsic f_T, indicating its potential for the low power and THz operation.

Keywords— InGaAs, InAs, InSb, HEMT, Strain, Band Calculation, Quantum-Corrected Monte Carlo Simulation, Delay Time, f_T, THz

I. INTRODUCTION

So far, performances of HEMTs have been improved by size scaling and channel engineering to increase electron velocity using Indium-rich channel. With regard to f_T, however, its progress has not been rapid recently: 562 GHz in 2002 by $In_{0.7}Ga_{0.3}As$ HEMT with L_g of 25 nm [1] and 688 GHz in 2011 by $In_{0.7}Ga_{0.3}As$ HEMT with L_g of 40 nm [2]. Therefore, a question about the frequency limits of the HEMTs has arisen [3] and an understanding about the mechanisms determining f_T has been strongly demanded [4,5]. In this paper, we investigate the frequency limits in f_T of the nanoscale HEMTs with various channel materials and discuss the essentials for the progress of f_T by using the quantum-corrected Monte Carlo (QC-MC) simulation.

II. SIMULATION METHOD

Strained band structures of the materials are calculated by empirical pseudopotential method with rigid ion approximation [6]. Band parameters are extracted for the QC-MC simulation, where the quantum mechanical and degeneracy effects are considered [7]. Three-valleys (Γ, L, X) non-parabolic spherical conduction band model is assumed for each material.

Scattering mechanisms that are taken into account are acoustic, polar, non-polar optical phonons, ionized impurity, alloy and impact ionization. Impact ionization threshold energy E_{th} is estimated considering energy and momentum conservations [6]. The delay time distribution along the channel $\tau(x)$ is calculated by using a combination of the delay time analysis and the QC-MC simulation [3,8]. The intrinsic f_T is calculated as $1/2\pi\tau_{total}$, where τ_{total} is the total delay time integrating $\tau(x)$ over the channel.

III. DEVICE MODELS

Three kinds of the devices with different layer structures are simulated, which are shown in Fig. 1. The combinations of channel/buffer•barrier are $In_{0.7}Ga_{0.3}As/In_{0.52}Al_{0.48}As$ (standard), $InAs/AlSb$, and $InSb/In_{0.75}Al_{0.25}Sb$ in the devices A, B, and C, respectively. The channels are 15 nm in thickness. The donor sheet densities are 2.0×10^{12}, 2.5×10^{12}, and 1.5×10^{12} cm^{-2}, respectively. The compressive or tensile strains are applied to the channels owing to the lattice mismatch for the buffers; the strain ratios $\varepsilon_{\parallel}$ are -1.1, +1.3, -1.3 for the devices, respectively. The gate length L_g is varied from 100 to 15 nm.

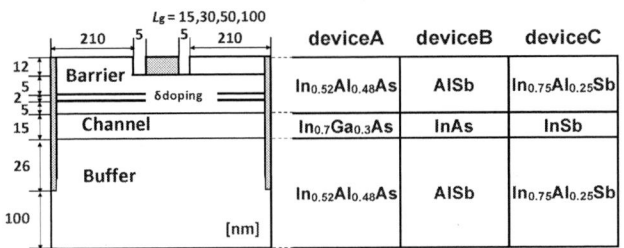

Figure 1 Schematic illustration of device geometry. Three kinds of devices with different layer structures are simulated.

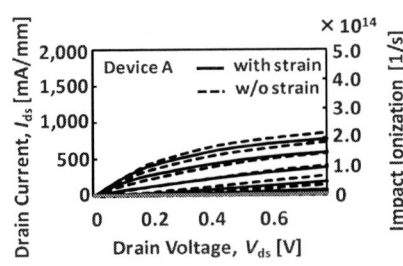
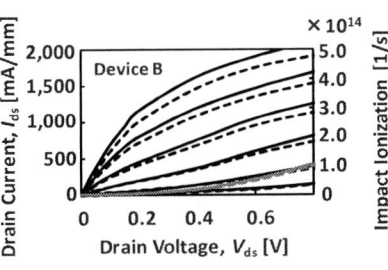
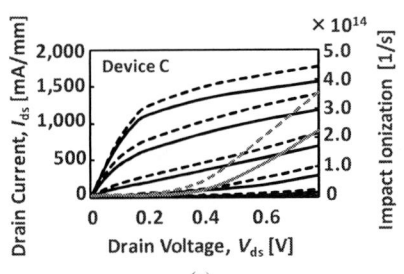

(a)   (b)   (c)

Figure 2 I_{ds} - V_{ds} characteristics for devices (a) A, (b) B, and (c) C. V_{gs} is varied from -0.8 to +0.2 V. Numbers of impact ionization events are also shown by red lines (V_{gs} is 0.0 V).

978-1-4673-6130-9/13 $31.00 © 2013 IEEE

ThD2-4 (Oral)
12:00 - 12:15

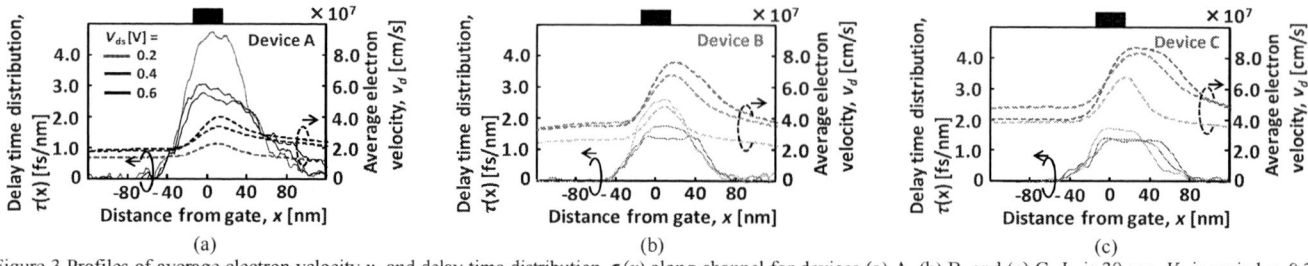

Figure 3 Profiles of average electron velocity v_d and delay time distribution $\tau(x)$ along channel for devices (a) A, (b) B, and (c) C. L_g is 30 nm. V_{ds} is varied as 0.2, 0.4, and 0.6 V. V_{gs} is biased as intrinsic f_T becomes maximum.

IV. SIMULATION RESULTS

A. I_{ds} - V_{ds} Characteristics

Fig. 2 shows the I_{ds} - V_{ds} characteristics for the devices A (a), B (b), and C (c), where the number of the impact ionization events are also shown. L_g is 30 nm. Fig. 3 shows the profile of the average electron velocity v_d along the channel when V_{ds} is varied as 0.2, 0.4, and 0.6 V. V_{gs} is biased to the maximum f_T conditions. The compressive strain makes the electron effective mass m^* larger in the devices A and C, whereas the tensile strain makes it smaller in the device B [6]. Nevertheless, v_d in the device C with the InSb channel is larger than the others from the low V_{ds}. Consequently, the device C exhibits the lower knee voltage of less than 0.2 V and the larger g_m. On the other hand, though the compressive strain makes E_{th} larger [6], the device C exhibits the severe impact ionization from the lower V_{ds} of about 0.4 V.

B. Estimation of Frequency Limits in intrinsic f_T

The profiles of $\tau(x)$ are also shown in Fig. 3. $\tau(x)$ almost distributes in the area under the gate, and has the tails for both sides. When V_{ds} increases, $\tau(x)$ under the gate decreases according to the increase of v_d. Meanwhile, the drain-side tail of $\tau(x)$ enlarges in turn, because of the extension of the high v_d area toward the drain. τ_{total} is divided into the source area τ_s, under the gate τ_g, and the drain area τ_d [3]. Owing to the larger v_d under the gate, the device C exhibits the smaller τ_g even at the low V_{ds} of 0.2 V: in which τ_d is also small. Here, we define the effective gate length $L_{g,eff}$ as corresponding length to $\tau(x)$ [3]. Fig. 4 shows the relationship between $L_{g,eff}$ and L_g. $L_{g,eff}$ is much longer than L_g, because of τ_s and τ_d; this indicates that the size scaling is finite practically. It is noteworthy to say that the difference between $L_{g,eff}$ and L_g depends on the channel structures: which will be discussed later. By extrapolating $L_{g,eff}$ to L_g of 0 nm, the lower limit of $L_{g,eff}$, $L_{g,eff}^{(0)}$, can be estimated. $L_{g,eff}^{(0)}$ becomes longer as V_{ds} increases owing to the increase of τ_d.

Fig. 5 shows the relationship between the intrinsic f_T and $1/L_{g,eff}$. Substantially the linear relationship can be seen. Then, the frequency limits in the intrinsic f_T, $f_T^{(0)}$, can be estimated by extrapolating f_T to $L_{g,eff}^{(0)}$, which are indicated by the arrows. The device C exhibits the higher intrinsic f_T from the lower V_{ds}. However, the intrinsic f_T does not increase even when V_{ds} increases, because the decrease of τ_g is canceled by the increase of τ_d. $f_T^{(0)}$ is limited practically by $L_{g,eff}^{(0)}$. Owing to the shorter $L_{g,eff}^{(0)}$, the device C exhibits the higher $f_T^{(0)}$ than the others at V_{ds} of 0.2 V, indicating its potential for the low power and THz operation.

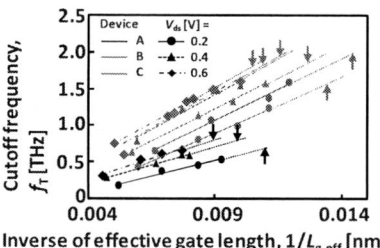

Figure 5 Dependence of intrinsic f_T on $1/L_{g,eff}$ for devices. V_{ds} is varied as 0.2, 0.4, and 0.6 V. V_{gs} is biased to maximum f_T conditions.

V. CONCLUSION

The frequency limits in the intrinsic f_T of the nanoscale HEMTs with various channel materials have been investigated by using the QC-MC simulation. The device C with the InSb channel has exhibited the higher intrinsic f_T from the lower V_{ds} because of the larger v_d. Owing to the shorter $L_{g,eff}^{(0)}$ at the lower V_{ds}, the device C has also exhibited the higher $f_T^{(0)}$, indicating its potential for the low power and THz operation.

ACKNOWLEDGMENT

This work is partly supported by Advanced Device Laboratories, Tokyo University of Science, and Grant-in-Aid for Scientific Research (C) (22560346).

REFERENCES

[1] Y. Yamashita et al., IEEE Electron Device Lett, Vol. 23, pp. 573-575, Oct. 2002.

[2] D.-H. Kim et al., IEDM Tech. Dig., pp. 13.6.1-13.6.4, Dec. 2011.

[3] T.Takegishi et al., 2010 IEICE Trans. Electron., vol. E93-C, no. 8, pp. 1258-1265, Aug. 2010.

[4] F. Machida et al., 2011 IPRM Proceedings, pp. 437-440, May. 2011.

[5] J. Sato et al., 2012 IPRM Proceedings, pp. We-2E.4, Aug. 2012.

[6] H. Nishino et al., 2010 IPRM Proceedings, pp. 156-159, May. 2010.

[7] H. I. Fujishiro et al., ISCS Abstracts, pp. 2795-2798, Oct. 2007.

[8] Y. Kwon and D. Pavlidis, IEEE Trans. Electron Devices., vol. 43, pp. 228-237, Feb. 1996.

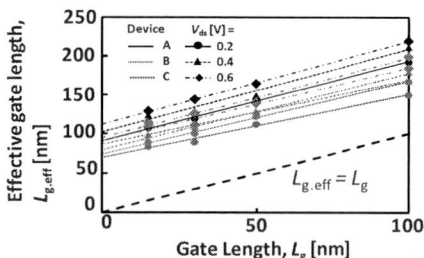

Figure 4 Relationship between $L_{g,eff}$ and L_g for devices. V_{ds} is varied as 0.2, 0.4, and 0.6 V. V_{gs} is biased to maximum f_T conditions.

978-1-4673-6130-9/13 $31.00 © 2013 IEEE

ThD2-5 (Oral)
12:15 - 12:30

InP/InGaAs DHBT Technology Using SiN/SiO$_2$ Sidewall Spacers

Norihide Kashio, Kenji Kurishima, Minoru Ida, and Hideaki Matsuzaki

NTT Photonics Laboratories, NTT Corporation
3-1, Morinosato Wakamiya, Atsugi-shi, Kanagawa Pref., 243-0198, Japan
E-mail: kashio.norihide@lab.ntt.co.jp

Abstract—This paper describes 0.25-μm-emitter InP/InGaAs DHBT technology that uses SiN/SiO$_2$ sidewall spacers. The technology enables the fabrication of HBTs with a passivation ledge (0.10-μm width) and narrow base metal (< 0.25 μm). The fabricated HBT exhibits a high current gain of over 50 and an f_t of 491 GHz at a collector current density of 18 mA/μm^2.

Keywords—InP DHBT, ledge passivation, sidewall spacer

I. INTRODUCTION

In the last several years, there have been rapid advances in the high-frequency performance of InP double heterojunction bipolar transistors (InP DHBTs) with a narrow emitter and base metal (< 0.25 μm) [1]. However, such aggressive lateral scaling degrades the current gain, which may seriously influence device reliability. An effective way to suppress the current gain degradation is to form a passivation ledge. Although InP HBTs with a passivation ledge exhibit high current gain, the high-frequency performance is limited by the relatively large total collector capacitance [2]. To improve the high-frequency performance while maintaining high current gain, we have newly developed InP DHBT technology that utilizes SiN/SiO$_2$ sidewall spacers for the formation of a narrow base metal and a passivation ledge with i-line lithography.

II. HBT STURUCRURES AND FABRICATION

We used double HBT structures grown on InP substrates by molecular beam epitaxy. The HBT consists of a degenerately doped n$^+$-InGaAs emitter contact, 10-nm-thick n-doped InP emitter, 15-nm-thick psedomorphic In$_x$Ga$_{1-x}$As base, 75-nm-thick InGaAs/InAlGaAs/InP collector, and InGaAs/InP subcollector. Here, the base layer is doped with carbon to a concentration of 6 x 10^{19} cm^{-3}. The indium mole fraction, x, is 0.44. The base sheet resistance was estimated to be about 1200 Ω/sq. from transmission-line-model measurements. Figure 1 shows a cross-sectional view of the fabricated HBT. The device fabrication starts with the deposition of W-based emitter metal. After a photoresist mask had been formed using i-line lithography, the W-based emitter was etched by dry etching. The combination of dry etching and selective wet etching was used to form the n$^+$-InGaAs emitter mesa. Then, SiN and SiO$_2$ films were deposited. The SiO$_2$ film was patterned, and the SiN and InP emitter were selectively etched. During this process, an undercut was made in the SiN and an InP ledge was also

formed. A base metal was then formed by a lift-off technique. Note that the undercut between the SiO$_2$ and SiN enables the formation of the narrow base metal without any emitter-base short circuits. The base-collector mesa, emitter and collector metals were formed. Each device was isolated and a BCB film was spin-coated on them. It was cured and then etched back to the emitter metal. Next, base and collector contact holes were formed and the interconnect metal was deposited. Figure 2 shows a SEM image of the fabricated HBT. A 0.25-μm-emitter HBT with a passivation ledge and a narrow base metal was successfully fabricated with i-line lithography. The width of the base metal is 0.22 μm and the emitter-base spacing is 0.15 μm.

III. DEVICE CHARACTERISTICS

Figure 3 shows typical *I-V* characteristics for a fabricated HBT. The HBT exhibits excellent turn-on characteristics and high current density of 20 mA/μm^2. The differential current gain is about 60 at a collector current density (J_c) of over 6 mA/μm^2. Figure 4 shows the Gummel plots for the HBT with an emitter-base (EB) spacing of 0.15 μm. The HBT exhibits a high current gain of 55 at a J_c of 18 mA/μm^2. To investigate the effectiveness of the ledge passivation, we measured Gummel plots of 0.25-μm-emitter HBTs with various EB spacing ranging from 0.10 μm to 0.95 μm. Figure 5 shows the current gain as a function of J_c. The current gain gradually decreases with decreasing EB spacing. This is attributed to an increase in the leakage current through the ledge layer. Note that a current gain of over 50 is maintained even in the HBT with an EB spacing of 0.10 μm. This indicates that the 0.10-μm passivation ledge firmly suppresses the base surface recombination current. Finally, we performed on-wafer measurements of *S*-parameters from 0.5 to 50 GHz with an HP8510C network analyzer. Figure 6 shows the current gain (h_{21}), Mason's unilateral power gain (*UG*), and maximum stable gain (*MSG*) as a function of frequency at a J_c of 18 mA/μm^2. The f_t and f_{max} were obtained by extrapolation of h_{21} and *UG* with a -20 dB/decade slope line. The HBT with an EB spacing of 0.10 μm exhibits an f_t of 491 GHz and an f_{max} of 225 GHz, respectively. The extracted total collector capacitance (C_{Tc}) is about 8 fF, which is lower by 30% than that of our previous work [2]. The reduction of C_{Tc} is due to our developed technology which enables the successful formation of the narrow base metal.

978-1-4673-6130-9/13 $31.00 © 2013 IEEE

IV. SUMMARY

We have successfully fabricated 0.25-μm-emitter InP DHBTs with a 0.10-μm passivation ledge and 0.22-μm base metal. The fabricated HBT exhibits a current gain of over 50 and an f_t of 491 GHz at a J_c of 18 mA/μm². This results demonstrate that the SiN/SiO₂ sidewall process is very useful for the fabrication of scaled-down InP HBTs with high reliability and high-frequency performance.

REFERENCES

[1] V. Jain, E. Lobisser, A. Baraskar, B. J. Thibeault, M. J. W. Rodwell, M. Urteaga, D. Loubychev, A. Snyder, Y. Wu, J. M. Fastenau, W. K. Liu, "InGaAs/InP DHBTs demonstrating simultaneous $f_t/f_{max} \sim 460/850$ GHz in a refractory emitter process," in Proc. IPRM 2011, Mo.1.2.2, pp. 1-4.

[2] N. Kashio, K. Kurishima, Y. K. Fukai, M. Ida, and S. Yamahata, "0.25-μm-Emitter InP Heterojunction Bipolar Transistors with a Thin Ledge Structure," Jpn. J. Appl. Phys., vol. 49, pp. 04DF02-1-5, 2010.

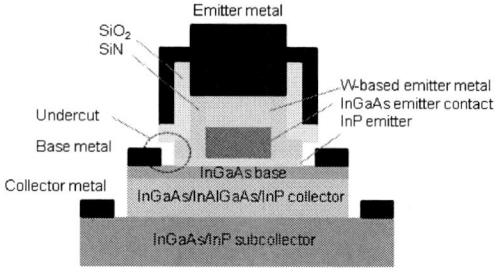

Figure 1. Cross-sectional view of an InP DHBT with a passivation ledge.

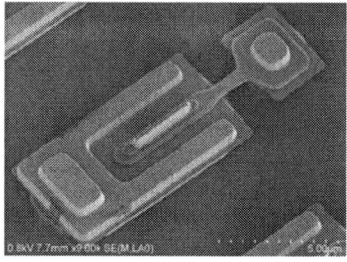

Figure 2. SEM image of a fabricated InP DHBT with an emitter size of 0.25 μm × 4 μm.

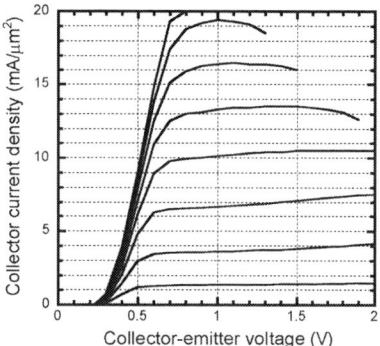

Figure 3. Common-emitter collector I-V charactersitcis for the HBT with an emitter size of 0.25 μm × 4 μm. The I_B step is 50 μA

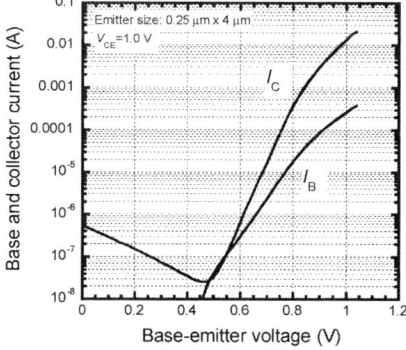

Figure 4. Gummel plots of the HBT with an emitter size of 0.25 μm × 4 μm

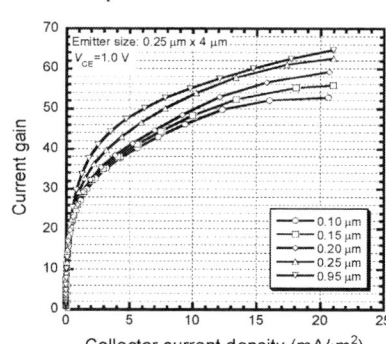

Figure 5. Current gain characteristics for 0.25-μm-emitter HBTs with various emitter-base spacings.

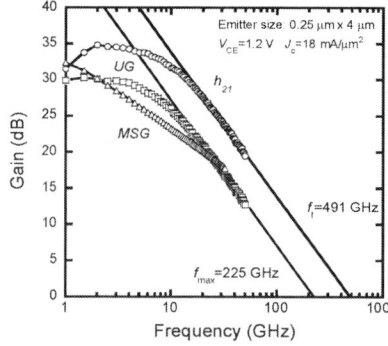

Figure 6. Current gain (h_{21}), Mason's unilateral gain (UG), and maximum stable gain (MSG) as a function of frequency at a collector current density of 18 mA/μm²